VOLUME SEVEN HUNDRED AND THIRTY FOUR

METHODS IN ENZYMOLOGY

Histone Deacetylases in Neurodegenerative Disease - Part B

METHODS IN ENZYMOLOGY

VOLUME SEVEN HUNDRED AND THIRTY FOUR

METHODS IN ENZYMOLOGY

Histone Deacetylases in Neurodegenerative Disease - Part B

Edited by

SUBASHCHANDRABOSE CHINNATHAMBI
Additional Professor Department of Neurochemistry
National Institute of Mental Health and Neurosciences Hospital
Bangalore, Karnataka, India

ELSEVIER

ACADEMIC PRESS
An imprint of Elsevier

Academic Press is an imprint of Elsevier
50 Hampshire Street, 5th Floor, Cambridge, MA 02139, United States
125 London Wall, London, EC2Y 5AS, United Kingdom

First edition 2026

ISBN: 978-0-443-47480-4
ISSN: 0076-6879

For information on all Academic Press publications
visit our website at https://www.elsevier.com/books-and-journals

Publisher: Zoe Kruze
Editorial Project Manager: Saloni Vohra
Production Project Manager: James Selvam
Cover Designer: Muthukumaran Thangaraj

Typeset by MPS Limited, India

Contents

Contributors

Niharika
Biochemistry and Molecular Biology Group, Epigenetics and Cancer Research Laboratory, Department of Life Science; Department of Life Science, ICMR Collaborating Centre of Excellence (ICMR-CCoE), National Institute of Technology Rourkela, Sundergarh, Odisha, India

Vidya Jyothi Alli
Academy of Scientific and Innovative Research (AcSIR), Ghaziabad; Department of Natural Products and Medicinal Chemistry, CSIR-Indian Institute of Chemical Technology, Hyderabad, India

Shalini Aricthota
Laboratory of Chromatin Biology and Epigenetics, BRIC-Centre for DNA Fingerprinting and Diagnostics, Uppal, Hyderabad, India

Abhishek Ankur Balmik
Department of Pharmacology and Chemical Biology, Emory University School of Medicine, Rollins Research Center, Atlanta, GA, United States

Jyotirmoy Banerjee
Department of Biophysics, All India Institute of Medical Sciences, New Delhi, India

Trayambak Basak
School of Biosciences and Bioengineering, Indian Institute of Technology Mandi, Kamand, Himachal Pradesh, India

Prasun Kumar Bhunia
School of Biosciences and Bioengineering, Indian Institute of Technology Mandi, Kamand, Himachal Pradesh, India

Karthick Chennakesavan
Department of Pharmaceutical Sciences, Irma Lerma Rangel College of Pharmacy, Texas A&M, University, College Station, TX, United States

Subashchandrabose Chinnathambi
Department of Neurochemistry, National Institute of Mental Health and Neuro Sciences Hospital (NIMHANS), Institute of National Importance, Bangalore, Karnataka; Neurobiology Group, Division of Biochemical Sciences, CSIR-National Chemical Laboratory, Pune, India

Sribas Chowdhury
CSIR-Institute of Genomics and Integrative Biology, Mathura Road, New Delhi, India

James P. Clement
Faculty of Clinical and Biomedical Science, University of Exeter, Exeter, United Kingdom

Chandrima Das
Biophysics and Structural Genomics Division, Saha Institute of Nuclear Physics, Kolkata; Homi Bhabha National Institute, Mumbai, India

Indrasis Dasgupta
Laboratory of Drug Design and Discovery, Department of Pharmaceutical Technology, Jadavpur University, Kolkata, India

Gayan Heruka De Zoysa
The MacDiarmid Institute for Advanced Materials and Nanotechnology, Victoria University of Wellington, Wellington, New Zealand

Aparna Banerjee Dixit
Dr. B. R. Ambedkar Centre for Biomedical Research, University of Delhi, Delhi, India

Humaira Farooq
Department of Molecular & Human Genetics, Institute of Science, Banaras Hindu University, Varanasi, UP, India

Shovanlal Gayen
Laboratory of Drug Design and Discovery, Department of Pharmaceutical Technology, Jadavpur University, Kolkata, India

Kuppan Gokulakrishnan
Department of Neurochemistry, National Institute of Mental Health & Neuro Sciences (NIMHANS), Bangalore, India

Nalini Vijay Gorantla
Department of Pathology, University of Chicago, Chicago, IL, United States

Nitesh Kumar Gupta
Department of Biochemistry, Institute of Science, Banaras Hindu University, Varanasi, UP, India

Rajaneesh Kumar Gupta
Department of Molecular & Human Genetics, Institute of Science, Banaras Hindu University, Varanasi, UP, India

Devyani Haldar
Laboratory of Chromatin Biology and Epigenetics, BRIC-Centre for DNA Fingerprinting and Diagnostics, Uppal, Hyderabad, India

Surender Singh Jadav
Academy of Scientific and Innovative Research (AcSIR), Ghaziabad; Department of Natural Products and Medicinal Chemistry, CSIR-Indian Institute of Chemical Technology, Hyderabad, India

Varghese Merin Jerusha
Department of Neurochemistry, National Institute of Mental Health & Neuro Sciences (NIMHANS), Bangalore, India

Shasi V. Kalivendi
Department of Applied Biology; Department of Organic Synthesis and Process Chemistry, CSIR-Indian Institute of Chemical Technology, Hyderabad, India

Prasad Kasturi
School of Biosciences and Bioengineering, Indian Institute of Technology Mandi, Kamand, Himachal Pradesh, India

Sonali Kumar
Department of Biotechnology, Lingaya's Vidyapeeth, Faridabad, Haryana; Dr. B. R. Ambedkar Centre for Biomedical Research, University of Delhi, Delhi, India

Tapas K. Kundu
Molecular Biology and Genetics Unit, Transcription and Disease Laboratory, Jawaharlal Nehru Centre for Advanced Scientific Research, Jakkur, Bengaluru; Academy of Scientific and Innovative Research (AcSIR), Ghaziabad, India

Jeff Kuret
Department of Biological Chemistry & Pharmacology, The Ohio State University, Columbus, OH, United States

Prajjval Mishra
Molecular Biology and Genetics Unit, Transcription and Disease Laboratory, Jawaharlal Nehru Centre for Advanced Scientific Research, Jakkur, Bengaluru, India

Abhijit Mohapatra
Department of Applied Biology, CSIR-Indian Institute of Chemical Technology, Hyderabad; Academy of Scientific and Innovative Research (AcSIR), Ghaziabad, India

Atanu Mondal
Biophysics and Structural Genomics Division, Saha Institute of Nuclear Physics, Kolkata; Homi Bhabha National Institute, Mumbai, India

Niharika Nag
Molecular and Structural Biophysics Laboratory, Department of Biochemistry, School of Life Sciences, North-Eastern Hill University, Shillong, India

Anam Naseer
Division of Toxicology and Experimental Medicine, CSIR-Central Drug Research Institute (CDRI), Lucknow, India

Ayesha Nasreen
CSIR-Institute of Genomics and Integrative Biology, Mathura Road, New Delhi; Academy of Scientific and Innovative Research (AcSIR), Ghaziabad, Uttar Pradesh, India

Vivek T. Natarajan
CSIR-Institute of Genomics and Integrative Biology, Mathura Road, New Delhi; Academy of Scientific and Innovative Research (AcSIR), Ghaziabad, Uttar Pradesh, India

Aamir Nazir
Division of Toxicology and Experimental Medicine, CSIR-Central Drug Research Institute (CDRI), Lucknow, India

Ayush Nigam
School of Biosciences and Bioengineering, Indian Institute of Technology Mandi, Kamand, Himachal Pradesh, India

Srihari Pabbaraja
Department of Natural Products and Medicinal Chemistry; Department of Organic Synthesis and Process Chemistry, CSIR-Indian Institute of Chemical Technology, Hyderabad, India

Priyanka Pandey
Molecular Biology and Genetics Unit, Transcription and Disease Laboratory, Jawaharlal Nehru Centre for Advanced Scientific Research, Jakkur, Bengaluru, India

Chitrakshi Pant
Department of Applied Biology, CSIR-Indian Institute of Chemical Technology, Hyderabad; Academy of Scientific and Innovative Research (AcSIR), Ghaziabad, India

Kamal D. Patel
School of Chemical Sciences and the Centre for Green Chemical Science, The University of Auckland, Auckland, New Zealand

Samir Kumar Patra
Biochemistry and Molecular Biology Group, Epigenetics and Cancer Research Laboratory, Department of Life Science; Department of Life Science, ICMR Collaborating Centre of Excellence (ICMR-CCoE), National Institute of Technology Rourkela, Sundergarh, Odisha, India

Tazeen Qureshi
Neurobiology Group, Division of Biochemical Sciences, CSIR-National Chemical Laboratory, Pune; Department of Neurochemistry, National Institute of Mental Health and Neuro Sciences Hospital (NIMHANS), Institute of National Importance, Bangalore, Karnataka, India

Amrish Rai
Division of Neuroscience and Ageing Biology, CSIR-Central Drug Research Institute, Lucknow; Academy of Scientific and Innovative Research (AcSIR), Ghaziabad, India

Thangavel Samikkannu
Department of Pharmaceutical Sciences, Irma Lerma Rangel College of Pharmacy, Texas A&M, University, College Station, TX, United States

Marisel Sanchez
Neuroscience Graduate Program, The Ohio State University, Columbus, OH, United States

Vijayalekshmi Sarojini
School of Chemical Sciences and the Centre for Green Chemical Science, The University of Auckland, Auckland, New Zealand

Singh Ankit Satyaprakash
Department of Biochemistry, Institute of Science, Banaras Hindu University, Varanasi, UP, India

Ozasvi R. Shanker
Dr. B. R. Ambedkar Centre for Biomedical Research, University of Delhi, Delhi; All India Institute of Ayurveda (AIIA), Government of India, New Delhi, India

Akash Kumar Singh
Molecular Biology and Genetics Unit, Transcription and Disease Laboratory, Jawaharlal Nehru Centre for Advanced Scientific Research, Jakkur, Bengaluru; Neuroscience Unit, Jawaharlal Nehru Centre for Advanced Scientific Research, Bangalore, India

Surya Pratap Singh
Department of Biochemistry, Institute of Science, Banaras Hindu University, Varanasi, UP, India

Srijoni Sinha
Laboratory of Drug Design and Discovery, Department of Pharmaceutical Technology, Jadavpur University, Kolkata, India

Shweta Kishor Sonawane
Inst Neurophysiopathol, CNRS, INP, Aix-Marseille Univ, Marseille, France

Genji Sukumar
Department of Natural Products and Medicinal Chemistry; Department of Organic Synthesis and Process Chemistry, CSIR-Indian Institute of Chemical Technology, Hyderabad, India

Chinnasamy Thirumoorthy
Department of Neurochemistry, National Institute of Mental Health & Neuro Sciences (NIMHANS), Bangalore, India

Timir Tripathi
Molecular and Structural Biophysics Laboratory, Department of Zoology, School of Life Sciences, North-Eastern Hill University, Shillong, India

CHAPTER ONE

Lysine methylation is an endogenous post-translational modifications of tau protein in human brain and a modulator of aggregation propensity

Marisel Sanchez[a] and Jeff Kuret[b,*]
[a]Neuroscience Graduate Program, The Ohio State University, Columbus, OH, United States
[b]Department of Biological Chemistry & Pharmacology, The Ohio State University, Columbus, OH, United States
*Corresponding author. e-mail address: kuret.3@osu.edu

Contents

Methods in Enzymology, Volume 734
ISSN 0076-6879, https://doi.org/10.1016/bs.mie.2026.05.044

Abstract

Tau protein undergoes a broad range of post-translational modifications in the brain, influencing its structure, solubility, and propensity to aggregate. This chapter presents an integrated methodological framework for characterizing tau methylation and evaluating its impact on tau biology. We describe procedures for isolating soluble and filamentous tau from post-mortem human brain tissue while preserving modifications for proteomic analysis. These approaches support precise mapping of methylation sites alongside other co-occurring modifications. To model methylation under controlled conditions, we outline protocols for recombinant tau expression, purification, and chemical reductive methylation, including radiolabeled assays for determining modification stoichiometry. We then detail biophysical assays used to assess how methylation alters tau conformation and aggregation propensity. This methodological framework supports experimentation seeking insight into mechanisms relevant to Alzheimer's disease and related tauopathies.

Abbreviations

BCA	bicinchoninic Acid.
CAS	Chemical Abstract Service.
CD	circular dichroism.
DTT	dithiothreitol.
EDTA	ethylenediamine tetraacetic acid.
EGTA	ethyleneglycol-bis(β-aminoethyl)-N,N,N',N'-tetraacetic acid.
HEPES	4-(2-hydroxyethyl)-1-piperazineethanesulfonic acid.
HSHB	high salt homogenization buffer.
IMAC	immobilized metal affinity chromatography.
IPTG	isopropyl β-D-1-thiogalactopyranoside.
LB	lysogeny broth.
LC–MS/MS	liquid Chromatography-Tandem Mass Spectrometry.
LSHB	low salt Homogenization Buffer.
MES	2-(N-morpholino)ethanesulfonic acid.
ODS	octadecyl sulfate.
PBS	phosphate buffered saline.
PMSF	phenylmethanesulfonyl fluoride.
PTM	posttranslational modification.
rpm	revolutions per min.
SDS-PAGE	sodium dodecyl sulfate-polyacrylamide gel electrophoresis.
SP3	single-pot.solid-phase-enhanced sample-preparation.
TBSE	Tris-buffered saline with EDTA.
TCA	trichloroacetic acid.
TCEP	tris(2-carboxyethyl) phosphine hydrochloride.
TEAB	tetraethylammonium.
ThT	Thioflavin T.

1. Introduction

Tau is a microtubule-associated protein that normally localizes to the axon of mature neurons where it influences microtubule stability and supports long-distance axonal transport (Kevenaar & Hoogenraad, 2015). In Alzheimer's disease (AD) and other tauopathies, however, tau dissociates from microtubules and aggregates to form pathological hallmarks of these neurodegenerative disorders (Ando et al., 2016; Cohen et al., 2011). Because tau is subject to a complex network of post-translational modifications (PTMs) in human brain, ranging from phosphorylation and N-glycosylation on polar amino acid side chains to methylation, acetylation and ubiquitylation on Lys side chains (Alquezar et al., 2020; Kyalu Ngoie Zola et al., 2023), there is great interest in clarifying which PTMs regulate tau function and its conversion into pathological aggregates.

Lys methylation is a tau PTM implicated in the conversion. Proteomic analyses of human brain-derived tau have identified mostly mono- and di-methylated Lys residues distributed over at least twenty-seven sites (Fig. 1), including within the microtubule-binding repeat region that mediates aggregation, microtubule binding, and putative cofactor association (Bichmann et al., 2021; Huseby et al., 2019; Thomas et al., 2012; Wesseling et al., 2020). Site occupancies in soluble tau appear to be dynamically regulated, with dimethylation predominating in middle-aged individuals and shifting toward monomethylation with advancing age (Huseby et al., 2019). In AD brains, methylation is markedly reduced or absent, consistent with potential utility as a marker of disease as well as aging. These studies also revealed the potential for interaction with other

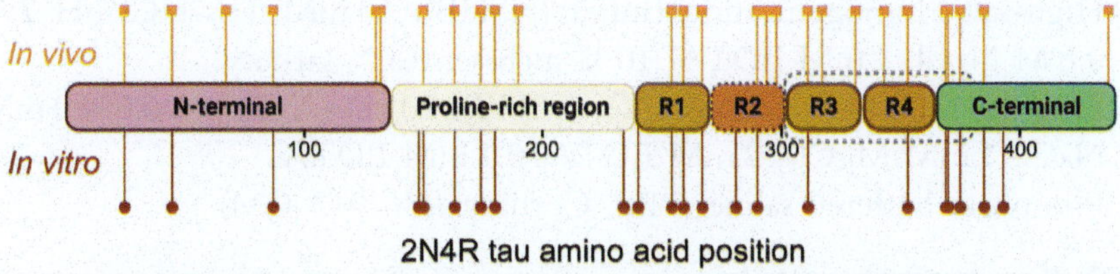

Fig. 1 Domain architecture of 2N4R tau and lysine methylation sites. Schematic representation of the human 2N4R tau isoform (441 aa) highlighting major domains, including the N-terminal, proline-rich, microtubule-binding repeat (R1–R4, where the R2 segment is encoded by alternatively spliced exon 10), and C-terminal regions. Methylation site distributions in human brain (in vivo) and in biochemical model systems (in vitro) are marked. Dashed line represents cross-β-sheet core region of AD-derived filaments.

PTMs ("crosstalk"), both directly with competing lysine modifications (e.g., ubiquitylation and acetylation) and indirectly with nearby phosphorylation sites (Balmik & Chinnathambi, 2021; Thomas et al., 2012).

In this chapter we provide experimental protocols to interrogate tau methylation biology. Because methylation effects on tau function depend first and foremost on sites of modification, we first provide protocols for isolation and proteomic characterization of human brain derived tau. Importantly, this approach captures the qualitative nature of methylation as it distributes among mono- and di-methylated Lys forms, and its spatial relationships with other PTMs. Second, we provide protocols for modeling tau methylation in vitro using recombinant full-length 2N4R tau and reductive methylation. Because PTM effects crucially depend on quantities of modification, a protocol for estimating methylation stoichiometry on recombinant preparations is included. Finally, we outline in vitro assay methods for quantifying effects of methylation on tau aggregation and microtubule assembly. This defined system enables direct comparison of methylated and unmodified tau, complementing recent cellular and structural studies examining residue-specific effects on microtubule binding and aggregation dynamics (Bichmann et al., 2021; Shams et al., 2022; Xia et al., 2023).

2. Proteomic mapping of brain-derived tau methylation sites

2.1 Before you begin

Timing: ~1.5 h.

- High-salt Homogenization Buffer (HSHB): 10 mM Tris-HCl pH 7.4, 0.8 M NaCl, 1 mM EDTA, 10 % sucrose, 0.1 % sarkosyl
- Low-salt Homogenization buffer (LSHB): 20 mM MES pH 6.8, 80 mM NaCl, 1 mM $MgCl_2$, 2 mM EGTA, 0.1 mM EDTA
- Phosphate buffered saline (PBS; Corning #46-013-CM)

2.2 Key resources table

Reagent or resource	Source	Identifier
Antibodies		
Tau-5 monoclonal antibody	Invitrogen	AHB0042
Biological Samples		

Fresh frozen post-mortem human frontal neocortex	Various sources	N/A
Chemicals, Peptides, and Recombinant Proteins		
Trypsin, mass spectrometry grade	Promega	V528A
Critical Commercial Assays		
BCA protein assay	ThermoFisher Scientific	23225
Other		
S-Trap kit	ProtiFi	K002-MICRO 0010KT
TiO_2 phosphopeptide enrichment kit	ThermoFisher Scientific	A32993

2.3 Materials and equipment

In addition to standard molecular biology and biochemistry laboratory equipment, this protocol leverages specialized instrumentation for protein purification and biophysical analysis. Reagents used in these protocols are of analytical grade or higher, and buffers are freshly prepared and filtered before use. All steps involving protein handling are performed at 4 °C unless otherwise noted.

2.3.1 Equipment

- Biosafety cabinet and personal protective equipment (lab coat, gloves, eye goggles, face shield and protective sleeves, etc.)
- Standard wet-laboratory equipment including superspeed-, ultra-, and benchtop centrifuges, spectrophotometer, balances, pH meter, −20 °C and −80 °C freezers, temperature-controlled water baths, vacuum concentrator, and gel electrophoresis equipment
- Equipment for tissue homogenization, including glass-Teflon homogenizers (5–55 mL capacities) and motor drive (1000 rpm capability)
- Strainer (stainless steel; 35 mesh, 0.5 mm pore diameter)
- Probe sonicator (100 W maximum power output, 20 kHz frequency; 1/8" probe)
- Standard column chromatography equipment including automated or manual low-pressure pumps and associated columns
- Major capital equipment accessed through core or other resources: Transmission electron microscope, LC/MS-MS equipment, circular dichroism spectropolarimeter

- Liquid scintillation counter
- Cotton swabs (Fisher Scientific #22-363-173)

2.3.2 Reagents

- 20 % sarkosyl solution (CAS #137-16-6)
- cOmplete Mini Protease Inhibitor Cocktail (Millipore Sigma #11836153001)
- PhosSTOP (Millipore Sigma #4906837001)
- Deacetylase inhibitors trichostatin A (CAS #58880-19-6) and nicotinamide (CAS #98-92-0)
- Perchloric acid (CAS #7601-90-3)
- TCA (CAS #76-03-9)
- Dithiothreitol (DTT; CAS #3483-12-3)
- High performance liquid chromatography-grade solvents including formic acid and acetonitrile
- $(NH_4)HCO_3$ (50–100 mM, pH 7.8)
- Iodoacetamide (CAS #144-48-9)

2.4 Step-by-step method details

2.4.1 Preparation of human brain-derived tau filaments

Filamentous tau aggregates are isolated from human brain gray matter using high-salt, sarkosyl-based extraction for enrichment (Fig. 2) (Guo et al., 2016).

Timing: ~18–22 h.

Note: All procedures were performed on ice or at 4 °C unless otherwise specified.

1. **Day 1:** Dissect gray matter from frozen human brain tissue on a chilled metal plate kept on ice.
   Note: Gray matter typically constitutes ~65–70 % of the total tissue slice weight.
2. Weigh tissue portions (~9 g each) using a tared weigh boat and record sample weights.
3. Place dissected tissue into labeled sandwich bags, seal, and store at −80 °C until homogenization.
4. Clean glassware and metal tools with bleach between samples and dispose of all disposables as biohazard waste.
5. **Day 2:** Prepare High-Salt Homogenization Buffer (HSHB).
   a. Supplement HSHB immediately before use with DTT (2 mM final concentration), and with one cOmplete mini protease inhibitor tablet and one PhosSTOP phosphatase inhibitor tablet per 10 mL buffer.

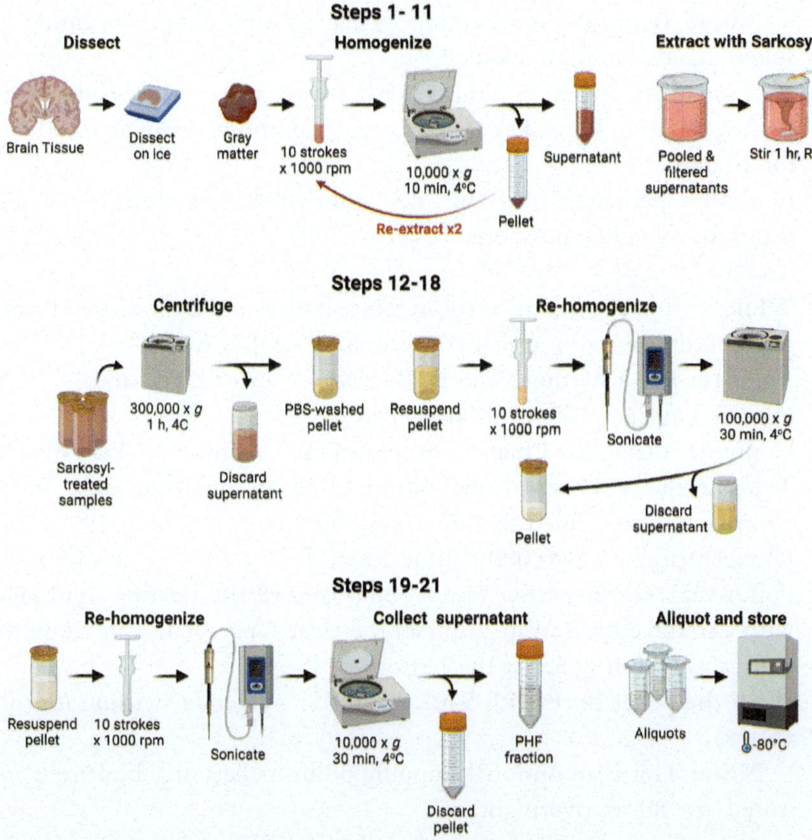

Fig. 2 Workflow for isolation of sarkosyl-insoluble tau aggregates from human brain tissue (Section 2.4.1). Tissue is dissected and homogenized in HSHB with 0.1 % Sarkosyl, followed by clarification by centrifugation. Supernatants are pooled, adjusted to 1 % Sarkosyl, and incubated for 1 h at room temperature. Sarkosyl-insoluble material is pelleted by ultracentrifugation, washed in PBS, and subjected to repeated homogenization and sonication steps. Final centrifugation yields a supernatant enriched in filamentous tau. *Figure created with BioRender.com.*

Note: Prepare sufficient HSHB volume for at least 5 mL/g tissue wet weight.

6. Homogenize gray matter in 5 volumes of supplemented HSHB (e.g., 45 mL buffer per 9 g tissue) using a glass–Teflon homogenizer (10 strokes at 1000 rpm).
7. Transfer homogenates to centrifuge tubes and centrifuge at 10,000 × *g* for 10 min at 4 °C. Carefully collect the supernatant using a serological pipette and retain the pellet for re-extraction.

Note: Because the resulting pellet is soft, direct decanting of supernatant should be avoided.

8. Resuspend the pellet in 5 mL HSHB per gram of tissue using gentle homogenization (5 strokes at 1000 rpm). Centrifuge again at 10,000 × *g* for 10 min at 4 °C.
9. Pool all supernatants (~10 mL per gram of starting tissue) and filter through a strainer positioned over a chilled beaker to remove large debris.
10. While stirring the filtrate at room temperature, add 20 % sarkosyl stock solution dropwise to a final concentration of 1 % (w/v).

    Note: For a starting volume V_1, add $V_1/19$ of 20 % sarkosyl stock (e.g., 5.3 mL per 100 mL homogenate).
11. Continue stirring for 1 h at room temperature with the beaker covered.
12. Transfer the sarkosyl-treated homogenate to ultracentrifuge tubes, ensuring they are filled to ≥ 90 % capacity.
13. Ultracentrifuge at 300,000 × *g* for 1 h at 4 °C.
14. Following centrifugation, remove and discard the floating lipid layer with a sterile cotton swab. Aspirate the clear supernatant and retain the compact reddish pellet at the bottom of the tube.
15. Wash the pellet once with 5 mL cold PBS by gentle swirling (do not vortex).

    Note: This is an optional stopping point; pellets may be frozen and stored at −80 °C overnight.
16. Day 3: Resuspend the washed pellets in PBS supplemented with cOmplete and PhosSTOP inhibitors (~200 μL per gram of starting tissue).
17. Homogenize using a 30 mL glass–Teflon homogenizer (10 strokes at 1000 rpm), followed by sonication (20 pulses, 0.5 s/pulse, 4 W power) to disperse aggregates.
18. Centrifuge at 100,000 × *g* for 30 min at 4 °C. Discard the supernatant and retain the pellet.
19. Resuspend the pellet again in PBS supplemented with cOmplete and PhosSTOP inhibitors (200 μL/g tissue) and homogenize and sonicate as before.
20. Centrifuge at 10,000 × *g* for 30 min at 4 °C. Collect the supernatant, which contains the purified filamentous tau fraction.
21. Aliquot the final filament preparation (50–500 μL) into labeled microcentrifuge tubes and store at −80 °C until use.

2.4.2 *Preparation of human brain-derived soluble tau*

In contrast to filamentous tau, monomeric tau is enriched from homogenates by sequential extraction under denaturing conditions, leveraging its high solubility and resistance to heat.

Timing: ~10–12 h.

1. Prepare buffer containing 20 mM MES pH 6.8, 80 mM NaCl, 1 mM $MgCl_2$, 2 mM EGTA, 0.1 mM EDTA, and 1 mM PMSF.
2. Supplement homogenization buffer with cOmplete, PhosSTOP and deacetylase inhibitors (2 μM trichostatin A, 10 mM nicotinamide) to preserve endogenous modifications.
3. Homogenize gray matter tissue (5 volumes of homogenization buffer per tissue weight) using a glass-Teflon homogenizer (ten strokes at 1000 rpm).
4. Centrifuge homogenates at 27,000 × *g* for 20 min at 4 °C. Collect the supernatant for subsequent purification steps.
   Note: Cytosolic tau is a highly soluble protein that can be efficiently extracted under mild conditions that preserve PTMs.
5. Adjust the supernatant to 0.5 M NaCl and 2 % (v/v) 2-mercaptoethanol.
6. Boil the sample for 10 min to denature non–heat-stable proteins.
7. Centrifuge again at 27,000 × *g* for 20 min at 4 °C and collect the heat-stable supernatant.
8. Add perchloric acid to a final concentration of 2.5 % (v/v) to precipitate remaining non-tau proteins.
9. Centrifuge for 20 min at 27,000 × *g* at 4 °C. Collect the acid-soluble supernatant containing enriched tau protein.
10. Precipitate proteins from the acid-soluble supernatant by adding TCA to a final concentration of 20 % (w/v). Incubate on ice for 30 min
11. Centrifuge for 20 min at 27,000 × *g* and wash the resulting protein pellet twice with cold acetone to remove residual TCA.
12. Air-dry the washed pellets briefly and store dry at −20 °C until solubilization for LC–MS/MS analysis.

Note: Avoid over-drying TCA pellets, as this can make resuspension difficult; brief air-drying until acetone dissipates is sufficient.

2.4.3 *Proteomic mapping of tau methylation sites*

Although detailed protocols for LC-MS/MS acquisition and methylated peptide validation are available (Thomas & Yang, 2017), identification of tau methylation sites in brain-derived materials continues to face challenges. First is the low occupancy nature of tau methylation, which averages only ~0.1 mol methyl group per mol modification site in cell culture models (Huseby et al., 2019). Second is the depression in methylation occupancy with age and disease status (Huseby et al., 2019). Third is the proteolytic loss of tau sequences in the "fuzzy coat" N-terminal and C-terminal segments (Fig. 1) relative to tau monomers in AD brain-derived aggregates (Zhou et al., 2018), which acts to decrease tau methylation occupancies in these regions still further. Consequently, modified peptides are available only in low abundance and may exhibit substantial variability in human AD brain-derived samples. For this reason, sample preparation is paramount for detecting tau methylation. Multiple solid-phase techniques, including filter-aided sample preparation, suspension trapping (HaileMariam et al., 2018), and SP3, support large-scale proteomic workflows and are commercially available. Among these, filter-aided sample preparation, where samples are bound to filters before digestion, has achieved > 90 % sequence coverage when applied brain-derived tau aggregates (Wesseling et al., 2020). Because suspension trapping pairs well with trypsin-mediated digestion (Ludwig et al., 2018), we provide a protocol for its application to brain-derived tau aggregates below.

Timing: ~8–10 h.

1. This protocol leverages S-trap-based solid phase capture (HaileMariam et al., 2018) for sample preparation using the commercial ProtiFi kit. Begin by resuspending tau samples (100 μg) in lysis buffer containing 10 % (w/v) SDS and 100 mM TEAB.

   Note: Samples tend to suspend well but can be subjected to bath sonication if necessary.
2. Centrifuge samples at high speed to pellet insoluble debris and transfer the clarified supernatant to a clean tube.
3. Reduce disulfide bonds by adding DTT to a final concentration of 20 mM and incubate samples for 15 min at 95 °C.
4. Cool samples to room temperature, then alkylate free sulfhydryl groups by adding iodoacetamide and incubating for 30 min at room temperature in the dark. Remove excess iodoacetamide during subsequent S-Trap binding and wash steps according to manufacturer's instructions.

   Note: Iodoacetamide (IAA) is used at a concentration at least twice that of DTT to guarantee full alkylation of reduced cysteines. Because IAA is light-sensitive, protect it from light throughout the procedure.

Ensure complete removal of IAA before digestion to prevent trypsin inactivation (Inagami, 1965).

5. Acidify samples to pH ≤ 1 using phosphoric acid immediately prior to addition of S-Trap binding buffer to induce protein precipitation.
6. Dilute each sample with S-Trap binding buffer (100 mM TEAB in 90 % methanol) at a 1:6 (sample: binding buffer) ratio. Load the protein suspension onto an S-Trap micro spin column and centrifuge until all material has passed through and precipitated proteins are retained within the matrix.

   Note: The time spent in the presence of acidified methanol should be minimized to avoid Fischer esterification.
7. Wash S-Trap columns three times with binding/wash solution, centrifuging after each wash to remove residual detergents and contaminants. Perform a final centrifugation step to remove residual wash buffer prior to digestion.
8. For standard proteomic analysis, digest 100 μg of S-Trap–bound protein by adding Trypsin Gold, mass spectrometry grade (Promega, V528A). Reconstitute trypsin in TEAB at 1 mg/mL and add at a 1:10 enzyme-to-protein ratio. Incubate the digestion overnight at 37 °C.
9. Elute peptides sequentially by centrifugation using the following solutions in order:
   - 50 mM TEAB
   - 0.2 % (v/v) formic acid in water
   - 0.2 % (v/v) formic acid in 50:50 water/acetonitrile
10. Collect all eluates, combine if necessary, and dry samples under vacuum.
11. Resuspend dried peptides in 0.1 % formic acid prior to LC–MS/MS analysis.

2.5 Expected outcomes

Frontal neocortex dissected from late-stage AD cases typically yields ~40–60 μg filamentous tau per g brain gray matter at 20–40 % purity. In contrast, normal control and AD cases typically similar amounts of soluble tau per g brain gray matter at ~20 % purity. As a result, even small-scale preparations are sufficient to map the spatial distribution of mono-, di-, and tri-methyl lysine sites on tau (Funk et al., 2014; Thomas et al., 2012; Wesseling et al., 2020).

2.6 Advantages and limitations

Purification methods are applicable up to 100 g wet weight starting material, and so can be adapted to different brain regions and scales. Furthermore, estimation of total protein yield by BCA assay is broadly applicable to brain-derived samples. However, quantification of tau content is limited by reliance on immunoassays conducted with anti-tau antibodies. Although outstanding monoclonal antibodies for this purpose are commercially available (Ellis et al., 2024), tau epitopes located outside the microtubule binding repeat region are underrepresented in filamentous relative to soluble tau preparations (Zhou et al., 2018). Therefore, when quantifying tau content using immunoassays, it is best to employ antibodies that bind epitopes close to the microtubule binding repeat region that makes up the misfolded core region of tau filaments. Tau-5 antibody (LoPresti et al., 1995) is recommended for this purpose. Nonetheless, it may underestimate filamentous tau levels depending on the extent of endogenous proteolysis.

2.7 Safety considerations and standards

This approach calls for deidentified human brain tissue samples obtained from brain banks operating under institutional review board oversight. It also requires bloodborne pathogens training by personnel and oversight by a local Institutional Biosafety Committee.

2.8 Alternative methods/procedures

2.8.1 In gel digestion

While solid-phase techniques perform well for generating peptides, they can be substituted with in-gel digestion (Fig. 3) (Thomas & Yang, 2017). In addition to requiring lower amounts of sample, the approach can bias capture toward full-length tau isoforms so that modifications outside the cross-β-sheet core region are well represented (Thomas et al., 2012).

Timing: ~12–14 h.

1. Resuspend brain-derived tau samples (≥20 μg per sample) in SDS sample buffer and fractionate by SDS–PAGE using 4–20 % Tris–glycine gels (Bio-Rad).
2. Excise gel regions corresponding to full-length tau isoforms (50–75 kDa).

   Note: SDS-PAGE must be performed with caution owing to risk of Fischer esterification, where free carboxyl groups react with methanol in an acid catalyzed reaction (Mannu & Mele, 2024) (Fig. 4). Accordingly, staining and destaining of gels in methanol/acetic acid should be avoided

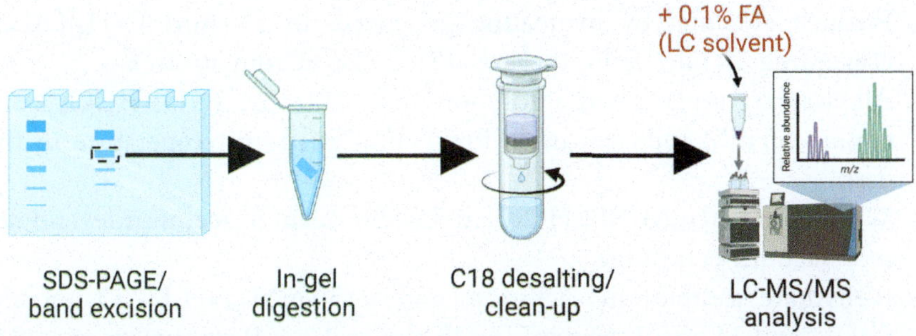

Fig. 3 Gel-based workflow for in-gel tryptic digestion and LC–MS/MS analysis of full-length tau. Tau protein samples are resolved by SDS–PAGE, and gel regions corresponding to full-length tau isoforms are excised. Excised gel pieces are subjected to in-gel reduction, alkylation, and tryptic digestion, followed by peptide extraction and reverse-phase desalting. Purified peptides are dried, reconstituted in 0.1 % formic acid, and analyzed by LC–MS/MS to enable identification of tau-derived peptides and post-translational modifications. *Figure created with BioRender.com.*

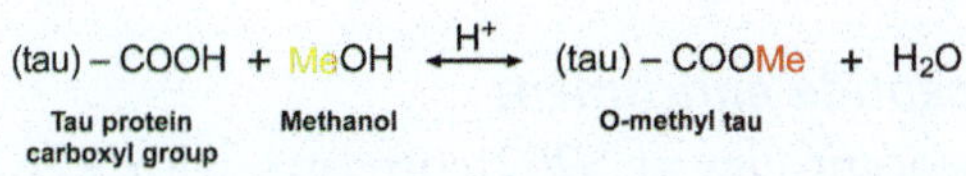

Fig. 4 Fischer esterification reaction. Free carboxyl groups in tau protein react with methanol to form O-methylated adducts. The reaction is acid catalyzed.

during sample preparation (Jung et al., 2008). Furthermore, methanol should be avoided as a co-solvent during liquid chromatography separation of peptides. The problem can be minimized by avoiding gel staining altogether and using pre-stained markers spanning the range ~45–65 kDa (Thomas & Yang, 2017). If sample visualization is critical, then replacing methanol with isopropanol can minimize spurious incorporation of methyl groups during staining/destaining of gels. Because tau lacks tryptophan residues, stain-free protein detection of proteins with trichloroethanol (Holzmüller & Kulozik, 2016) cannot resolve tau content in samples.

3. Cut excised bands into 1 mm^3 pieces and wash with 100 mM $(NH_4)HCO_3$, followed by a 5 min dehydration in acetonitrile. Dry samples completely in a vacuum centrifuge.

   Note: Volumes of all washes and treatments should be adjusted as needed to fully cover the 1 mm^3 gel pieces. This size ensures optimal trypsin access and complete digestion of the sample.

4. Reduce cysteines by incubating gel pieces in 100 mM $(NH_4)HCO_3$ containing 10 mM dithiothreitol (DTT) for 60 min at 60 °C.
5. Alkylate by replacing the solution with 100 mM $(NH_4)HCO_3$ containing 55 mM iodoacetamide for 45 min at room temperature in the dark.
6. Wash with 100 mM NH_4HCO_3, dehydrate again in acetonitrile, and dry in a vacuum centrifuge.
7. Rehydrate samples in digestion buffer (50 mM NH_4HCO_3, 5 mM $CaCl_2$, and 15 ng/μL trypsin) for 45 min at 4 °C. Remove excess buffer and incubate overnight at 37 °C to complete digestion.
8. Extract peptides sequentially: once with 50 mM $(NH_4)HCO_3$, once with 1:1 50 mM $(NH_4)HCO_3$/acetonitrile, and twice with 5 % formic acid/acetonitrile.
9. Desalt extracted peptides using C18 PepClean Spin Columns (Pierce/ThermoFisher Scientific) according to the manufacturer's protocol and dry by vacuum centrifugation.

2.8.2 Phosphopeptide enrichment

Phosphorylation occupancy at S262 correlates with K267 methylation in AD brain-derived tau aggregates, suggesting that methylation and phosphorylation may be coordinated at certain sites (Thomas et al., 2012). For such sites, selective enrichment of phosphopeptides may improve their capture, enabling a more complete assessment of modification crosstalk relevant to tau pathology. Titanium dioxide matrices (Thingholm & Larsen, 2016) provide a convenient approach for selectively enriching phosphopeptides prepared from brain-derived tau proteins (Tan et al., 2015). The following protocol integrates phosphopeptide enrichment with suspension trapping described above.

Timing: ~4–6 h.

1. Complete suspension trapping protocol (Section 2.4.3) with 300 μg brain-derived filamentous or soluble tau protein.
2. Resuspend dried peptides in 150 μL Binding/Equilibration Buffer supplied with TiO_2 enrichment kit by vortexing until fully dissolved. Verify that sample pH is < 3.0 using pH paper.
3. Wash and equilibrate TiO_2 spin columns following manufacturer's instructions.
4. Transfer the equilibrated TiO_2 spin tip into a 2 mL collection tube. Apply the peptide sample and centrifuge at $1000 \times g$ for 5 min. Reapply

the flow-through to the spin tip and centrifuge again at 1000 × *g* for 5 min to maximize phosphopeptide binding.

5. Wash the TiO_2 spin tip sequentially by transferring it back into the original collection tube. Add 20 μL Binding/Equilibration Buffer and centrifuge at 3000 g for 2 min, followed by 20 μL Wash Buffer and centrifugation at 3000 × *g* for 2 min. Repeat this wash sequence once. Perform a final wash with 20 μL LC–MS-grade water and centrifuge at 3000 × *g* for 2 min
6. Remove residual liquid by gently blotting the bottom of the spin tip on a clean laboratory tissue.
7. Elute bound phosphopeptides by placing the spin tip into a clean collection tube and adding 50 μL Phosphopeptide Elution Buffer. Centrifuge at 1000 × *g* for 5 min, then repeat the elution once and combine eluates.
8. Immediately dry the eluate under vacuum to remove the high-pH elution buffer and prevent phosphate loss.
9. Resuspend enriched phosphopeptides in 0.1 % (v/v) formic acid for LC–MS/MS analysis.

Note: Excessive drying of TiO_2 resin can reduce phosphopeptide recovery.

3. Tau methylation *in vitro*

3.1 Before you begin

Timing: ~1.5 h.

3.1.1 Recombinant tau expression and purification

- Sterile LB media containing ampicillin (0.15 mg/mL) and chloramphenicol (0.034 mg/mL)
- IMAC Lysis Buffer: 500 mM NaCl, 20 mM Tris-HCl pH 8.0, 0.5 mM PMSF, 0.1 % (v/v) 2-mercaptoethanol
- IMAC Elution Buffer: 500 mM NaCl, 20 mM Tris-HCl pH 8.0, 100 mM imidazole
- S300 Buffer: 10 mM HEPES-HCl pH 7.4, 250 mM NaCl, 0.1 mM EGTA
- Storage Buffer: 10 mM HEPES pH 7.4, 50 mM NaCl, 1 mM TCEP

3.1.2 *Chemical methylation and labeling*

- Sodium citrate buffer (0.1 M, pH 6.0)
- Desalting chromatography media (e.g., Bio-Gel P-6DG gel, BioRad #7322010 or Micro Bio-Spin P-6, BioRad #7326222, for work at mL or ~50 μL scales, respectively).
- Desalting Buffer (10 mM HEPES pH 7.4, 50 mM NaCl)
- CD Buffer: 100 mM $NaClO_4$, 20 mM H_3BO_3 buffer pH 7.4

3.1.3 *Aggregation assays*

- Assembly Buffer: 10 mM HEPES pH 7.4, 100 mM NaCl, 5 mM DTT

3.2 Key resources table

Reagent or resource	Source	Identifier
Bacterial and Virus Strains		
BL21CodonPlus (DE3)-RP cells	Agilent	Part # 230255
Chemicals, Peptides, and Recombinant Proteins		
[^{14}C] -formaldehyde (50–60 Ci/mol)	cliniSciences	CAS #3046-49-9
Critical Commercial Assays		
BCA	Thermo Fisher Scientific	23225
Recombinant DNA		
pT7II-2N4Rtau	Addgene	177653
pT7II-1N4Rtau	Addgene	249554
pT7II-0N4Rtau	Addgene	249555
pT7II-2N3Rtau	Addgene	249551
pT7II-1N3Rtau	Addgene	249552
pT7II-0N3Rtau	Addgene	249553
Other		
Sephacryl S300 size exclusion column and media	Cytiva	C26/70 and 17119601

3.3 Materials and equipment

3.3.1 Equipment

- Standard column chromatography equipment including automated or manual low-pressure pumps and associated columns
- Major capital equipment accessed through core or other resources: Transmission electron microscope, LC/MS-MS equipment, circular dichroism spectropolarimeter
- Liquid scintillation counter
- Flat-bottom 96-well plate (Corning #3651)
- Adhesive plate seals (Thermo Fisher Scientific #232699)
- Syringe filters: 5 μm (ThermoFisher #F2600-50) and 0.2 μm (Corning #431229) pore diameters
- Ultra filters: 30 kDa molecular weight cut-off (Millipore Sigma #UFC9030)

3.3.2 Reagents

- ODS (CAS #1120-04-3)
- Ovalbumin (CAS #9006-59-1)
- Thioflavin T (ThT) (CAS #2390-54-7)
- Formaldehyde (5 mM working solution)
- Sodium cyanoborohydride ($NaBH_3CN$, 0.1 M)
- Glycine (50 mM in water)
- Acetone (cold, for washing pellets)

3.4 Step-by-step method details

3.4.1 Production of full-length recombinant human tau isoforms

Recombinant tau proteins have long been prepared in soluble and fully active forms by leveraging heterologous expression in *E. Coli* and vectors derived from the T7 system (Studier et al., 1990). We have found that native full-length tau proteins can be purified by IMAC methods without the need for a polyhistidine tag (Pettis et al., 2026). The following protocol leverages our own T7-based expression vectors for doing so. Vectors encoding all six human central nervous system isoforms are available through the Addgene repository (Kamens, 2015).

Timing: ~32 h.

1. Transform pT7II-2N4R tau (or other full-length isoform) into an *E coli* BL21(DE3) cell line, followed by standard growth and IPTG-mediated or auto-induction of expression (Ayoub et al., 2023; Barghorn et al., 2005; Best et al., 2017). Store bacterial cell pellets at − 20 °C until purification.

2. Resuspend cell pellets in 5 volumes IMAC Lysis Buffer (v/w) containing freshly added 0.5 mM PMSF and 0.1% (v/v) 2-mercaptoethanol (final concentrations), then lyse by boiling on heated stir plate for 20 min (Fig. 5) (KrishnaKumar & Gupta, 2017).

   Note: Add a small stir bar to the suspension and heat in a boiling water bath for 30 min with continuous stirring to lyse cells and denature most *E. coli* proteins. Vortex briefly after 5 min to disperse the pellet, then continue boiling. Periodically vent the tube to release pressure.
3. After 5 min cooling in an ice bath, adjust lysate to pH 8.0, then subject to ultracentrifugation (400,000 × *g* for 1 h, 4 °C). Supernatants are collected and sequentially filtered through 5 μm and 0.2 μm syringe filters to remove remaining cellular debris.

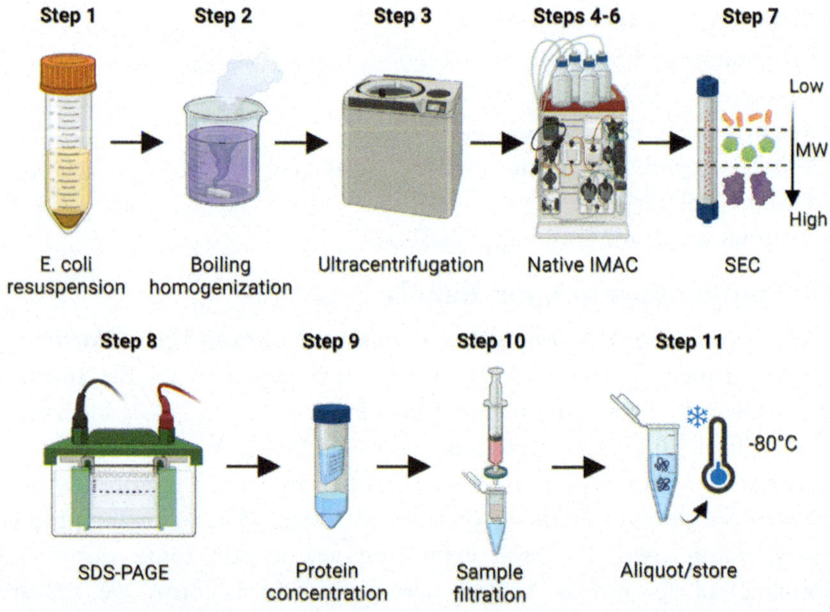

Fig. 5 Proteomic workflow for purification and characterization of recombinant tau. *E. coli* cells harboring overexpressed tau protein are homogenized by boiling to liberate heat-stable tau species. Homogenates are then centrifuged at 400,000 × *g* for 1 h at 4 °C, and soluble tau-containing fractions are recovered and purified by native immobilized metal affinity chromatography (IMAC) followed by size-exclusion chromatography. Purified fractions are assessed by SDS-PAGE, concentrated, filtered, and subjected to reductive stabilization prior to storage, yielding tau suitable for downstream biochemical and proteomic analyses. *Figure created with BioRender.com.*

4. Apply sample to HisTrap High Performance (HP) column (5 mL bed volume, Cytiva) pre-equilibrated with five bed volumes of ice-cold Lysis Buffer.

   Note: Apply to the column at a reduced flow rate of 0.5 mL/min to maximize binding
5. Wash the column with at least five column volumes of ice-cold Lysis Buffer at 1 mL/min.
6. Elute tau proteins with five column volumes of ice-cold Elution Buffer containing 0.5 mM PMSF at a flow rate of 1 mL/min. Monitor absorbance at 280 nm and pool fractions corresponding to the tau peak.
7. Add 80 % glycerol to the pooled eluate to a final concentration of 2 % (v/v), then subjected to size exclusion chromatography (SEC) on Sephacryl S300 media equilibrated and developed in S300 Buffer at ~20 mL/h flow rate. Collect sixty fractions per bed volume. Subject peak A280 fractions to SDS-PAGE.
8. Identify tau-containing fractions by SDS-PAGE (8–10 % Tris-glycine gels) and pool those corresponding to the monomeric tau peak.

   Note: All full-length human tau isoforms migrate anomalously on SDS-PAGE (Goedert & Jakes, 1990).
9. Concentrate the pooled sample by ultrafiltration (30 kDa molecular weight cut-off).
10. Filter proteins (0.2 μm pore diameter) before estimating protein concentrations by the 214 nm absorbance method (Kuipers & Gruppen, 2007) using S300 Buffer as the sample blank.
11. Add TCEP (2 mM final concentration) to depress cysteine oxidation. Aliquot and store at − 80 °C until used. Avoid repeated freeze–thaw cycles.

3.4.2 Reductive methylation of recombinant tau

Lysine methylation can be conveniently modeled by direct reductive methylation using formaldehyde substrate (Fig. 6), providing a controlled means to introduce methyl modifications into recombinant tau (Thomas et al., 2012). Replacement of formaldehyde with [^{14}C]-formaldehyde in parallel reactions enables precise estimation of methylation stoichiometry, allowing quantitative assessment of modification efficiency across preparations (Funk et al., 2014). The following procedures support modification, recovery, and structural validation of methylated tau while maintaining protein integrity.

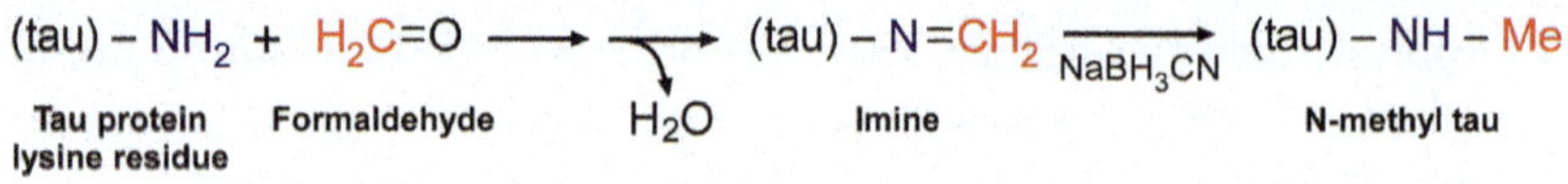

Fig. 6 Reductive methylation reaction. The ε-amine of lysine nucleophilically attacks the aldehyde to form a carbinolamine intermediate, which undergoes protonation and dehydration to generate an imine intermediate. Sodium cyanoborohydride selectively reduces this intermediate to yield N-methylated lysine while avoiding reduction of the starting carbonyl compound.

Timing: ~4 h.

1. Dilute purified recombinant 2N4R tau to a concentration of 1 mg/mL in 0.1 M sodium citrate buffer pH 6.0.
2. Add freshly prepared formaldehyde to 5 mM final concentration, followed addition of sodium cyanoborohydride ($NaBH_3CN$) to 0.1 M final concentration.

   Note: Handle $NaBH_3CN$ carefully in a fume hood, as it is toxic and releases HCN in acid.
3. Incubate the reaction mixture for up to 1 h at room temperature with gentle mixing.
4. Terminate the reaction by quenching unreacted formaldehyde with glycine at 50 mM final concentration.
5. Prepare a parallel non-methylated control sample under identical conditions except omitting formaldehyde.
6. Desalt methylated and control tau samples by chromatography in Desalting Buffer
7. Quantify tau concentration by absorbance at 214 nm, then aliquot and store at −20 °C until further use.

Note: If samples are to be subjected to LC–MS/MS analysis, replace Desalting Buffer with 50 mM ammonium bicarbonate ($(NH_4)HCO_3$ pH 7.8). For work at 50–75 μL scale, spin columns conveniently serve as the desalting media. Evaporate samples to dryness in a vacuum centrifuge and store at – 20 °C until further use.

3.4.3 [^{14}C]-formaldehyde incorporation and stoichiometry measurement

Timing: ~4–6 h.

1. For quantification of reductive methylation stoichiometry, substitute formaldehyde with 5 mM [^{14}C]-formaldehyde (final specific activity ~1 Ci/mol).

2. Perform reductive methylation (Section 3.4.2), maintaining identical incubation time and temperature as non-radioactive formaldehyde.
3. Remove unreacted [^{14}C]-formaldehyde by desalting chromatography.
4. Collect eluate fractions containing methylated tau and determine incorporated radioactivity by liquid scintillation spectroscopy.
5. Fit time-dependent methylation data to an exponential model: $y = y_{\max}(1 - e^{-kt})$, where k is the apparent rate constant and y_{max} is the maximal extent of methylation.

3.4.4 CD spectroscopy

Timing: ~4 h.

1. Prepare methylated and control tau samples by exchanging into far-UV transparent CD Buffer using desalting media.
2. Collect circular dichroism (CD) spectra at 20 °C using a 0.1 cm path-length quartz cuvette.
3. Record spectra from 187 to 245 nm with 0.5 nm step size and 1 nm bandwidth.
4. Average four consecutive scans, subtract buffer baseline, and apply no additional smoothing or filtering.
5. Convert raw ellipticity values (millidegrees) to mean residue molar ellipticity [θ]MRW for comparison between samples.
6. Evaluate any conformational changes resulting from methylation by comparing spectral profiles between methylated and control tau.

3.4.5 Effects of methylation on tau aggregation propensity

Because Lys methylation sites distribute throughout the tau molecule, including its microtubule binding repeat region (Fig. 1), it is positioned to modulate homotypic interactions critical for aggregation as well as heterotypic interactions with aggregation antagonists such as molecular chaperones. These relationships can be interrogated using in vitro aggregation assays employing purified full-length recombinant tau isoforms (Ayoub et al., 2023; Funk et al., 2014). Regardless of isoform selected for study, exogenous inducers are typically added to accelerate aggregation so that it proceeds within 1–2 days under near physiological conditions of pH, ionic strength and low micromolar tau concentration. Previously we reviewed considerations when choosing synthetic aggregation cofactor, and the advantages and disadvantages of different assays for quantifying tau aggregation (Nanavaty et al., 2017). Here we outline protocols for using aggregation assays to quantify effects of methylation on aggregation propensity.

Typical tau aggregation time series are sigmoidal, with distinct lag and plateau phases (Fig. 7A). Rising tau monomer concentrations tend to shorten the time taken for half of the available tau monomer to be converted into aggregates (half time, Fig. 7B) and raise the amounts of aggregates accumulating at plateau (Fig. 7C), reflecting increased aggregation rate and aggregate solubility, respectively. The former provides information on the key kinetic barriers to aggregation (including nucleation rate), whereas the latter reflects the minimal monomer concentration required to support aggregate nucleation. Because Lys methylation can modulate both aspects of aggregation (Funk et al., 2014), it is important to capture progress curves under conditions where plateau is proportional to input monomer concentration. The procedure below for doing so is compatible with data collected in arbitrary units without absolute quantification and is compatible with a broad range of cofactors and assay methods.

Timing: ~24–32 h.

1. Collect tau aggregation time series as a function of varying initial tau monomer concentrations (triplicate measurements). The resulting data will span lag and plateau phases of aggregation (Fig. 7A).
2. Quantify relative plateau aggregation.

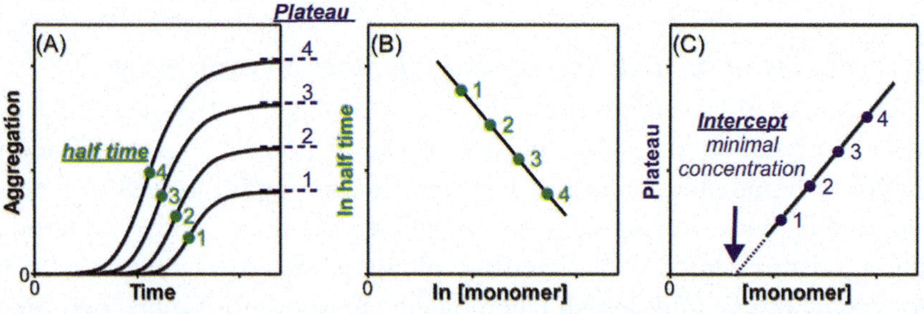

Fig. 7 Aggregation propensity analysis. (A), progress curves, where solid lines represent time-dependent appearance of aggregates at four input monomer concentrations (labeled 1 – 4). Each curve can be distinguished by a steady state (plateau) and a time to 50 % steady state (half time). (B) Replot of half time as a function of starting monomer concentration, where increasing monomer concentrations lead to faster aggregation (e.g., shorter half times). (C) Replot of plateau as a function of starting monomer concentration, showing greater amounts of aggregate at plateau. The abscissa intercept represents the minimal concentration of tau monomer needed to support aggregation.

3. Determine the minimal concentration by linear regression of these data as described previously (Huseby & Kuret, 2016).

   Note: Minimal concentration can be further disaggregated into its component association and dissociation rate constants as described previously (Funk et al., 2014; Huseby & Kuret, 2016).
4. Compare methylated- and non-methylated tau progress curves at identical supersaturation as described previously (Funk et al., 2014).

Note: Supersaturation is defined as the initial tau monomer concentration minus the minimal concentration estimated from step 3 above. Comparison of progress curves collected at equivalent supersaturations allows rigorous quantification of methylation-dependent effects on aggregation rate (Funk et al., 2014).

3.5 Expected outcomes

Tau purification and modification protocols should yield highly purified samples of non-methylated and lysine-methylated tau isoforms, with the latter calibrated for total methylation stoichiometry in units of mol methyl groups per mol tau protein. Because aggregation assays typically operate below 10 μM starting monomer concentration (<0.5 mg/mL, depending on tau isoform chosen for experimentation), final samples should be concentrated and aliquoted at ≥ 2 mg/mL concentration.

Aggregation assays should yield progress curves resembling those shown in Fig. 7A, with half times and plateau levels dependent on choice of cofactor and starting monomer concentrations. Replots of aggregation plateau as a function of starting tau monomer concentration yield estimates of the minimal tau concentration needed to support aggregation (Fig. 7C). Because minimal concentration approximates the equilibrium constant for aggregate extension, it provides a quantitative estimate of overall aggregate stability and therefore of tau supersaturation at any given starting monomer concentration. Progress curves collected at constant supersaturation allows direct comparison of native and methylated tau for differences in aggregation kinetics (Funk et al., 2014).

3.6 Quantification and statistical analysis

Minimal concentrations are estimated from replots of plateau aggregation as a function of starting monomer concentration (Fig. 7C). Because minimal concentration is captured from abscissa intercepts, inverse prediction methods are needed to estimate the statistical error. This can be done by applying Fieller's theorem after linear regression (Zhong et al., 2012). Pairwise comparison of non-methylated and methylated tau critical concentrations then can be assessed by z-test (Zhong et al., 2012).

Kinetic effects can be quantified from progress curves collected at constant supersaturation, defined as the difference between starting monomer and estimated minimal concentrations. Methylation dependent acceleration or deceleration of lag or half time can then be assessed by fitting with sigmoidal functions as described previously (Funk et al., 2014).

3.7 Advantages and limitations

This framework is compatible with data collected in arbitrary units without absolute quantification and so can be adapted to a broad range of cofactors and assay methods. However, it is limited to methylation effects on tau self-association. Because other PTMs have been reported to modulate heterotypic interactions with aggregation antagonists such as molecular chaperones (Moll et al., 2022), we outline a protocol that leverages ODS as cofactor and the Thioflavin T aggregation assay modality to capture this aspect of aggregation propensity (Ayoub et al., 2023; Jimenez-Harrison et al., 2023).

Timing: ~26 h.

1. Assemble reactions in a 96-well plate in the following order:
   - **a.** Assembly Buffer at 1x final concentration.
   - **b.** Tau samples (methylated- and non-methylated) at a final concentration of 2 μM.
   - **c.** ThT at 50 μM final concentration to enable monitoring of amyloid formation by fluorescence spectroscopy (excitation/emission: 440/485 nm).
   - **d.** Molecular chaperone of choice or a control protein at varying concentrations up to stoichiometric with tau monomer (i.e., up to 2 μM), corresponding to chaperone-to-tau molar ratios up to 1:1.

     Note: Because non-specific protein-protein interaction also can depress aggregation kinetics, it is important to include a non-specific protein control. Ovalbumin is a convenient reagent for this purpose (Ayoub et al., 2023; Jimenez-Harrison et al., 2023).
   - **e.** Initiate aggregation with addition of ODS cofactor to a final concentration of 30 μM (1:15 tau: ODS molar ratio).

     Note: Other colorless cofactors such as heparin are compatible with this assay.

Adjust the volume to 150 μL with water.

2. Seal the plate with an adhesive plate seal (Thermo Fisher Scientific, Cat. No. 232699) to prevent evaporation.
3. Incubate the plate at 37 °C for 24 h in a fluorescence plate reader.
4. Record ThT fluorescence continuously throughout the incubation period, agitating the plate at 100 rpm for 5 s prior to each measurement.
5. Quantify methylation-dependent effects on aggregation plateau as described previously (Ayoub et al., 2023; Jimenez-Harrison et al., 2023).

3.8 Optimization and troubleshooting

Aggregation plateaus tend to correlate linearly with starting monomer concentration within at least 2–3x of the minimal concentration. For this reason, appropriate initial tau monomer concentrations for yielding linearity should be identified empirically through initial screening before collecting a final data set. If linearity proves elusive, inverse prediction with Wald confidence intervals can be obtained after nonlinear fit using the investr R package (Greenwell & Kabban, 2014).

3.9 Alternative methods/procedures

Aggregation progress curves also can be interrogated using a kinetic modeling program. The online tool AmyloFit is specifically designed for analysis of nucleation-dependent kinetics as well as for characterizing rate-limiting secondary processes. Associated procedures have been described in detail (Meisl et al., 2016). As illustrated in Fig. 7, the approach can be combined with minimal concentration determination.

Methylation stoichiometry also can be estimated using LC/MS-MS methods (Cooper et al., 2018). This approach eliminates the need for radioactivity and so can be applied to brain-derived as well as synthetic samples.

4. Summary and conclusions

Like other tau PTMs, tau lysine methylation comprises multiple low-occupancy sites distributed throughout the tau molecule (Huseby et al., 2019). Our results demonstrate that endogenous lysine methylation patterns can be identified in human brain-derived specimens with high sensitivity and correlated with demographic variables including age and disease status (Funk et al., 2014). Moreover, recombinant expression and methylation workflows offer a controlled platform for generating well-defined tau species, enabling comparisons between structure and function that are not possible with brain-derived material alone. An example included here evaluated methylation-dependent effects on aggregation under defined conditions. Together these approaches support future efforts to map lysine methylation patterns in biological samples, characterize potential crosstalk with other PTMs, and identify effects on tau biological activity including aggregation propensity.

Acknowledgments

The authors thank members of the Interdisciplinary Research Network on Biologically Active Tau Aggregate Polymorphs from Alzheimer's Disease and Related Dementias for contributing to the development of protocols presented in this study. This work was supported by National Institutes of Health grants R01AG054018 and U24AG072458 (J.K.).

References

Alquezar, C., Arya, S., & Kao, A. W. (2020). Tau post-translational modifications: Dynamic transformers of tau function, degradation, and aggregation. *Frontiers in Neurology, 11*, 595532.

Ando, K., Maruko-Otake, A., Ohtake, Y., Hayashishita, M., Sekiya, M., & Iijima, K. M. (2016). Stabilization of microtubule-unbound tau via tau phosphorylation at Ser262/356 by Par-1/MARK contributes to augmentation of AD-related phosphorylation and Abeta42-induced tau toxicity. *PLoS Genetics, 12*(3), e1005917.

Ayoub, C. A., Wagner, C. S., & Kuret, J. (2023). Identification of gene networks mediating regional resistance to tauopathy in late-onset Alzheimer's disease. *PLoS Genetics, 19*(3), e1010681.

Balmik, A. A., & Chinnathambi, S. (2021). Methylation as a key regulator of Tau aggregation and neuronal health in Alzheimer's disease. *Cell Communication and Signaling, 19*(1), 51.

Barghorn, S., Biernat, J., & Mandelkow, E. (2005). Purification of recombinant tau protein and preparation of Alzheimer-paired helical filaments in vitro. *Methods in Molecular Biology, 299*, 35–51.

Best, R. L., Chung, P. J., Benbow, S. J., Savage, A., LaPointe, N. E., Safinya, C. R., et al. (2017). Expression and isolation of recombinant tau. *Methods in Cell Biology, 141*, 3–26.

Bichmann, M., Prat Oriol, N., Ercan-Herbst, E., Schondorf, D. C., Gomez Ramos, B., Schwarzler, V., et al. (2021). SETD7-mediated monomethylation is enriched on soluble Tau in Alzheimer's disease. *Molecular Neurodegeneration, 16*(1), 46.

Cohen, T. J., Guo, J. L., Hurtado, D. E., Kwong, L. K., Mills, I. P., Trojanowski, J. Q., et al. (2011). The acetylation of tau inhibits its function and promotes pathological tau aggregation. *Nature Communications, 2*, 252.

Cooper, G. L., Huseby, C. J., Chandler, C. N., Cocuron, J. C., Alonso, A. P., & Kuret, J. (2018). A liquid chromatography tandem mass spectroscopy approach for quantification of protein methylation stoichiometry. *Analytical Biochemistry, 545*, 72–77.

Ellis, M. J., Lekka, C., Holden, K. L., Tulmin, H., Seedat, F., O'Brien, D. P., et al. (2024). Identification of high-performing antibodies for the reliable detection of Tau proteoforms by Western blotting and immunohistochemistry. *Acta Neuropathologica, 147*(1), 87.

Funk, K. E., Thomas, S. N., Schafer, K. N., Cooper, G. L., Liao, Z., Clark, D. J., et al. (2014). Lysine methylation is an endogenous post-translational modification of tau

protein in human brain and a modulator of aggregation propensity. *Biochemical Journal, 462*(1), 77–88.
Goedert, M., & Jakes, R. (1990). Expression of separate isoforms of human tau protein: Correlation with the tau pattern in brain and effects on tubulin polymerization. *EMBO Journal, 9*(13), 4225–4230.
Greenwell, B. M., & Kabban, C. M. S. (2014). investr: An R package for inverse estimation. *R Journal, 6*(1), 90–100.
Guo, J. L., Narasimhan, S., Changolkar, L., He, Z., Stieber, A., Zhang, B., ... Lee, V. M. (2016). Unique pathological tau conformers from Alzheimer's brains transmit tau pathology in nontransgenic mice. *J Exp Med, 213*(12), 2635–2654.
HaileMariam, M., Eguez, R. V., Singh, H., Bekele, S., Ameni, G., Pieper, R., et al. (2018). S-Trap, an ultrafast sample-preparation approach for shotgun proteomics. *Journal of Proteome Research, 17*(9), 2917–2924.
Holzmüller, W., & Kulozik, U. (2016). Protein quantification by means of a stain-free SDS-PAGE technology without the need for analytical standards: Verification and validation of the method. *Journal of Food Composition and Analysis, 48*, 128–134.
Huseby, C. J., Hoffman, C. N., Cooper, G. L., Cocuron, J. C., Alonso, A. P., Thomas, S. N., et al. (2019). Quantification of tau protein lysine methylation in aging and Alzheimer's disease. *Journal of Alzheimers Disease, 71*(3), 979–991.
Huseby, C. J., & Kuret, J. (2016). Analyzing tau aggregation with electron microscopy. *Methods in Molecular Biology, 1345*, 101–112.
Inagami, T. (1965). The alkylation of the active site of trypsin with iodoacetamide in the presence of alkylguanidines. *Journal of Biological Chemistry, 240*(8), PC3453–PC3455.
Jimenez-Harrison, D., Huseby, C. J., Hoffman, C. N., Sher, S., Snyder, D., Seal, B., et al. (2023). DJ-1 molecular chaperone activity depresses tau aggregation propensity through interaction with monomers. *Biochemistry, 62*(5), 976–988.
Jung, S. Y., Li, Y., Wang, Y., Chen, Y., Zhao, Y., & Qin, J. (2008). Complications in the assignment of 14 and 28 Da mass shift detected by mass spectrometry as in vivo methylation from endogenous proteins. *Analytical Chemistry, 80*(5), 1721–1729.
Kamens, J. (2015). The Addgene repository: An international nonprofit plasmid and data resource. *Nucleic Acids Research, 43*, D1152–D1157 (Database issue).
Kevenaar, J. T., & Hoogenraad, C. C. (2015). The axonal cytoskeleton: From organization to function. *Frontiers in Molecular Neuroscience, 8*, 44.
KrishnaKumar, V. G., & Gupta, S. (2017). Simplified method to obtain enhanced expression of tau protein from E. coli and one-step purification by direct boiling. *Preparative Biochemistry and Biotechnology, 47*(5), 530–538.
Kuipers, B. J., & Gruppen, H. (2007). Prediction of molar extinction coefficients of proteins and peptides using UV absorption of the constituent amino acids at 214 nm to enable quantitative reverse phase high-performance liquid chromatography-mass spectrometry analysis. *Journal of agricultural & Food Chemistry, 55*(14), 5445–5451.
Kyalu Ngoie Zola, N., Balty, C., Pyr Dit Ruys, S., Vanparys, A. A. T., Huyghe, N. D. G., Herinckx, G., et al. (2023). Specific post-translational modifications of soluble tau protein distinguishes Alzheimer's disease and primary tauopathies. *Nature Communications, 14*(1), 3706.
LoPresti, P., Szuchet, S., Papasozomenos, S. C., Zinkowski, R. P., & Binder, L. I. (1995). Functional implications for the microtubule-associated protein tau: Localization in oligodendrocytes. *Proceedings of the National Academy of Sciences, 92*(22), 10369–10373.
Ludwig, K. R., Schroll, M. M., & Hummon, A. B. (2018). Comparison of in-solution, FASP, and S-trap based digestion methods for bottom-up proteomic studies. *Journal of Proteome Research, 17*(7), 2480–2490.
Mannu, A., & Mele, A. (2024). Fischer-speier esterification and beyond: Recent mechanicistic advances. *Catalysts, 14*(12), 16.

Meisl, G., Kirkegaard, J. B., Arosio, P., Michaels, T. C., Vendruscolo, M., Dobson, C. M., et al. (2016). Molecular mechanisms of protein aggregation from global fitting of kinetic models. *Nature Protocols, 11*(2), 252–272.

Moll, A., Ramirez, L. M., Ninov, M., Schwarz, J., Urlaub, H., & Zweckstetter, M. (2022). Hsp multichaperone complex buffers pathologically modified Tau. *Nature Communications, 13*(1), 3668.

Nanavaty, N., Lin, L., Hinckley, S. H., & Kuret, J. (2017). Detection and quantification methods for fibrillar products of in vitro tau aggregation assays. *Methods in Molecular Biology, 1523*, 101–111.

Pettis, J. A., Orshoski, M., Pal, S., Allen, A. M., Ortega Zepeda, M., Wysocki, W. H., et al. (2026). Efficient tag-less purification of recombinant human tau proteins. *Analytical Biochemistry* (in press.).

Shams, H., Matsunaga, A., Ma, Q., Mofrad, M. R. K., & Didonna, A. (2022). Methylation at a conserved lysine residue modulates tau assembly and cellular functions. *Molecular and Cellular Neuroscience, 120*, 103707.

Studier, F. W., Rosenberg, A. H., Dunn, J. J., & Dubendorff, J. W. (1990). Use of T7 RNA polymerase to direct expression of cloned genes. *Methods in Enzymology, 185*, 60–89.

Tan, H., Wu, Z., Wang, H., Bai, B., Li, Y., Wang, X., et al. (2015). Refined phosphopeptide enrichment by phosphate additive and the analysis of human brain phosphoproteome. *Proteomics, 15*(2-3), 500–507.

Thingholm, T. E., & Larsen, M. R. (2016). The use of titanium dioxide for selective enrichment of phosphorylated peptides. *Methods in Molecular Biology, 1355*, 135–146.

Thomas, S. N., Funk, K. E., Wan, Y., Liao, Z., Davies, P., Kuret, J., et al. (2012). Dual modification of Alzheimer's disease PHF-tau protein by lysine methylation and ubiquitylation: A mass spectrometry approach. *Acta Neuropathologica, 123*(1), 105–117.

Thomas, S. N., & Yang, A. J. (2017). Mass spectrometry analysis of lysine posttranslational modifications of tau protein from Alzheimer's disease brain. *Methods in Molecular Biology, 1523*, 161–177.

Wesseling, H., Mair, W., Kumar, M., Schlaffner, C. N., Tang, S., Beerepoot, P., et al. (2020). Tau PTM profiles identify patient heterogeneity and stages of Alzheimer's disease. *Cell, 183*(6), 1699–1713.

Xia, Y., Bell, B. M., & Giasson, B. I. (2023). Tau lysine pseudomethylation regulates microtubule binding and enhances prion-like tau aggregation. *International Journal of Molecular Sciences, 24*(9), 8286.

Zhong, Q., Congdon, E. E., Nagaraja, H. N., & Kuret, J. (2012). Tau isoform composition influences rate and extent of filament formation. *Journal of Biological Chemistry, 287*(24), 20711–20719.

Zhou, Y., Shi, J., Chu, D., Hu, W., Guan, Z., Gong, C. X., et al. (2018). Relevance of phosphorylation and truncation of tau to the etiopathogenesis of Alzheimer's disease. *Frontiers in Aging Neuroscience, 10*, 27.

CHAPTER TWO

Experimental approaches for investigating sirtuin-mediated mitochondrial function, cellular metabolism, and epigenetic regulation

Karthick Chennakesavan and Thangavel Samikkannu*
Department of Pharmaceutical Sciences, Irma Lerma Rangel College of Pharmacy, Texas A&M, University, College Station, TX, United States
*Corresponding author. e-mail address: thangavel@tamu.edu

Contents

Methods in Enzymology, Volume 734
ISSN 0076-6879, https://doi.org/10.1016/bs.mie.2026.06.023

Abstract

Cellular pathways for experimental discovery provide a comprehensive overview of sirtuin biology and its critical involvement in HIV-associated neurocognitive disorders (HAND) and related neurodegenerative diseases, highlighting the translational potential of sirtuin-targeted therapeutic strategies. As NAD^+-dependent deacetylases and ADP-ribosyl transferases, sirtuins regulate diverse cellular processes, including stem cell maintenance, cellular proliferation, metabolic homeostasis, apoptosis, autophagy, oxidative stress responses, and genomic stability, all of which contribute to neuronal dysfunction and disease progression. This chapter focuses on key mammalian sirtuins, including SIRT1 and SIRT2, which are primarily localized within the nucleus and cytosol; mitochondrial sirtuins SIRT3, SIRT4, and SIRT5; and nuclear/nucleolar sirtuins SIRT6 and SIRT7. Here, a method with a detailed protocol to isolate compartment-specific sirtuin expression and activity was used: subcellular fractionation was performed using a subcellular fractionation kit to obtain cytosolic and nuclear fractions, while mitochondrial isolation was carried out using Tom20 antibody-conjugated magnetic microbeads. These approaches were applied to brain tissues from HIV-positive individuals, as well as to HIV-Tat-treated human microglial (HMC3) cells and astrocytes. This experimental framework enables accurate assessment of compartment-resolved sirtuin regulation in disease-relevant models. Collectively, the chapter highlights the protective roles of sirtuins in mitigating key pathogenic mechanisms underlying HAND and related neurodegenerative diseases. These findings support the emerging concept that sirtuins represent promising pharmacological targets for the development of novel therapeutic interventions in neurodegeneration and HIV-associated brain disorders.

1. Introduction

Sirtuins (SIRT1–SIRT7) are NAD^+-dependent class III histone deacetylases that share a conserved catalytic core and NAD^+-binding domain but differ in their subcellular localization and functional roles (Wang et al., 2019). SIRT1 is predominantly localized in the nucleus and cytoplasm, whereas SIRT2 is mainly cytoplasmic but can translocate to the nucleus during the G2–M phase to regulate histone acetylation. SIRT3–SIRT5 are primarily mitochondrial proteins, while SIRT6 and SIRT7 shuttle between the nucleus and cytoplasm depending on cell-cycle dynamics and cellular context (Zou et al., 2018; Haorah et al., 2025). Through these compartment-specific localizations, SIRTs regulate diverse cellular processes, including oxidative stress responses, mitochondrial function, reactive oxygen species (ROS) production, epigenetic regulation, metabolic reprogramming, and inflammatory signaling (Haorah et al., 2025).

Functionally, SIRTs catalyze the NAD^+-dependent removal of acetyl and other acyl modifications from lysine residues on histone and non-histone proteins, thereby regulating transcription and metabolic pathways. Certain SIRT family members exhibit specialized enzymatic activities; for example, SIRT5 preferentially removes succinyl, glutaryl, and malonyl groups, whereas SIRT4 functions as an ADP-ribosyltransferase, and SIRT6 acts as a long-chain fatty-acid deacylase (Roessler et al., 2015; Yang et al., 2017; Park et al., 2013; Liu et al., 2017). Because SIRT activity is tightly linked to intracellular NAD^+ availability and the $NAD^+/NADH$ ratio, changes in cellular metabolism directly influence SIRT-mediated epigenetic regulation.

In the central nervous system, SIRTs regulate mitochondrial biogenesis, immune signaling, and inflammatory responses in glial cells, processes particularly relevant to HIV-associated neuroinflammation. HIV viral proteins, such as Tat, can induce mitochondrial dysfunction and sustained activation of microglia and astrocytes (Sivalingam et al., 2021a, 2021b; Doke et al., 2022, 2021).

This study investigates HIV-associated alterations in the epigenetic regulation of SIRT1–SIRT7 and their relationship to mitochondrial biogenesis in brain tissues from people with HIV (PWH) compared with age-matched uninfected controls. We further evaluated whether changes in the SIRT regulatory landscape correlate with key metabolic and mitochondrial regulators, including AMP-activated protein kinase (AMPK), the mitochondrial biogenesis master regulator PGC-1α, and mitochondrial transcription factors such as nuclear respiratory factors 1 and 2 (NRF-1/NRF-2), mitochondrial transcription factor A (TFAM), and oxidative phosphorylation (OXPHOS) components.

2. Rationale

Subcellular fractionation and mitochondrial isolation are critical for dissecting the cellular and molecular mechanisms underlying SIRT-mediated regulation of mitochondrial function and energy metabolism in the central nervous system. Sirtuins (SIRT1–SIRT7) regulate mitochondrial biogenesis and oxidative stress responses through the PGC-1α–NRF1/2–TFAM axis, which coordinates mitochondrial DNA replication, transcription, and oxidative phosphorylation. Accurate separation of nuclear, cytosolic, and mitochondrial fractions enables precise measurement of SIRT localization,

Table 1 Sirtuin family: location, function, and role in HIV-associated neuropathogenesis.

Sirtuin	Cellular location	Major biological function	Link to HIV / neuropathogenesis
SIRT1	Nucleus, cytoplasm	Histone deacetylation, regulation of transcription factors (p53, NF-κB, FOXO), inflammation control	HIV Tat protein inhibits SIRT1, leading to increased NF-κB activation, neuroinflammation, and neuronal damage in HIV-associated neurocognitive disorders (HAND) (Kwon et al., 2008)
SIRT2	Cytoplasm (also nucleus during mitosis)	Microtubule stability, cell cycle regulation, oligodendrocyte differentiation	Dysregulation may contribute to cytoskeletal instability and neuronal injury observed in HIV-induced neurodegeneration (Manjula et al., 2020)
SIRT3	Mitochondria	Regulates mitochondrial metabolism, antioxidant defense (SOD2), ROS detoxification	HIV infection and viral proteins increase mitochondrial ROS; reduced SIRT3 activity contributes to mitochondrial dysfunction and oxidative stress in neurons and glia (Thangaraj et al., 2021; Zhu et al., 2022).
SIRT4	Mitochondria	Controls metabolic enzymes, regulates fatty acid oxidation and ATP production	Altered mitochondrial metabolism during HIV infection may involve SIRT4-mediated metabolic imbalance in brain cells (Haorah et al., 2025)
SIRT5	Mitochondria	Desuccinylation, demalonylation, regulation of ammonia detoxification and metabolic enzymes	May influence mitochondrial metabolic stress and redox imbalance during HIV-induced neurotoxicity (Walter et al., 2022).

SIRT6	Nucleus	DNA repair, genomic stability, telomere maintenance, regulation of glucose metabolism	HIV infection increases DNA damage and inflammatory gene expression; SIRT6 deficiency may worsen genomic instability and neuroinflammation (Haorah et al., 2025)
SIRT7	Nucleolus	rRNA transcription, ribosomal biogenesis, cellular stress response	Potential involvement in cellular stress responses in HIV-infected neural cells, though less studied in NeuroHIV (Haorah et al., 2025)

enzymatic activity, and downstream effects on mitochondrial biogenesis and transcriptional regulation, while minimizing confounding signals arising from mixed cellular compartments.

In the context of HIV-associated neurocognitive disorders (HAND), viral proteins such as Tat, gp120, and Nef disrupt mitochondrial homeostasis and promote neuroinflammatory signaling in microglia and astrocytes. Fractionation-based approaches, therefore, allow assessment of cell-type- and compartment-specific alterations in SIRT expression, PGC-1α activation, and NRF1/2-mediated transcriptional programs, all of which collectively regulate mitochondrial integrity and bioenergetics. By enabling the isolation of mitochondria and other subcellular compartments, this methodology provides a mechanistic framework to link HIV-induced epigenetic dysregulation with mitochondrial dysfunction, oxidative stress, and neuroinflammation, thereby advancing our understanding of HIV neuropathogenesis and SIRT-dependent metabolic regulation (Table 1).

3. Methodology

3.1 Postmortem human brain tissues

Postmortem human brain tissues used in this study were obtained from the National NeuroAIDS Tissue Consortium (NNTC) and the National Neurological AIDS Bank (NNAB) at the University of California, Los Angeles (UCLA). These repositories provide well-characterized, high-quality biospecimens with detailed clinical histories and neuropathological evaluations, making them valuable resources for investigating the neuropathogenesis of HIV infection. All tissues were collected and distributed in accordance with institutional ethical guidelines and regulatory standards governing the use of human postmortem samples in biomedical research.

Brain samples were obtained from the frontal cortex, a region critically involved in higher-order cognitive functions such as executive control, working memory, and decision-making, and is frequently affected in HIV-associated neurocognitive disorders. The study cohort included brain tissues from HIV-positive individuals and HIV-seronegative control subjects. Clinical diagnosis of HIV infection and associated neurocognitive impairment was established according to criteria outlined in the Diagnostic and Statistical Manual of Mental Disorders, Fifth Edition (DSM-5). Within this framework, HIV-associated cognitive impairment is classified under neurocognitive disorders (NCDs), and only cases meeting these diagnostic criteria were included.

To minimize potential confounding factors that could independently affect brain pathology or mitochondrial function, strict exclusion criteria were applied. Individuals with a history of vascular disease, acute substance intoxication at the time of death, non–HIV related central nervous system infections, prion diseases, inflammatory neurological disorders, structural brain abnormalities, metabolic disorders, traumatic brain injury, delirium, or other genetic or systemic neurological conditions were excluded. This rigorous screening ensured that observed molecular alterations were primarily attributable to HIV infection rather than other neurological or systemic pathologies.

For stratification of HIV disease severity, samples were further categorized based on plasma viral load and $CD4^+$ T-cell counts, two widely used clinical indicators of immune status and disease progression. HIV-positive cases included in this study were characterized by viral load ≥ 200 copies/mL and $CD4^+$ counts ≤50 cells/μL, representing advanced immunosuppression. This classification enabled the evaluation of molecular changes associated with severe HIV disease in the central nervous system.

A total of ten postmortem brain samples were analyzed, including five HIV-seronegative controls and five HIV-positive individuals. From each sample, approximately 25 mg of frontal cortical tissue was carefully dissected under cold conditions to preserve protein integrity. The tissue was then subjected to subcellular fractionation to isolate mitochondrial, cytosolic, and nuclear compartments. Fractionation was performed using differential centrifugation protocols optimized to maintain the structural and functional integrity of subcellular organelles.

Following fractionation, protein expression analyses were performed to assess HIV-associated molecular alterations. Equal amounts of protein from mitochondrial, cytosolic, and nuclear fractions were analyzed by Western blotting to evaluate compartment-specific protein localization and expression. This approach enabled the assessment of mitochondrial dynamics, cellular signaling pathways, and nuclear regulatory mechanisms affected by HIV infection. Representatives' subcellular localization from these analyses are presented in Fig. 1.

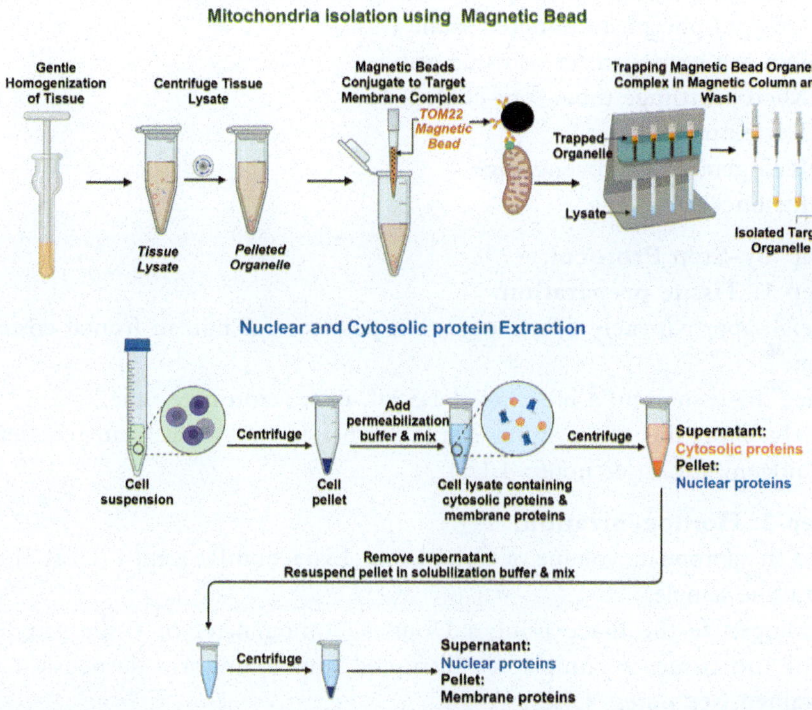

Fig. 1 Schematic workflow of mitochondrial isolation and nuclear/cytosolic protein extraction. Mitochondria are isolated using TOM22 magnetic beads and purified with a magnetic separation rack for downstream analyses. Nuclear and cytosolic proteins are sequentially extracted, with the supernatant containing the cytosolic protein fraction and the pellet subsequently processed to obtain the nuclear protein fraction.

4. Isolation of cytosolic, nuclear, and mitochondrial fractions from human brain tissues

4.1 Isolation of cytosolic and nuclear fractions from human brain tissues

Cytosolic and nuclear fractions from postmortem human brain tissues were isolated using the NE-PER Nuclear and Cytoplasmic Extraction Reagents (Catalog No. 78835) from Thermo Fisher Scientific, following the manufacturer's recommended protocol with minor modifications optimized for brain tissue samples (Haorah et al., 2025; Santos et al., 2023; Nabbi & Riabowol, 2015).

Materials required:

1. NE-PER Nuclear and Cytoplasmic Extraction Reagents
2. Ice-cold phosphate-buffered saline (PBS)
3. Dounce homogenizer or tissue grinder
4. Microcentrifuge tubes (pre-chilled)
5. Vortex mixer
6. Refrigerated microcentrifuge
7. Ice bucket

Step-by-Step Protocol.

Step 1: Tissue preparation.

Weigh approximately 25 mg of frozen postmortem human frontal cortex tissue.

Place the tissue into a pre-chilled 1.5 mL microcentrifuge tube.

Finely chop the tissue using sterile scalpel blades to facilitate efficient homogenization (see notes: 4.1.1).

Step 2: Homogenization.

Add an appropriate volume of Cytoplasmic Extraction Reagent I (CER I) to the tissue sample.

Homogenize the tissue using a Dounce homogenizer or tissue grinder until the tissue is completely disrupted and a uniform suspension is obtained (see notes: 4.1.2).

Step 3: Cytoplasmic extraction.

Vortex the homogenate vigorously for 15 s at the highest setting to fully suspend the cellular material.

Incubate the tube on ice for 10 min to allow cytoplasmic proteins to solubilize.

Step 4: Addition of CER II.
Add ice-cold Cytoplasmic Extraction Reagent II (CER II) to the homogenate. Vortex immediately for 5 s at the highest setting.
Incubate the mixture on ice for 1 min.

Step 5: Cytoplasmic fraction separation.
Centrifuge the sample at ~16,000 g for 10 min at 4 °C.
Carefully transfer the supernatant (cytoplasmic extract) to a new pre-chilled microcentrifuge tube without disturbing the pellet.
Keep the cytoplasmic extract on ice for immediate use or store at −80 °C (see notes: 4.1.3).

Step 6: Nuclear extraction.
The remaining insoluble pellet, containing nuclei and nuclear components, is retained for nuclear protein extraction.
Resuspend the pellet in ice-cold Nuclear Extraction Reagent (NER).

Step 7: Nuclear protein solubilization.
Vortex the suspension for 15 s at the highest setting.
Incubate the sample on ice for a total of 30 min, vortexing every 5 min for 15 s to enhance nuclear protein extraction.

Step 8: Nuclear fraction separation.
Centrifuge the sample at ~16,000 g for 10 min at 4 °C.
Carefully collect the supernatant containing the nuclear extract and transfer it to a fresh pre-chilled tube.
Remaining pellet stored at −80 °C for chromatin extraction (see notes: 4.1.4).

Step 9: Storage.
Both cytoplasmic and nuclear protein extracts were immediately stored at −80 °C until further analysis (see notes: 4.1.5).

Note:

4.1.1 Maintaining samples at low temperature throughout the procedure minimizes proteolytic degradation and preserves the native structure and function of proteins.

4.1.2 Complete homogenization is critical for efficient disruption of cellular membranes while maintaining nuclear integrity, thereby improving the purity of cytoplasmic and nuclear fractions.

4.1.3 CER I disrupts the cell membrane and releases cytoplasmic proteins while keeping the nuclear membrane intact. CER II improves the extraction of cytoplasmic proteins and helps achieve better separation between cytoplasmic and nuclear fractions. The supernatant should be carefully collected to prevent contamination from nuclear material and ensure the purity of the cytoplasmic fraction.

4.1.4 NER contains reagents that disrupt the nuclear envelope and solubilize nuclear proteins while maintaining compatibility with downstream applications such as immunoblotting and proteomic analyses. The residual pellet is enriched in chromatin-bound proteins, DNA, and other insoluble nuclear structures, making it suitable for chromatin fractionation and epigenetic analyses.

4.1.5 Repeated freeze–thaw cycles should be avoided because they can result in protein degradation and loss of biological activity. Aliquoting extracts before storage is recommended to preserve sample integrity.

4.2 Total mitochondrial isolation from human brain tissues

Mitochondria were isolated from postmortem human brain tissues using the Miltenyi Biotec Mitochondria Isolation Kit (Catalog No. 130–096−946) based on MACS Technology, according to the manufacturer's recommended protocol with minor modifications (Gross et al., 2011; Franko et al., 2013; Kandigian et al., 2022).

Materials required.

1. Miltenyi Biotec Mitochondria Isolation Kit
2. Anti-TOM22 MicroBeads
3. LS separation columns
4. 70 μm cell strainers
5. Dounce homogenizer
6. Refrigerated centrifuge
7. Storage buffer provided in the kit

Step-by-Step Protocol.

Step 1: Tissue homogenization.

- Equal weights of postmortem brain tissues from people with HIV (PWH) and uninfected controls were collected.
- Tissue samples were homogenized in ice-cold isolation buffer using a Dounce homogenizer to disrupt cells while preserving mitochondrial integrity.

Step 2: Filtration.

- The homogenized tissue lysates were filtered through a 70 μm separation filter to remove debris and large tissue fragments.

- The filtrate containing cellular components was collected and maintained on ice.
- Note: It is strongly recommended that the integrity of the mitochondria is examined after the lysis procedure, especially if the mitochondria are to be used in functional assays, e.g., by measuring citrate synthase activity.

Step 3: Magnetic labeling of mitochondria.

- The filtered lysates were incubated with 50 μL anti-TOM22 MicroBeads, which specifically bind to the mitochondria.
- Mix well and incubate for 1 h in the refrigerator (2–8°C) with gentle shaking, e.g., using the MACSmix Tube Rotator.

Step 4: Magnetic separation with LS columns.

- Place column in the magnetic field of a suitable MACS Separator (MidiMACS Separation Unit or QuadroMACS Separation Unit).
- Prepare column by rinsing with 3 mL of 1 × Separation Buffer.
- Place a Pre-Separation Filter (30 μm) on top of the LS Column
- After labeling incubation transfer the labelled lysates onto LS magnetic separation columns placed in a magnetic field and let the lysate run through.
- Wash filter and column with 3 × 3 mL of 1 × Separation Buffer.
- Note: Perform washing steps by adding buffer aliquots only when the column reservoir is empty.
- Remove column from the separator and place it on a suitable collection tube (15 mL conical tube).
- Pipette 1.5 mL of 1 × Separation Buffer onto the column.
- Immediately flush out the magnetically labeled mitochondria by firmly pushing the plunger into the column. Immediately proceed with downstream analysis
- Note: The 1 × Separation Buffer contains stabilizing agent. Before measuring mitochondria protein content, please carry out the following procedure, according to manufacturer's instruction, Miltenyi Biotec (#130-094-532):
  1. Centrifuge the isolated mitochondria at 13,000 g for 2 min at 4 °C. Aspirate the supernatant, it contains residual cytosolic protein.
  2. Resuspend the mitochondria pellet in 1000 μL of Storage Buffer.
  3. Centrifuge mitochondria suspension at 13,000 g for 2 min at 4 °C. Aspirate the supernatant.
  4. Resuspend the mitochondria pellet in 100 μL Storage Buffer.

5. *In vitro* study

5.1 Human astrocytes

Primary human astrocytes derived from the cerebral cortex were obtained from ScienCell Research Laboratories (Catalog No. 1800, Carlsbad, CA, USA). These cells provide a physiologically relevant in vitro model to investigate astrocyte-mediated mechanisms associated with neuroinflammation and mitochondrial dysfunction during HIV infection.

Astrocytes were cultured in astrocyte growth medium (Catalog No. 1801) supplied by the same manufacturer. The culture medium was supplemented with:

a) 5 % fetal bovine serum (FBS)
b) 1 % antibiotic–antimycotic solution

Cells were maintained in tissue culture flasks under standard incubator conditions at 37 °C in a humidified atmosphere containing 5 % CO_2. The culture medium was replaced every 2–3 days, and cells were passaged once they reached approximately 80–90% confluency. All experiments were conducted using cells within early passages to maintain cellular integrity and physiological responses.

5.2 Human microglia

Human microglial cells (HMC3) were obtained from the American Type Culture Collection (ATCC; Catalog No. CRL-3304, USA). HMC3 cells represent a well-characterized immortalized human microglial cell line widely used for studying neuroimmune responses in HIV-associated neurocognitive disorders.

Microglia were cultured in Eagle's Minimum Essential Medium (EMEM) supplemented with:

a) 10 % fetal bovine serum
b) 1 % Penicillin–Streptomycin (P/S) solution

Cells were maintained in a humidified incubator at 37 °C with 5 % CO_2. The culture medium was replaced every 2–3 days, and cells were sub-cultured upon reaching 80–90 % confluency to maintain optimal growth conditions.

6. HIV-Tat recombinant protein

HIV-1 Trans activator protein (Tat) is a small regulatory protein (~14–16 kDa) that plays a critical role in efficient HIV-1 replication and viral gene expression. It binds to the transactivation response (TAR) RNA stem–loop structure located at the 5′ end of nascent viral transcripts, thereby recruiting and activating the transcriptional machinery at the viral promoter and markedly enhancing transcriptional elongation. Through this mechanism, Tat significantly increases the efficiency of HIV-1 transcription. In addition to regulating viral gene expression, Tat also modulates host cellular genes and signaling pathways, creating a cellular environment favorable for viral replication and spread. Notably, Tat is expressed very early during infection, even before viral integration, and evidence suggests that it can be incorporated into HIV-1 virions to facilitate early reverse transcription and activation of viral gene expression (Thangaraj et al., 2021; Gandhi et al., 2009; Thangavel et al., 2018; Clark et al., 2024; Thangavel et al., 2009). Because of its central role in HIV transcription and replication, Tat represents an important target for therapeutic strategies aimed at limiting viral propagation (Cafaro & Ensoli, 2022; Cafaro et al., 2020; Ling et al., 2023) (see notes: 6.1).

Note:

6.1 The ability of Tat to cross cellular membranes and alter host gene expression makes it a widely used experimental tool for modeling HIV-induced cellular dysfunction in vitro and in vivo

The recombinant HIV-1 Tat protein used in this study was obtained through the NIH AIDS Research and Reference Reagent Program (Catalog No #2222, see notes: 6.2). The protein preparation was recombinant and purified to a purity level exceeding 95 %, ensuring minimal contamination by other viral or host proteins. The purified Tat protein was supplied in lyophilized form and was reconstituted in sterile molecular-grade water according to the supplier's instructions. Aliquots were prepared to avoid repeated freeze–thaw cycles and stored at −80 °C until use in vitro experiments. For cellular treatments, Tat protein was diluted in culture medium to the concentration of 50 ng/mL immediately before use (see notes: 6.3).

Note:

6.2 Aliquoting recombinant proteins prior to storage minimizes repeated freeze-thaw cycles, which can lead to protein degradation, aggregation, and reduced biological activity.

6.3 A concentration of 50 ng/mL is commonly employed in neuroHIV studies because it induces cellular and molecular responses associated with HIV neuropathogenesis, including oxidative stress, inflammatory signaling, mitochondrial dysfunction, and neuronal injury, while maintaining adequate cell viability under most experimental conditions.

7. Immunocytochemistry and microscopic imaging protocol

7.1 Immunocytochemistry for sirtuin (SIRT) expression in human microglia and astrocytes

Materials.

1. Human microglia and astrocytes
2. 8-well glass chamber slides
3. HIV-1 Tat protein (50 ng/mL)
4. Phosphate-buffered saline (PBS)
5. 4 % Paraformaldehyde (PFA)
6. Triton X-100
7. Bovine serum albumin (BSA)
8. Primary antibodies against SIRT proteins
9. Alexa Fluor® 488 or Alexa Fluor® 555 secondary antibodies
10. DAPI (4′,6′-diamidino-2-phenylindole)
11. Fluorescence microscope (Keyence or equivalent)

Procedure.

1. **Cell Seeding:** Seed human microglia (1×10^4 cells) and astrocytes (3×10^4 cells) onto sterile 8-well glass chamber slides and incubate under standard culture conditions (37 °C, 5 % CO_2).
2. **Cell Growth:** Allow cells to grow until they reach approximately 70–80 % confluency.
3. **Treatment:** Treat cells with HIV-1 Tat protein (50 ng/mL) in 2% serum free media and incubate for 24 h.
4. **Cell Fixation:** Remove culture media and gently wash cells with PBS (2–3 times). Fix cells with 4 % paraformaldehyde for 20 min at room temperature.
5. **Permeabilization:** Wash cells with PBS and permeabilize with 0.01 % Triton X-100 in PBS for 15 min at room temperature.
6. **Blocking:** Incubate cells with 2 % BSA in PBS for 1 h at room temperature to block nonspecific antibody binding.

7. **Primary Antibody Incubation:** Incubate cells with primary antibodies against SIRT proteins (1:500 dilution) overnight at 4 °C.
8. **Washing:** Wash cells 3 times with PBS (5 min each wash) to remove unbound primary antibodies.
9. **Secondary Antibody Incubation:** Incubate cells with Alexa Fluor® 488 or Alexa Fluor® 555 conjugated secondary antibodies (1:2000 dilution) for 1 h at room temperature in the dark.
10. **Nuclear Staining:** Counterstain cell nuclei with DAPI (0.2 μg/mL) for 5 min
11. **Final Wash:** Wash cells 2–3 times with PBS to remove excess stain.
12. **Imaging:** Capture fluorescent images using a Keyence fluorescence microscope or equivalent imaging system.

7.2 Mitochondrial membrane potential analysis using JC-1 dye

JC-1 is a membrane-permeable fluorescent dye commonly used to assess mitochondrial membrane potential in apoptosis studies. The dye accumulates in mitochondria in a potential-dependent manner: at low membrane potential it remains in a monomeric form, emitting green fluorescence (~529 nm), whereas at higher membrane potential it forms red-fluorescent J-aggregates within the mitochondria. Mitochondrial depolarization is therefore indicated by a decrease in the red-to-green fluorescence intensity ratio. Because this ratio depends primarily on membrane potential rather than mitochondrial size, shape, or density, JC-1 fluorescence provides a reliable method for comparing mitochondrial health and quantifying mitochondrial responses to cellular stimuli (Sivandzade et al., 2019).

Materials.

1. Live cell cultures (microglia or astrocytes)
2. Opti-MEM medium
3. JC-1 dye (5 μM; Invitrogen)
4. Fluorescence microscope (Keyence or equivalent)

Procedure.

Preparation and setup.

Note: As a first step, allow the JC-1 powder and DMSO solutions to come to 25 °C before use.

a. Prepare a fresh 200 μM JC-1 dye stock solution immediately prior to use by reconstituting the lyophilized JC-1 dye with DMSO to obtain a 100x stock solution.

b. Mix the preparation until the solution is clear of aggregates and the dye powder is completely dissolved.

Staining protocol for adherent cells.

The following protocol describes the process of JC-1 staining in cultured adherent cells and following analysis using either fluorescence microscopy.

a. Cells can be cultured on pre-coated 8-well glass chamber slides cover-slips. Make sure not to exceed a max cell density of ~1×10^4 cells/cm^2.
b. Treat cells with HIV-1 Tat protein (50 ng/mL) in 2 % serum free media and incubate for 24 h.
c. Remove the cell culture medium and then replace it with a warm medium (~37 °C), PBS, or Opti-MEM media (see notes: 7.2.1).
d. Add 5 μM (final concentration) of JC-1 dye and incubate the cells at 37 °C, 5 % CO2 for 30–60 min (see notes: 7.2.2)
e. Remove the culture medium of all samples and wash the cells once by adding warm PBS or Opti-MEM media (~37 °C).
f. Immediately capture fluorescence images using a Keyence fluorescence microscope.

Green fluorescence (~529 nm) indicates JC-1 monomers and reflects low mitochondrial membrane potential. Red fluorescence (J-aggregates) indicates high mitochondrial membrane potential (see notes: 7.2.3).

Data Analysis: Determine mitochondrial depolarization by calculating the ratio of red to green fluorescence intensity. A decrease in this ratio indicates mitochondrial membrane potential loss.

Note:

7.2.1 Wash cells once by adding 2 mL of warm phosphate-buffered saline (PBS) or other buffer to each tube of cells.
7.2.2 Protect the reagent from light at all times.
7.2.3 A decrease in the red/green fluorescence ratio indicates mitochondrial depolarization, which is commonly associated with oxidative stress, mitochondrial dysfunction, apoptosis, and cellular injury.

8. Subcellular fractionation from cultured cells

Isolation of Cytosolic, Nuclear, and Mitochondrial Fractions.

Subcellular fractions from cultured cells were isolated using the Subcellular Protein Fractionation Kit (Catalog No. ab65320) supplied by Abcam, following the manufacturer's recommended protocol (Nabbi & Riabowol, 2015; Kandigian et al., 2022; Huynh et al., 2024).

Step-by-Step Procedure.
Cell harvesting.

- Cultured astrocytes or microglia were first washed once with ice-cold phosphate-buffered saline (PBS) to remove residual culture medium.
- Cells were then detached using trypsin and transferred into centrifuge tubes.

Cell pellet preparation.

- The collected cells were centrifuged at 300–400 g for 5 min at 4 °C.
- The supernatant was discarded, leaving a compact cell pellet.

Cytosolic extraction.

- The cell pellet was resuspended in cytosolic extraction buffer (CEB) provided in the kit.
- Approximately 200 μL of lysis buffer per 10^6 cells was used.
- The cell suspension was gently homogenized to disrupt the plasma membrane while preserving organelle integrity.
- The homogenized sample was incubated on ice for 10 min, with gentle mixing every few minutes.

Separation of nuclear fraction.

- The homogenate was transferred into a 1.5 mL microcentrifuge tube.
- The sample was centrifuged at 1000 *g* for 10 min at 4 °C.
- The supernatant, containing cytosolic and mitochondrial components, was carefully collected.
- The pellet, containing nuclear material, was retained.

Nuclear protein extraction.

- The nuclear pellet was resuspended in nuclear extraction reagent (NER).
- The suspension was incubated at 4 °C for 30 min with intermittent gentle mixing.
- The sample was centrifuged at 500 *g* for 5 min.
- The supernatant containing nuclear proteins was transferred to a fresh microcentrifuge tube.

Mitochondrial isolation.

- The previously collected cytosolic supernatant was centrifuged at 10,000 *g* for 30 min at 4 °C. The supernatant contained cytosolic proteins and the pellet remains mitochondria.

Collection of fractions

- The cytosolic fraction was transferred to a fresh tube and stored at −80 °C.
- The mitochondrial pellet was resuspended in 100 μL of mitochondrial extraction buffer containing DTT and protease inhibitors.
- The sample was vortexed briefly for 10 s and stored at −80 °C for downstream analyses.

9. Protein determination

The Bicinchoninic Acid (BCA) protein assay is a colorimetric method widely used to determine protein concentration in biological samples. The assay is based on the reduction of Cu^{2+} to Cu^{1+} by proteins in an alkaline environment (biuret reaction). The reduced copper ion then forms a purple-colored complex with bicinchoninic acid, which can be quantified by measuring absorbance at 562 nm. The intensity of the color produced is proportional to the protein concentration in the sample. The protocol is followed by the manufacturer's instructions, Thermo Scientific (Pierce BCA protein assay kit# 23227).

Materials:

BCA working reagent (W*R*):

- BCA reagent A: 50 mg/mL of bicinchoninic acid in 2 % sodium carbonate solution
- BCA reagent B: 4 % cupric sulfate solution in 0.4 N sodium hydroxide
- Mix 50 parts of BCA reagent A with 1 part of BCA reagent B. Mix well and incubate at room temperature for 30 min before use.

BSA standard.

a) Prepare diluted albumin (BSA) standards in the working range from 20 to 2000 μg/mL, as per manufacturer's instructions.
b) Protein sample
c) 96-well microplate
d) Microplate reader

Procedure:

1. Pipette 25 μL of each standard or unknown sample replicate into a microplate well (working range = 20–2000 μg/mL).
   - Note: If sample size is limited, 10 μL of each unknown sample and standard can be used (sample to W*R* ratio = 1:20), however, the working range is limited to 125–2000 μg/mL.

2. Add 200 μL of the *WR* to each well and mix plate thoroughly on a plate shaker for 30 s
3. Cover the plate and incubate at 37 °C for 30 min.
4. Equilibrate the plate to room temperature. Measure the absorbance at or near 562 nm on a plate reader.
   - Note: Wavelengths from 540 to 590 nm has been used successfully with this method.
5. Plate readers use a shorter light path length than cuvette spectrophotometers, hence the microplate procedure requires a greater sample to *WR* ratio to obtain the same sensitivity as the standard test tube procedure. If higher 562 nm measurements are desired, increase the incubation time to 2 h.
6. Increasing the incubation time or ratio of sample volume to *WR* increases the net 562 nm measurement for each well and lowers both the minimum detection level of the reagent and the working range of the assay. As long as all standards and unknowns are treated identically, such modifications are useful.
7. Subtract the average 562 nm absorbance measurement of the blank standard replicates from the 562 nm measurements of all other individual standard and unknown sample replicates.
8. Prepare a standard curve by plotting the average blank–corrected 562 nm measurement for each BSA standard vs. its concentration in μg/mL.
9. Use the standard curve to determine the protein concentration of each unknown sample.
   - Note: Tips for a successful Bicinchoninic Acid (BCA) Protein Assay:
   - Ensure that the BCA reagents are mixed well before use.
   - Use a blank well with no protein sample to calibrate the microplate reader.

- Use BSA as a standard protein because it has a well-defined structure and is stable in solution.
- Use distilled water to prepare all solutions and dilute the protein sample.
- Avoid high concentrations of detergents or reducing agents in the protein sample, as these can interfere with the assay.

10. Western blot analysis

Protein expression analysis was performed using Western blotting to evaluate mitochondrial, cytosolic and nuclear proteins, and determine the volume of protein extract to ensure 20 μg in each well. Add 4 μL (4X) Laemmli buffer to the sample and mix well, kept at 95 °C for 5 min and bring to room temperature. (Tip: Total volume of 15 μL per lane is suggested).

Loading and running the gel.

The 4–20 % gel (Biorad, #4561096) is immersed in 1X Tris/Glycine/SDS buffer, the protein samples are loaded, and an electrical current is applied to the gel, which causes proteins to migrate from one end of the gel (negative electrode) to the other (positive electrode). Proteins are separated by size; smaller proteins travel more quickly through the gel, so appear further down.

To confirm the size of each protein in your sample, they are run alongside molecular weight ladders (Table 2).

Run the gel according to the manufacturer's instructions.

- Optimize running times and voltages according to the machine you're using and the target protein. Ideal running times and voltages can vary according to the manufacturer, the gel composition, and the protein of interest. Larger proteins should be run at a higher voltage for a longer time.

  Remove the gel from the running apparatus when ready to transfer.
- Use a gel knife to carefully pry open the apparatus and remove the gel.

Transferring from the gel to the membrane.

After performing electrophoresis, proteins are then transferred (or 'blotted') onto a membrane, ready for antibody incubation. This membrane can be made of PVDF. As with electrophoresis, transfer to the membrane is achieved by applying an electrical charge, which causes the proteins to migrate. The proteins travel away from the gel near the negative electrode and towards the positive electrode, where they bind to the membrane.

Table 2 Recommended gradient gel chemistries for different protein sizes. Our lab use gradient gels, but gels with fixed acrylamide concentration can also be used.

Protein size	Recommended gel and buffer system
10–30 kDa	Any-kDa acrylamide gradient Bis-Tris gel Tris/Glycine/SDS running buffer
31–150 kDa	4–20 % acrylamide gradient Bis-Tris gel Tris/Glycine/SDS running buffer

Semi-dry transfer.

Semi-dry transfer requires additional equipment but has a much shorter transfer time with easier setup.

Materials required.

- Protein separated SDS-PAGE gel
- Transfer apparatus (Trans-Blot Turbo Transfer System, Biorad #1704150)
- Prepare ice cold Transfer Buffer
- PVDF membrane (Biorad, #1620174)
- Ethanol (if using PVDF)

Procedure:

- Soak the membrane in ethanol, if using a PVDF membrane. (This step is essential for PVDF membranes but not needed for nitrocellulose membranes. As PVDF is naturally hydrophobic, it requires activation with ethanol to allow the buffer to pass through effectively).
- Soak the membrane in ice cold transfer buffer for 10 min at room temperature. Assemble the SDS-PAGE gel and the membrane in the transfer cassette.
- The gel should be closest to the negative electrode.
- The membrane should be closest to the positive electrode.
- Press down on the stack with a small roller to remove any bubbles.
- For detailed guidance on assembling the transfer apparatus, refer to the manufacturer's instructions (Biorad, 1704150).
- Run the transfer according to the manufacturer's instructions.

Blocking and antibody incubation.

Use the procedures below for antibody incubations. If using loading control antibodies in chemiluminescent western blot, the procedure below can be repeated on the same membrane after stripping.

Materials required.

- Every blot- Blocking buffer (Biorad, 12010020)
- Wash buffer (example TBST)
- Transferred PVDF membrane
- Conjugated primary antibody

Steps.

- Incubate the protein transferred PVDF membrane into Every blot-Blocking buffer for 10 min at room temperature with gentle rocking.

- Dilute the primary antibodies listed in the table below with Every blot -Blocking buffer for overnight at 4 °C with gentle rocking.
- After primary antibody incubation (Table 3), wash the membrane three times with 1 × TBST (Tris-buffered saline with Tween-20) for 5 min per wash with gentle agitation.
- Note: If high background signal is observed with certain antibodies, the duration of the wash steps may be increased; however, in most cases three 5-min washes provide an optimal signal-to-noise ratio. When preparing TBST, verify the composition of the TBS buffer, as formulations may vary between manufacturers.
- In particular, ensure that the NaCl concentration is approximately 150 mM, since higher salt concentrations (e.g., 500 mM NaCl) may interfere with Western blot performance.
- Following the wash steps, incubate the membrane with the appropriate secondary antibody diluted in blocking buffer for 1 h at room temperature with gentle agitation. For chemiluminescent detection, the secondary antibody should be horseradish peroxidase (HRP)-conjugated. After incubation, wash the membrane again three times with TBST for 5 min each to remove excess secondary antibody.
- For signal detection, prepare the Clarity Max Western ECL substrate according to the manufacturer's instructions (Bio-Rad, #1705062) by mixing solutions A and B in the recommended proportions. Apply sufficient ECL reagent to fully cover the membrane and incubate for 2–5 min. A 1000 μL pipette may be used to distribute the reagent evenly across the membrane surface. The chemiluminescent signal can then be detected using an Azure Biosystems c300 imaging system, typically with the Chemi Hi Sensitivity setting and an initial 10 s exposure time. Images are subsequently analyzed using ImageJ software (National Institutes of Health, Bethesda, MD, USA).
- Note: If the signal intensity is weak, the membrane can be incubated with the ECL substrate for slightly longer; however, prolonged incubation beyond 5 min may reduce the duration of peak signal intensity. Therefore, it is recommended to remove excess ECL reagent and limit incubation time to ≤ 5 min, although optimal conditions may vary depending on the specific ECL reagent used.
- Protein expression levels (band intensities) are normalized to loading controls such as Histone H3, β-actin, or VDAC, and results are expressed as fold change relative to the control group.

Table 3 List of antibodies.

Antibody	Company	Cat #	Molecular weight (kDa)	Dilution for WB/ICC
SIRT1	Protein Tech	13,161-1-AP	110 kDa	1:1000/1:500
SIRT2	Protein Tech	19,655-1-AP	40 kDa	1:1000
SIRT3	Protein Tech	1000-99-1-AP	45 kDa	1:1000/1:500
SIRT4	Protein Tech	66,543-1-AP	35 kDa	1:1000/1:500
SIRT5	Protein Tech	15,122-I-AP	32 kDa	1:1000
SIRT6	Protein Tech	13,572-1-AP	37 kDa	1:1000
SIRT7	Protein Tech	12,994-1-AP	45 kDa	1:1000/1:500
Phospho-AMPK (Thr183/172)	Elabscience	E-AB-21121	63 kDa	1:500
AMPK-a	Elabscience	E/AB/30490	63 kDa	1:1000
NRF1	Protein Tech	2482-I-AP	70 kDa	1:1000
NRF2	Protein Tech	2482-I-AP	110 kDa	1:1000
PGC-1a	Abcam	ab72230	91 kDa	1:1000/1:500
OXPHOS	Abcam	ab110413	20-, 30-, 40-, 48-, 55 kDa	1:1000
VDAC	Protein Tech	10,866–1-AP	30 kDa	1:1000
β-actin	Protein Tech	60,008-I-AP	42 kDa	1:2500
Histone H3	Protein Tech	17,168-I-AP	15−17 kDa	1:2000
Lamin B	Abclonal	A1910	60–66 kDa	1:2000
NF-kB P65	Protein Tech	10,745–1-AP	65 kDa	1:1000

11. Concluding remarks

Compartment-specific analysis of sirtuins provides key mechanistic insight into HIV-associated neurocognitive disorders (HAND) and related neurodegenerative diseases, as their functions are distinctly regulated within the nucleus, cytosol, and mitochondria, making subcellular fractionation essential for accurate spatial resolution. Brain tissues from HIV-positive individuals, along with HIV-Tat-treated HMC3 cells and astrocytes, provide a relevant translational model to study disease-associated alterations in sirtuin signaling. Using validated fractionation approaches consistent with Methods in Enzymology, reliable assessment of compartment-resolved sirtuin expression and activity can be achieved. Collectively, these analyses link sirtuin dysregulation to mitochondrial dysfunction, oxidative stress, neuroinflammation, and epigenetic instability, all of which are central to HAND pathology. Importantly, these findings reinforce the emerging view that sirtuins play protective roles in maintaining cellular homeostasis. Accordingly, sirtuins represent promising therapeutic targets for the development of precision interventions in HIV-associated and neurodegenerative disorders.

Acknowledgments

The National Institutes of Health (NIH)/National Institute on Drug Abuse (NIDA) supported this work under grant DA044872.

References

Cafaro, A., & Ensoli, B. (2022). HIV-1 therapeutic vaccines in clinical development to intensify or replace antiretroviral therapy: The promising results of the Tat vaccine. *Expert Review of Vaccines, 21*(9), 1243–1253.

Cafaro, A., et al. (2020). HIV-1 Tat protein enters dysfunctional endothelial cells via integrins and renders them permissive to virus replication. *International Journal of Molecular Sciences, 22*(1).

Clark, A., et al. (2024). Inflammatory biomarker reduction with fostemsavir over 96 weeks in heavily treatment-experienced adults with multidrug-resistant HIV-1 in the BRIGHTE study. *Open Forum Infectious Diseases, 11*(9), ofae469.

Doke, M., et al. (2022). HIV-1 Tat and cocaine coexposure impacts piRNAs to affect astrocyte energy metabolism. *Epigenomics, 14*(5), 261–278.

Doke, M., Pendyala, G., & Samikkannu, T. (2021). Psychostimulants and opioids differentially influence the epigenetic modification of histone acetyltransferase and histone deacetylase in astrocytes. *PLoS One, 16*(6), e0252895.

Franko, A., et al. (2013). Efficient isolation of pure and functional mitochondria from mouse tissues using automated tissue disruption and enrichment with anti-TOM22 magnetic beads. *PLoS One, 8*(12), e82392.

Gandhi, R. T., et al. (2009). A randomized therapeutic vaccine trial of canarypox-HIV-pulsed dendritic cells vs. canarypox-HIV alone in HIV-1-infected patients on antiretroviral therapy. *Vaccine, 27*(43), 6088–6094.

Gross, V. S., et al. (2011). Isolation of functional mitochondria from rat kidney and skeletal muscle without manual homogenization. *Analytical Biochemistry, 418*(2), 213–223.

Haorah, J., et al. (2025). Epigenetics and mitochondrial biogenesis: The role of sirtuins in HIV neuropathogenesis. *Molecular Neurobiology, 62*(8), 10333–10348.

Huynh, H. T., et al. (2024). Biochemical separation of cytoplasmic and nuclear fraction for downstream molecular analysis. *Current Protocols, 4*(5), e1042.

Kandigian, S. E., et al. (2022). Proteomic characterization of post-mortem human brain tissue following ultracentrifugation-based subcellular fractionation. *Brain Communications, 4*(3).

Kwon, H. S., et al. (2008). Human immunodeficiency virus type 1 Tat protein inhibits the SIRT1 deacetylase and induces T cell hyperactivation. *Cell Host & Microbe, 3*(3), 158–167.

Ling, L., et al. (2023). Transient CD4+ T cell depletion during suppressive ART reduces the HIV reservoir in humanized mice. *PLoS Pathogens, 19*(12), e1011824.

Liu, G., et al. (2017). Loss of NAD-dependent protein deacetylase sirtuin-2 alters mitochondrial protein acetylation and dysregulates mitophagy. *Antioxidants & Redox Signaling, 26*(15), 849–863.

Manjula, R., Anuja, K., & Alcain, F. J. (2020). SIRT1 and SIRT2 activity control in neurodegenerative diseases. *Frontiers in Pharmacology, 11*, 585821.

Nabbi, A., & Riabowol, K. (2015). Rapid isolation of nuclei from cells in vitro. *Cold Spring Harbor Protocols, 2015*(8), 769–772.

Park, J., et al. (2013). SIRT5-mediated lysine desuccinylation impacts diverse metabolic pathways. *Molecular Cell, 50*(6), 919–930.

Roessler, C., et al. (2015). A novel continuous assay for the deacylase sirtuin 5 and other deacetylases. *Journal of Medicinal Chemistry, 58*(18), 7217–7223.

Santos, L., et al. (2023). Sirtuins: The NAD(+)-dependent multifaceted modulators of inflammation. *Antioxidants & Redox Signaling, 39*(16-18), 1185–1208.

Sivalingam, K., et al. (2021a). HIV-Tat and cocaine impact brain energy metabolism: Redox modification and mitochondrial biogenesis influence NRF transcription-mediated neurodegeneration. *Molecular Neurobiology, 58*(2), 490–504.

Sivalingam, K., et al. (2021b). Influence of psychostimulants and opioids on epigenetic modification of class III histone deacetylase (HDAC)-sirtuins in glial cells. *Scientific Reports, 11*(1), 21335.

Sivandzade, F., Bhalerao, A., & Cucullo, L. (2019). Analysis of the mitochondrial membrane potential using the cationic JC-1 dye as a sensitive fluorescent probe. *Bio-protocol, 9*(1).

Thangaraj, A., et al. (2021). HIV TAT-mediated microglial senescence: Role of SIRT3-dependent mitochondrial oxidative stress. *Redox Biol, 40*, 101843.

Thangavel, S., et al. (2009). HIV-1 clade B and C Tat differentially induce indoleamine-2, 3-dioxygenase (IDO) and serotonin in immature dendritic cells (IDC): Implications for neuro-AIDS. *Journal of Neurovirology*.

Thangavel, S., et al. (2018). Oxidative stress in HIV infection and alcohol use: Role of redox signals in modulation of lipid rafts and ATP-binding cassette transporters. *Antioxidants & Redox Signaling, 28*(4), 324–337.

Walter, M., et al. (2022). SIRT5 is a proviral factor that interacts with SARS-CoV-2 Nsp14 protein. *bioRxiv*.

Wang, G., et al. (2019). Regulation of UCP1 and mitochondrial metabolism in brown adipose tissue by reversible succinylation. *Molecular Cell, 74*(4), 844–857.e7.

Yang, L., et al. (2017). Sirtuin 5: A review of structure, known inhibitors and clues for developing new inhibitors. *Science China Life Sciences, 60*(3), 249–256.

Zhu, S., et al. (2022). Sirt3 promotes chondrogenesis, chondrocyte mitochondrial respiration and the development of high-fat diet-induced osteoarthritis in mice. *Journal of Bone and Mineral Research: the Official Journal of the American Society for Bone and Mineral Research, 37*(12), 2531–2547.

Zou, P., et al. (2018). Resveratrol pretreatment attenuates traumatic brain injury in rats by suppressing NLRP3 inflammasome activation via SIRT1. *Molecular Medicine Reports, 17*(2), 3212–3217.

CHAPTER THREE

Methodological frameworks to probe histone acetylation–driven circuit plasticity in SYNGAP1–associated neurodevelopmental disorders

Akash Kumar Singh[a,b], Prajjval Mishra[a], Amrish Rai[c,d], Priyanka Pandey[a], James P. Clement[e], and Tapas K. Kundu[a,b,*]

[a]Molecular Biology and Genetics Unit, Transcription and Disease Laboratory, Jawaharlal Nehru Centre for Advanced Scientific Research, Jakkur, Bengaluru, India
[b]Neuroscience Unit, Jawaharlal Nehru Centre for Advanced Scientific Research, Bangalore, India
[c]Division of Neuroscience and Ageing Biology, CSIR-Central Drug Research Institute, Lucknow, India
[d]Academy of Scientific and Innovative Research (AcSIR), Ghaziabad, India
[e]Faculty of Clinical and Biomedical Science, University of Exeter, Exeter, United Kingdom
*Corresponding author. e-mail address: tapas@jncasr.ac.in

Contents

Abstract

SYNGAP1 encodes a Ras GTPase-activating protein essential for neurodevelopment and synaptic plasticity, and heterozygous loss-of-function mutations are strongly associated with intellectual disability (ID) and autism spectrum disorder (ASD). While synaptic and circuit-level impairments resulting from SYNGAP1 haploinsufficiency

Methods in Enzymology, Volume 734
ISSN 0076-6879, https://doi.org/10.1016/bs.mie.2026.05.055

have been well characterized, the contribution of epigenetic dysregulation remains poorly understood. In this chapter, we describe experimental approaches to investigate histone acetylation–dependent mechanisms underlying SYNGAP1-associated neurodevelopmental phenotypes. Using *Syngap1*$^{+/-}$ mice, we demonstrate a robust reduction in p300/CBP-specific histone acetylation marks in the adolescent hippocampus, accompanied by impaired dendritic arborization of adult-born doublecortin-positive (DCX^{+}) neurons, indicative of disrupted adult hippocampal neurogenesis. To establish causality between altered chromatin state and behavioural and circuit dysfunction, we employed a glucose-derived carbon nanosphere–conjugated small-molecule activator of p300/CBP (CSP–TTK21) in young adult *Syngap1*$^{+/-}$ mice (2–4 months). Pharmacological enhancement of p300/CBP activity restored histone acetylation, rescued synaptic and structural plasticity, normalized experience-dependent cortical circuit reorganization, and significantly improved behavioural performance to levels comparable with wild-type littermates. Transcriptomic profiling by hippocampal RNA sequencing revealed reversal of dysregulated gene expression programs, including key regulators of synaptic plasticity and neurogenesis. Collectively, the methods described in this chapter provide a framework for dissecting epigenetic contributions to SYNGAP1-associated neurodevelopmental disorders and establish chromatin-targeted modulation of p300/CBP as a reversible and therapeutically actionable mechanism for ID/ASD-related circuit dysfunction.

1. Introduction

Intellectual disability, or ID, shows up early in life, before age 18, and sticks around. People with ID struggle with intellectual tasks and everyday life skills, and most have an IQ below 70 (Schalock et al., 2021) (Fig. 1). Clinicians categorize ID as mild, moderate, severe, or profound, and then further classify it as syndromic or non-syndromic based on whether other physical differences or health issues are present (Van Bokhoven, 2011). Sure, some cases are due to environmental factors, but genetics plays a huge role. So far, scientists have linked over 450 genes to ID. Still, for most of these genes, we don't really know how the mutations end up messing with the brain's ability to adapt and learn.

Take non-syndromic ID, for example. One gene keeps popping up: SYNGAP1. Mutations in this gene account for about 2–3 % of cases (Hamdan et al., 2011, 2009). SYNGAP1 makes SynGAP1, a protein that works as a Ras and Rap GTPase-activating protein (GAP) and accumulates in the postsynaptic density of excitatory synapses (Fig. 2). Basically, SynGAP1 puts the brakes on specific signalling pathways that shape synapse development, regulate AMPA receptor trafficking, and control how synapses change with experience (Komiyama et al., 2002; Rumbaugh

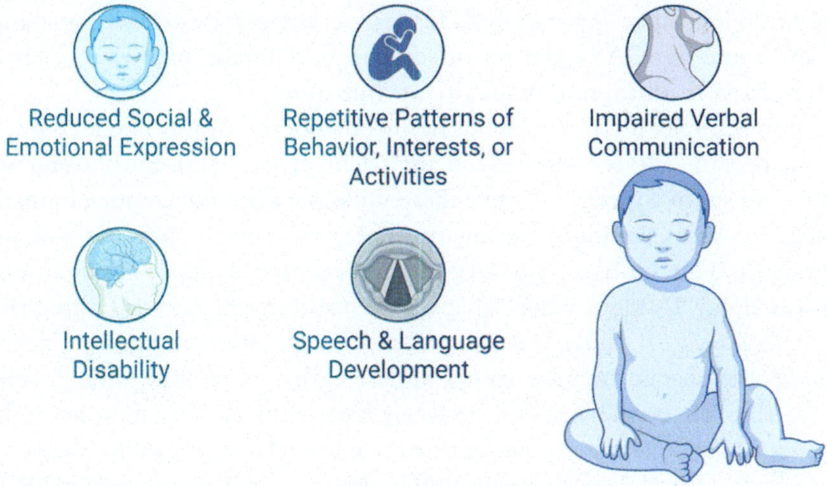

Fig. 1 Intellectual disability. Behavioural characteristics.

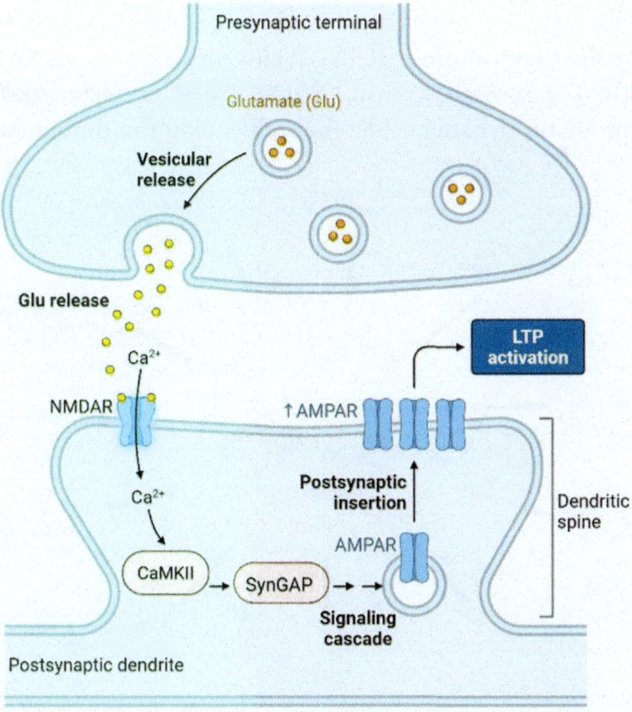

Fig. 2 NMDAR-dependent synaptic plasticity. SynGAP1 is a post-synaptic protein that acts downstream of NMDAR-mediated synaptic plasticity.

et al., 2006). Since Ras-family GTPases are key players in cell signalling, losing some SynGAP1 throws off downstream kinase pathways, such as ERK/MAPK, disrupting brain circuit formation.

For example, if you knock out both copies of the *Syngap1* gene in mice, the pups don't survive past early infancy (Kim et al., 2003). Even losing just one copy, as in *Syngap1*$^{+/-}$ mice that mimic the gene dosage seen in many patients, causes significant problems. These mice have clear issues at the synapse and circuit level. SynGAP1 levels peak in the hippocampus around postnatal day 14, right when synapses are maturing the fastest. During that period, *Syngap1*$^{+/-}$ mice show their dendritic spines maturing too quickly: mushroom-shaped, mature spines appear earlier than they should. This early shift alters the balance between excitation and inhibition in the hippocampus, making the network overly excitable and causing the brain to close its critical developmental windows too soon (Clement et al., 2012). In the end, these changes leave the mice with lasting problems in thinking, social interaction, and learning (Guo et al., 2009) (Fig. 3).

Recent studies indicate that epigenetic mechanisms are disrupted in neurodevelopmental conditions, including Kabuki syndrome and Rubinstein-Taybi syndrome (RTS) (Schwenty-Lara et al., 2020; Singh et al., 2018; Lopez-Atalaya et al., 2014). These modifications primarily regulate chromatin dynamics, thereby influencing the timing and location

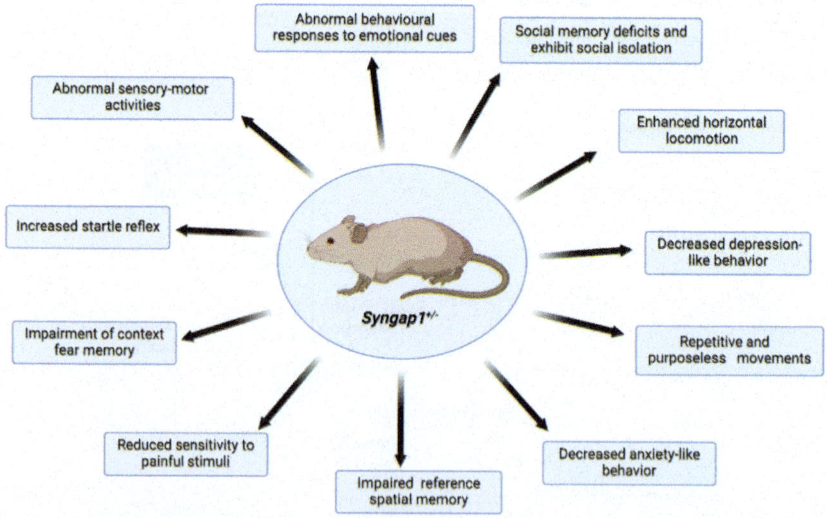

Fig. 3 *Syngap1*$^{+/-}$ mouse. Schematic depicting behavioural abnormalities associated with the *Syngap1* heterozygous mouse.

of gene expression associated with synapse formation, maintainence, and pruning from the earliest developmental phases. The reversible acetylation of histones within nucleosomes, along with Lysine acetyltransferases (KATs) and Lysine deacetyltransferases (KDACs), represents a significant epigenetic process that is implicated in various physiological and pathophysiological conditions, including intellectual disabilities and autism spectrum disorders. The primary epigenetic enzymes, p300 and CBP, have been identified as crucial KATs required for neuronal development, with KAT activity vital for neuronal differentiation and maturation (Sheikh, 2014; Smitha et al., 2024). Additionally, p300 and CBP are also instrumental in cognitive functions and memory (Alarcon et al., 2004; Korzus et al., 2004; Fernandez-Albert et al., 2019; Maddox et al., 2013; Mitchnick et al., 2016). Research utilizing transgenic mice expressing a truncated p300 variant or with conditional knockouts of specific p300 regions revealed impairments in cognition and the ability to retain long-term memories (Oliveira et al., 2011), resembling features found in Rubinstein-Taybi syndrome (RTS) (Coupry et al., 2002; Tsai et al., 2011). Experiments using zebrafish, iPSC-neurons from RTS patients, and mouse models of RTS show that restoring acetylation, either by genetic reintroduction of p300/CBP or by administration of KDAC inhibitors such as TSA and SAHA, significantly ameliorates deficits in morphology, cognition, and synaptic plasticity (Babu et al., 2018; Alari et al., 2021). Likewise, focusing on these epigenetic changes rather than just synaptic targets is yielding encouraging results across various neurodevelopmental disorders (Nakayama et al., 2024; Yan, 2024).

2. Rationale

p300/CBP-mediated acetylation plays a key role during brain development, learning, and cognition, and thereby boosting KAT activity could be a real therapeutic option for several neurological disorders (Singh et al., 2018; Singh & Kundu, 2026). Still, using KDAC inhibitors in patients hasn't worked out well due to their broad, nonspecific effects. This makes it clear we need alternatives that target KATs more directly. One promising approach is to develop small-molecule activators of KATs. Some p300/CBP activators exist, but most don't enter cells efficiently, limiting their usefulness in living animals. To get around this, we developed and studied CSP-TTK21, a small molecule that selectively enhances p300/CBP KAT activity. CSP-TTK21 enters cells, crosses the blood–brain barrier, and increases histone acetylation without apparent toxicity. When given systemically, it boosts adult neurogenesis and extends long-term spatial memory (Chatterjee et al., 2013). More recent work shows that oral gavage delivers CSP-TTK21 efficiently to the brain, inducing histone acetylation and long-term potentiation similar to those seen with intraperitoneal injection (Singh et al., 2024).

In SYNGAP1-related intellectual disability (ID), there's a convergence of problems: synaptic signaling defects and altered p300/CBP-mediated transcription. So, modulating p300/CBP activity could be a strong therapeutic strategy (Singh et al., 2025).

In this chapter, we describe procedures to investigate the therapeutic effects of CSP-TTK21 in a *Syngap1*$^{+/-}$ mouse model of ID upon oral administration. We highlight a set of biochemical, epigenetic, and functional assays to measure KAT activation, histone acetylation, synaptic plasticity, and cognitive performance. These methods should shed light on how targeting chromatin-regulating enzymes can shape phenotypes associated with neurodevelopmental diseases.

3. Materials and methods

3.1 Investigating histone acetylation and adult neurogenesis in *Syngap1*$^{+/-}$ mice

3.1.1 Equipment

1. Dounce homogenizer (Sigma-Aldrich, #D9938)
2. Refrigerated centrifuge (12,000 rpm)
3. 1.5 mL microcentrifuge tubes
4. Cryotome (Leica, #CM3050 S)
5. Freezers: one at −80 °C, one at −20 °C
6. Temperature-controlled incubator
7. Humidified staining chamber
8. Light and confocal fluorescence microscope
9. Ice bucket and cold dissection tools

3.1.2 Reagents

1. Buffers and Chemicals: Phosphate-buffered saline (PBS) pH 7.4, Triton X-100, 4 % paraformaldehyde (PFA), Sucrose, Sodium azide, Optimal cutting temperature (OCT) compound.

2. Inhibitors: Protease inhibitor cocktail (Roche, #11836170001), Phosphatase inhibitor cocktail (Sigma-Aldrich, #P0044).
3. Blocking Reagents: Normal horse serum or BSA.
4. Primary Antibodies: Anti-H4K12ac, Anti-H3K14ac, Anti-H2Bac, and DCX.
5. Secondary Antibodies and Detection: Goat anti-rabbit, Fluorophore-conjugated secondary antibodies (species-specific, standard suppliers), Nuclear counterstain (DAPI or similar)

3.1.3 Procedure

A. Preparation of Brain Lysates

1. Euthanize mice using cervical dislocation, following the institution's animal ethics rules.
2. Rapidly remove the brain, rinse it with ice-cold 1 X PBS (pH 7.4) to wash off any blood.
3. Keep everything cold. Dissect the brain regions using clean, chilled tools, then drop the tissue straight into pre-cooled tubes.
4. Use the Dounce homogenizer (Sigma #D9938) to break up the tissue in ice-cold RIPA lysis buffer. RIPA buffer composition: 150 mM NaCl, 50 mM Tris-Cl (pH 7.4), 5 mM EDTA, 0.25 % sodium deoxycholate, 0.1 % SDS, 0.1 % Triton X-100, 10 μl of protease inhibitor cocktail per 100 mL, and 2 μl phosphatase inhibitor per 100 mL.
5. Let the homogenates sit on ice for 30 min. Give them a gentle swirl now and then so the cells break open entirely.
6. Spin the samples down at 12,000 rpm for 20 min at 4 °C.
7. Carefully collect the supernatant without disturbing the pellet—aliquot lysates into 20 μl volumes and store at −80 °C until further use.

B. Brain Fixation, Cryoprotection, and Sectioning

1. Euthanize the mice using cervical dislocation, then quickly remove the brain. Wash the brain in 1 × PBS (pH 7.4).
2. Drop the whole brain into 4 % paraformaldehyde (PFA) and leave it overnight at 4 °C.
3. Wash the brains three times with 1 × PBS, 5 min each, at room temperature.
4. Transfer the brains to a 30 % (w/v) sucrose solution in 1 × PBS with 0.02% sodium azide. Keep at 4 °C for 48 h, or until they sink.
5. Embed the brains in optimal cutting temperature (OCT) compound and freeze at −20 °C for 20 min.

6. Cut coronal sections through the dorsal hippocampus at 20 μm thickness using a cryotome (Leica, #CM3050 S).
7. Place the sections onto coated glass slides and store them at −20 °C until you're ready to do immunostaining.

C. Immunohistochemistry

1. Let the frozen sections warm up to room temperature for 10 min.
2. Incubate the sections in 1 × PBS with 0.3 % Triton X-100 (0.3 % PBTx) for 15 min.
3. Block nonspecific binding by incubating the sections in 5 % normal horse serum (NHS) in 1 × PBS with 0.1 % Triton X-100 (0.1 % PBTx) for 30 min at 37 °C.
4. Wash the sections three times with 0.1 % PBTx, 10 min each time.
5. Incubate the sections overnight at 4 °C with primary antibodies diluted in blocking buffer: Anti-H4K12ac (1:250, lab-raised), Anti-H3K14ac (1:250, lab-raised), and Anti-H2Bac (1:500, Sigma-Aldrich, #07-373).
6. For secondary antibody incubation, nuclear counterstaining, and imaging, follow the usual immunofluorescence procedures.

3.1.4 Notes

Tissue integrity and postmortem delay: Brain extraction must be completed within one minute following sacrifice. Any delays may result in postmortem deterioration and inconsistent histone acetylation signals. It is crucial to pre-chill all instruments and buffers, and to process each animal individually instead of in groups.

Lysis and homogenization efficiency: Achieving complete tissue disruption in RIPA buffer is vital for consistent protein yield. Insufficient homogenization can lead to weak or uneven immunoblot signals. Employ 10–15 strokes with a tight Dounce pestle and visually verify the uniformity of the homogenate.

Preservation of acetylation marks: Protease and phosphatase inhibitors should be freshly incorporated into lysis buffers. A decrease in acetylation signal generally indicates enzymatic degradation during processing. It is essential to keep samples on ice throughout the entire procedure.

Cryoprotection and section quality: Brains must undergo complete cryoprotection before freezing. Incomplete sucrose infiltration may result in tissue cracking or distorted cryosections. Only proceed to embedding once the tissues have fully sunk in the sucrose solution.

Antibody penetration and staining consistency: Sufficient permeabilization and blocking are necessary for consistent immunostaining. Weak or inconsistent signals frequently arise from inadequate Triton X-100 permeabilization or excessive background. Adjust detergent concentration and blocking duration as required.

Inter-group reproducibility: All experimental groups should be processed simultaneously. Variability in fixation durations, antibody incubation, or imaging parameters can lead to batch effects. Consistent conditions must be upheld across all samples.

3.2 Performing oral treatment of KAT activator CSP-TTK21 in Syngap1$^{+/-}$ mouse

General Considerations

All procedures should be performed in accordance with institutional and national animal ethics guidelines.

Animals should be acclimatized to handling before drug administration to minimize stress.

Ensure solutions are freshly prepared, sterile, and at room temperature before administration.

3.2.1 Materials

1. Stainless steel curved gavage needle with a rounded (ball) tip, appropriate for mouse size
2. Sterile syringe
3. Test compound or vehicle solution

3.2.2 Reagents

1. Saline
2. CSP
3. CSP-TTK21

3.2.3 Methodology

1. Draw the appropriate volume of solution into a syringe attached to a stainless-steel curved gavage needle. Ensure the absence of air bubbles.
2. Scruff the mouse gently such that the forelimbs are extended laterally, and the head is tilted backward, maintaining alignment with the body axis. This positioning straightens the esophagus and facilitates safe gavage.
3. Introduce the gavage needle into the oral cavity and advance it gently through the pharynx into the esophagus. Advancement should occur smoothly, guided by gravity, without any force.

4. The absence of resistance, coughing, or respiratory distress indicates proper insertion. If resistance is encountered, withdraw the needle and reattempt insertion.
5. Administer the solution slowly to prevent reflux or aspiration.
6. Withdraw the gavage needle carefully along the same trajectory used for insertion.
7. Monitor mice for 5–10 min following gavage for signs of discomfort, regurgitation, labored breathing, or other complications.

3.2.4 Notes

Injection speed: Rapid delivery can cause discomfort or leakage; slow, controlled administration is essential for both IP injection and gavage.

Gavage resistance: Never force the gavage needle; resistance may indicate tracheal entry or esophageal obstruction.

Animal welfare: Any mouse showing persistent distress after administration should be excluded and evaluated in accordance with ethical guidelines.

3.3 Investigating the effect of CSP-TTK21 treatment on synaptic plasticity

3.3.1 Assessment of synaptic plasticity in acute hippocampal slices

A. Preparation of Acute Hippocampal Slices

1. Sacrifice adult mice (>3 months of age) of the indicated experimental groups by cervical dislocation in accordance with institutional animal ethics guidelines.
2. Rapidly remove brain and immediately immerse in ice-cold, carbogenated (95 % O_2/5 % CO_2) sucrose-based cutting solution to preserve tissue viability. The sucrose cutting solution composition (in mM): 124 sucrose, 3 KCl, 24 $NaHCO_3$, 2 $CaCl_2$, 1.25 NaH_2PO_4, 1 $MgSO_4$, and 10 D-glucose.
3. Following initial cooling, remove the cerebellum and mount the brain onto the vibratome stage.
4. Prepare transverse hippocampal slices (350 μm thickness) using a vibratome in ice-cold sucrose cutting solution under continuous carbogen supply.
5. Immediately after sectioning, transfer slices to a holding chamber containing artificial cerebrospinal fluid (aCSF) maintained at 37 °C and continuously bubbled with carbogen for 1 h. The aCSF composition (in mM): 124 NaCl, 3 KCl, 24 $NaHCO_3$, 2 $CaCl_2$, 1.25 NaH_2PO_4, 1 $MgSO_4$, and 10 D-glucose.

6. Place slices on a nylon mesh to allow optimal perfusion. After the initial recovery period, maintain slices at room temperature in carbogenated aCSF until electrophysiological recordings.

B. Electrophysiological Recording Setup

1. For recordings, transfer a single hippocampal slice to a submerged recording chamber and continuously perfuse it with warmed (34 °C), carbogenated aCSF at a flow rate of 400–550 μl/min. Maintain remaining slices at room temperature in carbogenated aCSF during recordings.
2. Record extracellular field excitatory postsynaptic potentials (fEPSPs) from the CA region using low-resistance glass microelectrodes (3–5 MΩ). Prepare recording electrodes from borosilicate glass capillaries (inner diameter 0.6 mm, outer diameter 1.2 mm; Harvard Apparatus) using a horizontal micropipette puller (Flaming-Brown P-97, Sutter Instruments, USA) and fill it with aCSF (pH 7.3).
3. Position the stimulating and recording electrodes using a three-axis micromanipulator to allow precise placement within the slice. Orient both electrodes at approximately 45° relative to the slice's vertical axis. Evoke synaptic responses at a stimulation frequency of 0.1 Hz throughout baseline and post-induction recordings (Fig. 4).

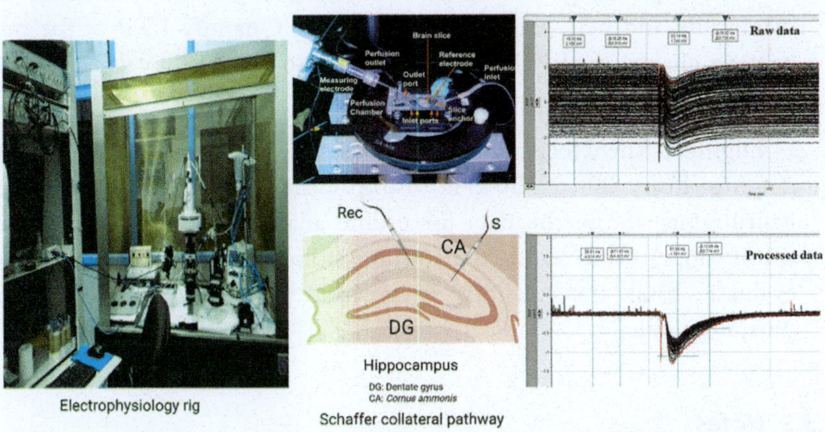

Fig. 4 Electrophysiological analysis. A schematic showing the rig used for recordings. A zoomed image of the slide holder showing the orientation of electrodes (reference and recording), perfusion chamber, slice anchor, and Brain slice. The Schaffer collateral pathway from where the recordings were performed, and the raw and processed data from the recordings.

C. Assessment of Basal Synaptic Transmission

Create input–output (I/O) curves to evaluate basal synaptic transmission and to identify a suitable stimulus intensity for subsequent experiments. Vary the stimulation pulse widths from 20 to 40 μs. Incrementally increase the stimulus intensity by 20 μA per sweep, ranging from 0 to 160 μA. Plot the resulting fEPSP slopes against the stimulus intensity to assess synaptic responsiveness.

D. Assessment of Presynaptic Function

Perform paired-pulse facilitation (PPF) to evaluate alterations in presynaptic neurotransmitter release probability. Administer two consecutive stimuli at different inter-pulse intervals, separated by quarter-log increments, with intervals ranging from 10 to 1000 ms. Determine the paired-pulse ratio as the ratio of the slope of the second fEPSP to that of the first. Interpret the variations in PPR as changes in presynaptic release dynamics.

E. Induction and Recording of Long-Term Potentiation (LTP)

After establishing a stable baseline, record synaptic responses for 15–20 min. Discard recordings if baseline responses showed instability or drift. Induce Long-term potentiation (LTP) through a single train of high-frequency stimulation (100 Hz for 1 s). Following the induction, record synaptic responses for an additional 55–60 min to evaluate the magnitude and persistence of potentiation. Quantify LTP as the percentage change in fEPSP slope compared to baseline levels.

F. Data Acquisition and Analysis

Digitized the electrophysiological signals at sampling frequencies ranging from 20 to 50 kHz using Clampex 10 software (Axon Instruments). Store the data for offline analysis. Measure and analyse fEPSP slopes using standard electrophysiological analysis workflows to evaluate basal transmission, presynaptic function, and long-term synaptic plasticity.

3.3.2 Notes

All solutions should be prepared fresh and continuously carbogenated during the experiment. Standardize the conditions for slice preparation, recovery, and recording across all experimental groups to reduce variability. Only slices that exhibit stable baseline recordings and strong synaptic responses should be incorporated into the final analysis.

3.4 Studying the effect of CSP-TTK21 treatment on behavioural deficits in adult *Syngap1*$^{+/-}$ mouse (social and learning deficits)

3.4.1 Materials

1. Dedicated behavioral testing room (sound-attenuated, low-traffic area)
2. Overhead LED lighting system (adjustable intensity)
3. Video recording system (high-resolution camera; top-mounted)
4. Computer with behavioral tracking software (e.g., ANY-maze, EthoVision, SMART)
5. Infrared lighting (if testing during dark cycle)
6. Stopwatch (backup timing)
7. Digital weighing balance
8. Cleaning station
9. Cage holding racks (for habituation before testing)
10. Temperature and humidity monitor
11. White noise generator (optional)

3.4.2 Reagents

1. 70 % ethanol (apparatus cleaning)
2. Paper towels
3. Disposable gloves
4. Distilled water
5. Animal marking tools (non-toxic marker or tail marking pen)

3.4.3 Behavioral assessments procedures

1. Experimental Design and Animal Handling

   Organize mice into groups and treat for 3 weeks (one dosage per week) with Saline, CSP, or CSP-TTK21 before the behavioral assays. Perform behavioural experiments after a week following the final treatment, to negate any stress due to the injection (Fig. 5). Handle each mouse for a period of 1–2 min daily over the course of 3–5 days before proceeding to the Socio-Behavior experiments. For all behavioral experiments, transport the mice to the behavior room 30 min before the experiment. Upon completion of the experiment, return each mouse to their home cage. Sanitize the chamber with 70% alcohol and allow it to dry between tests. At the end of the testing day, return the mice to the holding room.
2. **Open Field:** Position each mouse in one corner of an acrylic open field arena measuring 52 × 50 × 45 cm (a custom-made white plywood box

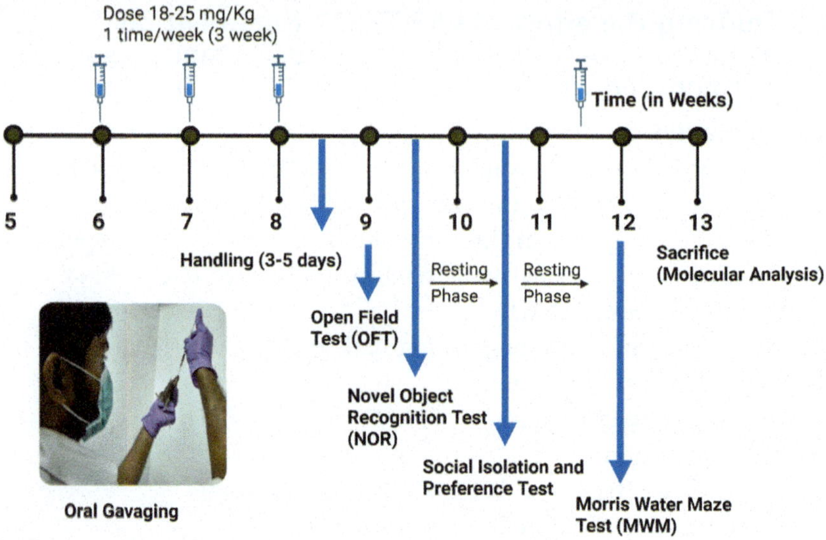

Fig. 5 Treatment timeline and behavioural testing.

at JNCASR), illuminated at 80–100 lx. Allow the mouse to explore the apparatus, and immediately record its activity using a CCTV camera (Panlab Harvard Apparatus #DCBA5-E83) for 10 min. Analyze the total distance traveled, distance in the center (20 × 20), number of entries into the center, and time spent in the center using Smart V 3.0 Software (Panlab Harvard Apparatus #DCBA5-E83).

3. **Novel Object Recognition (NOR):** On the day following Open Field Test (OFT), place each mouse in one corner of the same open field arena and allow it to acclimate to two similar objects positioned at opposite corners of the arena. Familiarize the mouse with two identical objects for 2 consecutive days, each session for 10 min. Perform the test after 24 h following the last familiarization session. During testing, substitute one of the familiar objects with a novel object and record the session for 10 min using a CCTV mounted at the top of the apparatus. Manually analyze all the videos, and measure the interaction time (active contacts and the mice facing the object within a 2 cm periphery) with each object using stopwatches. Calculate the recognition index (time spent exploring the novel object divided by total exploration time) and discrimination index ((novel − familiar)/total exploration time). Animals with less than 20 s of total object interaction time should be excluded from analysis.

4. Three-Chamber Social Interaction Test

Social Isolation: A custom-made rectangular transparent plexiglass box measuring 40.6 × 21 × 35 cm was utilized—the central chamber featured rectangular openings, which permitted access to the side chambers.

Place each mouse in the central chamber for 5 min, with both the right and left side entrances secured using heavy-weight cardboard sheets to allow acclimatization to the apparatus. For social Isolation, position a C57BL/6 J male mouse (Stranger 1), which had not previously encountered the test mouse, inside a wire cage in one of the side chambers, and an empty wire cage in the opposite side chamber. The wire cage should be small and cylindrical, facilitating direct contact through the bars while preventing aggressive interactions. Analyze the time spent and the number of body contacts by the test mouse with Stranger 1 and the empty wire cage. To assess social preference, introduce a second novel C57BL/6 J male mouse (Stranger 2) into the previously empty wire cage, and allow the test mouse to explore both Stranger 1 and Stranger 2 for an additional 10 min. Analyze the time spent and the number of direct body contacts between the test mouse and both Stranger 1 and Stranger 2.

5. **Morris Water Maze (MWZ):** A white-water tank with a diameter of 94 cm and a depth of 50 cm was used in the experiment.

Fill the tank with water to a depth of 38 cm and add non-toxic white tempera paint to make it opaque. Maintain the water temperature at 25 ± 1 °C by adding warm water. Submerge an escape platform (10 cm in diameter) 1 cm below the water surface in the center of one of the quadrants (the target quadrant). Keep the platform location constant throughout the entire training session; remove the platform during the probe testing phase. Several distal extra-maze cues should be used to guide the mouse. We used a low-lux bulb, a door, an empty wall, and a standing experimenter, which were strategically placed around the tank, with their positions remaining unchanged throughout the training and testing sessions (Fig. 6).

Commence a trial by positioning the mouse in one quadrant, and allow it to locate the platform for 60 s.

1. Let the mice acclimate for 15 s to the distal extra-maze cues upon reaching the platform. If the mouse fails to find the platform within 60 s, the experimenter should manually guide it to the platform and allow 15 s to adjust to the extra-maze cues.

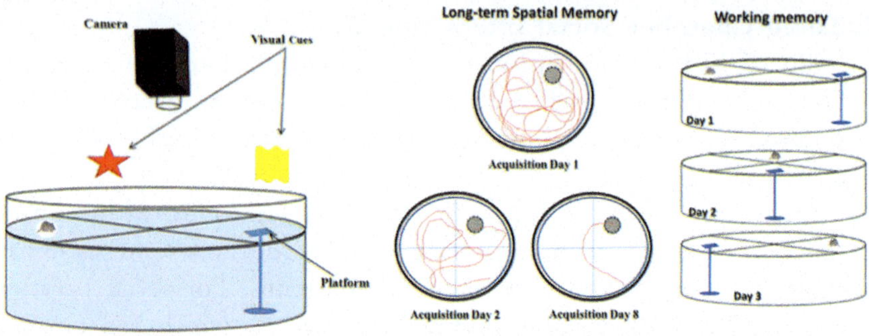

Fig. 6 Morris water maze. Representative experimental setup used for long-term spatial memory and working memory.

2. Each day, conduct four trials. Initiate the first and fourth trials from the opposite quadrant, and the second and third trials from the right and left quadrants, relative to the target quadrant.
3. Following the trials, remove and dry the mouse using tissue paper, followed by a terrycloth towel, and subsequently transfer them to an empty cage positioned under a lamp for a minimum of 10 min to avert hypothermia.
4. Visually inspect each mouse for dryness and other potential issues before being placed back in the home cage.
5. Conduct all sessions at approximately the same time each day to minimize variability associated with the time of day.
6. Document the trials using a smart digital tracking system (Panlab) and analyze with Smart v.3 software.

For working memory assessment, perform three trials per day over three consecutive days for each mouse, with the escape platform location rotated clockwise daily. The first two trials each day will serve as training sessions for the new escape platform position, while the third trial will serve as a test of working memory. Measure the latency to reach the target platform during the testing trials across both days, average, and subsequently plot the graph.

3.5 Experimental controls and data analysis

All behavioral experiments should be conducted by investigators blinded to treatment groups. Testing conditions, lighting, apparatus dimensions, and trial parameters should be kept constant across all experimental cohorts. Analyze data using appropriate statistical methods.

3.6 Analyzing the effect of CSP-TTK21 treatment on neuronal rewiring (developmental plasticity)

3.6.1 Equipment

1. Fine curved iris scissors or blunt micro-scissors (sterilized)
2. Clean gloves (powder-free)
3. Soft absorbent pad or clean paper towel
4. Warming surface (e.g., pre-warmed pad at ~30–32 °C)
5. Timer
6. Cage card or log sheet for documentation
7. Glass slides and bottles
8. Cryostat, microscope, and Neurolucida software

3.6.2 Reagents

1. Golgi-Cox solutions
2. Alcohol,
3. Ammonia
4. DPx mounting media

3.6.3 Procedure

A. **Whisker Trimming:** Perform whisker-trimming on WT and *Syngap1*$^{+/-}$ pups beginning at postnatal day 7 (PND7) and concluding at PND21.

1. **Gentle Removal of the Litter**
   - Calmly move the home cage to a quiet workspace.
   - Gently transfer the dam to a temporary holding cage with nesting material.
   - Move the litter together onto a soft, warm surface to maintain social contact.
2. **Handling the Pup**
   - Pick up one pup at a time using a gentle scoop with gloved hands.
   - Support the pup's body fully to prevent struggling.
   - Keep handling time as short as possible.
3. **Whisker Trimming**
   - Using fine scissors, trim the whiskers close to the base without touching the skin.
   - Do not pull whiskers or cut into the follicle.
   - Avoid contact with the eyes or facial skin.
   - Inspect briefly to confirm clean trimming and absence of injury.

4. **Return to the Litter**
   - Immediately return the pup to its littermates.
   - Repeat for the remaining pups efficiently and calmly.
5. **Reuniting with the Dam**
   - Return the entire litter to the home cage.
   - Reintroduce the dam promptly.
   - Observe from a distance for 5–10 min to confirm normal maternal behavior (retrieval, grooming, nursing).
6. **Post-Procedure Monitoring**
   - Check pups daily for:
     - Normal nursing and weight gain
     - Skin integrity around the snout
     - Signs of maternal neglect or aggression
   - Record each trimming session and any observations.

B. **Oral administration**: Perform oral administration of CSP/CSP-TTK21 (20 mg/kg) following the whisker trimming at PND30, PND36, and PND42. Euthanize animals at PND51 and subsequently place the brain into Golgi-Cox staining solution (Fig. 7).

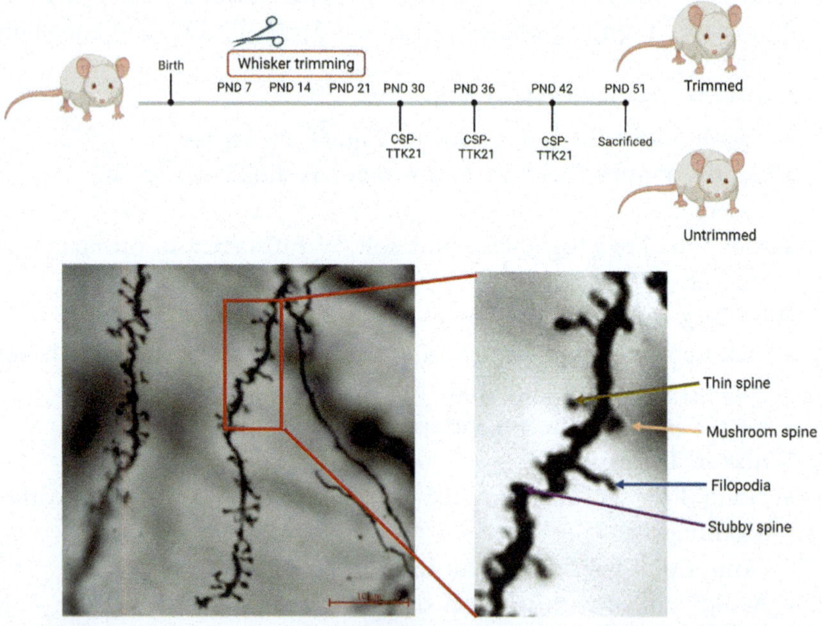

Fig. 7 Whisker trimming and Golgi staining. Schematic depicting the whisker trimming and treatment timeline. Golgi-stained neurons with dendritic spines.

C. Golgi-Cox Staining of Neurons

Composition of Golgi–Cox Solution (Stock).

Solution A.

5 % Potassium dichromate ($K_2Cr_2O_7$).

Dissolve 5 g of potassium dichromate in 100 mL of distilled water.

Solution B.

5 % Mercuric chloride ($HgCl_2$).

Dissolve 5 g of mercuric chloride in 100 mL of distilled water.

Warm gently and stir continuously to aid dissolution.

Solution C.

5 % Potassium chromate (K_2CrO_4).

Dissolve 5 g of potassium chromate in 100 mL of distilled water.

Preparation of Working Golgi–Cox Solution.

Mix 5 parts Solution A + 5 parts Solution B + 4 parts Solution C.

Example: 50 mL A + 50 mL B + 40 C.

Add 10 parts distilled water.

Example: add 100 mL of water.

1. Mix thoroughly in a dark bottle
2. Allow the solution to stand for 24–48 h in the dark
3. Filter the solution to remove precipitates
4. Store in amber or foil-wrapped glass bottles at room temperature (Protect from light)
5. Stable for several weeks to months if uncontaminated
6. Re-filter before use if precipitate forms

Golgi-Staining and sample processing.

1. Clean glass slides using 0.1 N HCl, and coat with 3 % Gelatin.
2. Euthanize the mouse at PND51, dissect the brain, wash with dd·H_2O, and bisect into two halves to enhance impregnation.
3. Place each hemisphere into separate small glass bottles filled with Golgi-Cox solution and store in the dark at room temperature (RT).
4. After 24 h, transfer each brain sample to a new glass bottle containing Golgi-Cox solution by either pouring the solution with the brain sample onto histological cassettes or utilizing plastic forceps.
5. Incubate the small bottles with brains in the dark at RT for a duration of 7 days.
6. Subsequently, rinse the brain sample with dd·H_2O and transfer it into a glass bottle containing a tissue-protectant solution, store in the dark at 4 °C.

7. After 24 h, transfer the brain sample to a fresh tissue-protectant solution and store at 4 °C in the dark for 4–7 days.
8. Prepare 80 μm thick sections using a cryostat set at an object temperature (OT) of −20 to −22 °C and a chamber temperature (CT) of −22 to −24 °C.
9. Place the sections onto the gelatin-coated slides.
10. The Golgi-stained sections were developed and mounted using D.P.X mountant.
11. Capture images in layer IV of the barrel cortex at 100X zoom using a Zeiss LSM-510 META confocal laser scanning microscope.
12. Assess spine and filopodia density using Neurolucida software, approximately 8−14 dendritic segments (20–50 μm in length) per mouse.

4. Notes

1. **Tissue Collection and Handling**
    - Use fresh, non-perfused brains. Perfusion with PFA significantly reduces impregnation efficiency. Rapid dissection (within 2–3 min post-euthanasia) improves staining consistency.
    - Avoid mechanical damage during dissection; even mild compression can interfere with uniform impregnation.
    - Rinse briefly in distilled water before immersion in Golgi–Cox solution to remove blood residues.
2. **Impregnation Phase**
    - Ensure complete immersion of the brain in Golgi–Cox solution; tissue should not float or contact the air.
    - Use amber glass bottles or wrap containers in aluminum foil to prevent light exposure.
    - Do not agitate excessively during incubation; gentle handling prevents mechanical artifacts.
    - Incubation time typically ranges from 10 to 14 days (can extend to 21 days depending on strain and age).
    - Replace solution after 24 h if heavy precipitate forms.
    - Temperature should remain stable at room temperature (20–25 °C); avoid refrigeration.

3. **Precipitate Formation**
   - Precipitation is common due to chromium-mercury reactions.
   - Always filter solution before use.
   - If excessive precipitate adheres to tissue, gently rinse in distilled water before transferring to sucrose.
   - Do not scrape tissue to remove crystals.
4. **Cryoprotection and Sectioning**
   - After impregnation, transfer tissue to 30 % sucrose until it sinks.
   - Section thickness should be 100–150 μm for optimal dendritic spine visualization.
   - Thinner sections (<80 μm) may lose complete dendritic arborization.
   - Use sharp blades and minimal vibration to avoid tearing.
   - Collect sections in distilled water rather than PBS to prevent chromate reactions.
5. **Development and Mounting**
   - Avoid prolonged exposure to ammonium hydroxide during development; overdevelopment increases background.
   - Dehydrate quickly through graded ethanol to prevent fading.
   - Use non-aqueous mounting medium.
   - Ensure complete drying before coverslipping to avoid artifact formation.
6. **Safety Considerations** ⚠
   - Golgi–Cox reagents contain mercuric chloride and chromium salts, both highly toxic.
   - Always prepare and discard solutions in a certified chemical fume hood.
   - Dispose of waste as hazardous heavy metal waste.
   - Avoid skin contact and inhalation.
   - Use designated glassware for heavy metal reagents.

5. Summary and discussion

In this chapter, a conceptual and methodological link between epigenetic regulation of neural plasticity and synaptic gene mutations is established. SYNGAP1 has been traditionally studied as a postsynaptic signaling regulator downstream of NMDA receptor pathways. However, the methods presented expand this framework to include chromatin-level mechanisms. Synaptic signaling deficits may converge on transcriptional coactivator systems that control long-term gene expression programs necessary for neurogenesis and circuit maturation, as evidenced by the observed decrease in p300/CBP-dependent histone acetylation in the *Syngap1*$^{+/-}$ hippocampus. The methodology's multiscale integration is one of its main advantages. The framework facilitates causal inference at the molecular, cellular, and systems levels by integrating histone acetylation assays, adult neurogenesis measurements, electrophysiological analyses, circuit mapping, and behavioral testing. As a selective p300/CBP activator, CSP-TTK21 offers a potent pharmacological tool for temporally precise chromatin-state manipulation in vivo. Notably, the rescue of phenotypes in young adult animals indicates that SYNGAP1-associated dysfunction retains a degree of reversibility beyond early developmental critical periods. This finding challenges the long-standing assumption that neurodevelopmental disorders are largely irreversible in adulthood and highlights epigenetic plasticity as a therapeutic entry point. From a translational standpoint, p300/CBP targeting is a more targeted substitute for broad-spectrum histone deacetylase inhibitors, which have limited clinical applicability and pleiotropic effects. Nevertheless, these approaches raise a few issues. First, even though CSP-TTK21 shows promise in a mouse model, long-term safety, dosage schedules, and cell-type specificity continue to be crucial factors for further improvement. Second, because SYNGAP1-related disorders vary widely, more models will be needed to assess whether chromatin-based therapies are broadly applicable. Third, a thorough examination of the duration and time frame of therapeutic rescue is necessary, especially in light of experience-dependent plasticity and developmental stage. All things considered, the chapter offers a solid experimental framework for examining the role of epigenetics in neurodevelopmental disorders by proving that p300/CBP activation is chromatin-targeted.

Acknowledgments

This work was supported by the Jawaharlal Nehru Center for Advanced Scientific Research (JNCASR), J C Bose Fellowship, Department of Science and Technology (DST), India (grant no. JBR/2020/000029) and core grant of SERB (grant no. CRG/2021/000356) to TKK.

References

Alarcon, J. M., Malleret, G., Touzani, K., Vronskaya, S., Ishii, S., Kandel, E. R., & Barco, A. (2004). Chromatin acetylation, memory, and LTP are impaired in CBP+/− mice: A model for the cognitive deficit in Rubinstein-Taybi syndrome and its amelioration. *Neuron, 42*, 947–959.

Alari, V., Scalmani, P., Ajmone, P. F., Perego, S., Avignone, S., Catusi, I., ... Larizza, L. (2021). Histone deacetylase inhibitors ameliorate morphological defects and hypoexcitability of iPSC-neurons from rubinstein-taybi patients. *International Journal of Molecular Sciences, 22*, 5777.

Babu, A., Kamaraj, M., Basu, M., Mukherjee, D., Kapoor, S., Ranjan, S., ... Sachidanandan, C. (2018). Chemical and genetic rescue of an ep300 knockdown model for Rubinstein Taybi Syndrome in zebrafish. *Biochimica et Biophysica Acta (BBA) - Molecular Basis of Disease, 1864*(4 Pt A), 1203–1215.

Chatterjee, S., Mizar, P., Cassel, R., Neidl, R., Selvi, B. R., Mohankrishna, D. V., ... Boutillier, A. L. (2013). A novel activator of CBP/p300 acetyltransferases promotes neurogenesis and extends memory duration in adult mice. *The Journal of Neuroscience, 33*, 10698–10712.

Clement, J. P., Aceti, M., Creson, T. K., Ozkan, E. D., Shi, Y., Reish, N. J., ... Rumbaugh, G. (2012). Pathogenic SYNGAP1 mutations impair cognitive development by disrupting maturation of dendritic spine synapses. *Cell, 151*, 709–723.

Coupry, I., Roudaut, C., Stef, M., Delrue, M. A., Marche, M., Burgelin, I., ... Arveiler, B. (2002). Molecular analysis of the CBP gene in 60 patients with Rubinstein-Taybi syndrome. *Journal of Medical Genetics, 39*, 415–421.

Fernandez-Albert, J., Lipinski, M., Lopez-Cascales, M. T., Rowley, M. J., Martin-Gonzalez, A. M., Del Blanco, B., ... Barco, A. (2019). Immediate and deferred epigenomic signatures of in vivo neuronal activation in mouse hippocampus. *Nature Neuroscience, 22*, 1718–1730.

Guo, X., Hamilton, P. J., Reish, N. J., Sweatt, J. D., Miller, C. A., & Rumbaugh, G. (2009). Reduced expression of the NMDA receptor-interacting protein SynGAP causes behavioral abnormalities that model symptoms of Schizophrenia. *Neuropsychopharmacology: Official Publication of the American College of Neuropsychopharmacology, 34*, 1659–1672.

Hamdan, F. F., Daoud, H., Piton, A., Gauthier, J., Dobrzeniecka, S., Krebs, M. O., ... Michaud, J. L. (2011). De novo SYNGAP1 mutations in nonsyndromic intellectual disability and autism. *Biological Psychiatry, 69*, 898–901.

Hamdan, F. F., Gauthier, J., Spiegelman, D., Noreau, A., Yang, Y., Pellerin, S., ... Michaud, J. L. (2009). Synapse to disease, G., mutations in SYNGAP1 in autosomal nonsyndromic mental retardation. *The New England Journal of Medicine, 360*, 599–605.

Kim, J. H., Lee, H. K., Takamiya, K., & Huganir, R. L. (2003). The role of synaptic GTPase-activating protein in neuronal development and synaptic plasticity. *The Journal of Neuroscience, 23*, 1119–1124.

Komiyama, N. H., Watabe, A. M., Carlisle, H. J., Porter, K., Charlesworth, P., Monti, J., ... Grant, S. G. (2002). SynGAP regulates ERK/MAPK signaling, synaptic plasticity, and learning in the complex with postsynaptic density 95 and NMDA receptor. *The Journal of Neuroscience, 22*, 9721–9732.

Korzus, E., Rosenfeld, M. G., & Mayford, M. (2004). CBP histone acetyltransferase activity is a critical component of memory consolidation. *Neuron, 42*, 961–972.

Lopez-Atalaya, J. P., Valor, L. M., & Barco, A. (2014). Epigenetic factors in intellectual disability: The Rubinstein-Taybi syndrome as a paradigm of neurodevelopmental disorder with epigenetic origin. *Progress in Molecular Biology and Translational Science, 128*, 139–176.

Maddox, S. A., Watts, C. S., Doyere, V., & Schafe, G. E. (2013). A naturally-occurring histone acetyltransferase inhibitor derived from Garcinia indica impairs newly acquired and reactivated fear memories. *PLoS One, 8*, e54463.

Mitchnick, K. A., Creighton, S. D., Cloke, J. M., Wolter, M., Zaika, O., Christen, B., ... Winters, B. D. (2016). Dissociable roles for histone acetyltransferases p300 and PCAF in hippocampus and perirhinal cortex-mediated object memory. *Genes, Brain and Behavior, 15*(6), 542–557.

Nakayama, T., Singh, A. K., Fukutomi, T., Uchida, N., Terao, Y., Hamada, H., ... Akagawa, K. (2024). Activator of KAT3 histone acetyltransferase family ameliorates a neurodevelopmental disorder phenotype in the syntaxin 1A ablated mouse model. *Cell Reports, 43*, 114101.

Oliveira, A. M., Estevez, M. A., Hawk, J. D., Grimes, S., Brindle, P. K., & Abel, T. (2011). Subregion-specific p300 conditional knock-out mice exhibit long-term memory impairments. *Learning & Memory (Cold Spring Harbor, N. Y.), 18*, 161–169.

Rumbaugh, G., Adams, J. P., Kim, J. H., & Huganir, R. L. (2006). SynGAP regulates synaptic strength and mitogen-activated protein kinases in cultured neurons. *Proceedings of the National Academy of Sciences of the United States of America, 103*, 4344–4351.

Schalock, R. L., Luckasson, R., & Tasse, M. J. (2021). An overview of intellectual disability: Definition, diagnosis, classification, and systems of supports. *American Journal on Intellectual and Developmental Disabilities, 126*(6), 439–442.

Schwenty-Lara, J., Nehl, D., & Borchers, A. (2020). The histone methyltransferase KMT2D, mutated in Kabuki syndrome patients, is required for neural crest cell formation and migration. *Human Molecular Genetics, 29*, 305–319.

Sheikh, B. N. (2014). Crafting the brain - role of histone acetyltransferases in neural development and disease. *Cell and Tissue Research, 356*, 553–573.

Singh, A. K., & Kundu, T. K. (2026). Landscape of protein post-translational modifications in age-related neurodegenerative disorders: Implications in protein aggregation, altered cellular processes, and therapeutics. *ACS Omega*.

Singh, A. K., Halder-Sinha, S., Clement, J. P., & Kundu, T. K. (2018). Epigenetic modulation by small molecule compounds for neurodegenerative disorders. *Pharmacological Research: the Official Journal of the Italian Pharmacological Society, 132*, 135–148.

Singh, A. K., Joshi, I., Reddy, N. M. N., Purushotham, S. S., Eswaramoorthy, M., Vasudevan, M., ... Kundu, T. K. (2025). Epigenetic modulation rescues neurodevelopmental deficits in Syngap1(+/-) mice. *Aging Cell, 24*, e14408.

Singh, A. K., Rai, A., Joshi, I., Reddy, D. N., Guha, R., Alka, K., ... Kundu, T. K. (2024). Oral administration of a specific p300/CBP lysine acetyltransferase activator induces synaptic plasticity and repairs spinal cord injury. *ACS Chemical Neuroscience, 15*, 2741–2755.

Smitha, A. S., Singh, A. K., R, P., Bhatt, R., ... Kundu, T. K. (2024). p300/CBP KATs are critical for maturation and differentiation of adult neural progenitors. *ACS Chemical Biology, 19*(11), 2345–2358.

Tsai, A. C., Dossett, C. J., Walton, C. S., Cramer, A. E., Eng, P. A., Nowakowska, B. A., ... Cheung, S. W. (2011). Exon deletions of the EP300 and CREBBP genes in two children with Rubinstein-Taybi syndrome detected by aCGH. *European Journal of Human Genetics: EJHG, 19*, 43–49.

Van Bokhoven, H. (2011). Genetic and epigenetic networks in intellectual disabilities. *Annual Review of Genetics, 45*, 81–104.

Yan, Z. (2024). Targeting epigenetic enzymes for autism treatment. *Trends in Pharmacological Sciences, 45*, 764–767.

CHAPTER FOUR

Mechanistic insights on HDAC6 ZnF UBP–Tau interaction through biophysical and computational approaches

Abhishek Ankur Balmik[a,*], Shweta Kishor Sonawane[b], Nalini Vijay Gorantla[c], and Subashchandrabose Chinnathambi[d,*]

[a]Department of Pharmacology and Chemical Biology, Emory University School of Medicine, Rollins Research Center, Atlanta, GA, United States
[b]Inst Neurophysiopathol, CNRS, INP, Aix-Marseille Univ, Marseille, France
[c]Department of Pathology, University of Chicago, Chicago, IL, United States
[d]Department of Neurochemistry, National Institute of Mental Health and Neuro Sciences Hospital (NIMHANS), Institute of National Importance, Bangalore, Karnataka, India
*Corresponding authors. e-mail address: abhi.balmik@gmail.com; subashneuro@nimhans.ac.in

Contents

Abstract

Histone deacetylase 6 is a unique cytoplasmic deacetylase implicated in cellular functions such as microtubule dynamics, protein quality control, ubiquitin-mediated degradation and neurodegenerative disorders. The ZnF UBP (zinc finger ubiquitin binding protein) domain of HDAC6 is known to be directly modulate several cellular processes linked to neurodegeneration such as sequestering polyubiquitinated aggregates and regulating protein aggregate clearance mechanisms in neurons. Microtubule associated protein Tau (MAP Tau) undergoes aggregation in neurodegenerative conditions like Alzheimer's disease (AD) and several other tauopathies. Tau is a natively disordered protein which is functionally regulated by wide array of post-translational modifications (PTMs) as well as by interacting with several proteins. This methodological study aims to understand the molecular interaction between HDAC6 ZnF UBP domain and Tau protein in order to elucidate the role of HDAC6 ZnF UBP domain in Tau aggregation and stability. We employed an integrated biochemical,

Methods in Enzymology, Volume 734
ISSN 0076-6879, https://doi.org/10.1016/bs.mie.2026.07.009

biophysical and computational workflow to characterize the interaction between HDAC6 ZnF UBP and Tau. NMR spectroscopy, isothermal titration calorimetry and pull-down assay with purified HDAC6 ZnF UBP and Tau proteins demonstrated direct interaction between the two, with interaction associated structural perturbations and favourable binding kinetics as observed in NMR and ITC respectively. Computational analyses further suggest the formation of Tau-HDAC6 ZnF UBP complex and provided underlying molecular interactions involved in the binding of these two proteins. The findings in this study helps to advance the current understanding of regulatory role of HDAC6 specifically in Tau biology and further provides a useful framework for investigating the modulation of aggregation prone proteins via protein-protein interaction in neurodegenerative diseases.

1. Introduction

MAP Tau is the common denominator in set of diseases collectively termed as tauopathies which includes progressive supranuclear palsy (PSP), corticobasal degeneration (CBD), frontotemporal dementia, Frontotemporal Lobar Degeneration with MAPT mutations, Argyrophilic Grain Disease (AGD), Pick's disease (PiD), Chronic Traumatic Encephalopathy (CTE) and Alzheimer's disease (AD) (Serrano-Pozo et al., 2026; Wang & Mandelkow, 2016; Zheng et al., 2024; Sonawane et al., 2022). Tau functions in the regulation of microtubule stability and axonal transport under physiological state while accumulates as aggregates as neurofibrillary tangles (NFTs) under pathological state (Weingarten et al., 1975; Stelzl et al., 2022; Goedert et al., 2017; Rawat et al., 2022). The formation of Tau aggregates can be triggered by numerous factors that can affect the conformational stability of Tau thereby leading to its detachment from microtubules and self-association to form aggregates (Zheng et al., 2024; Stelzl et al., 2022; Vaquer-Alicea & Diamond, 2019). Pathological changes in neuron during AD condition results in aberrant function of various key proteins, including Tau. Tau is highly regulated through the dynamics of its signature PTM profile as well as its interaction with various key proteins modulating the local conformations crucial for its function (Fig. 1) (Martin et al., 2011; Arakhamia et al., 2020; Di Lorenzo, 2024; Hanger et al., 2007). Changes in this PTM and interaction profile disrupts Tau physiological state and function while production of various aggregated state of Tau (soluble/insoluble oligomers, protofibrils and mature aggregates) may create novel interactome profile of Tau (Wright & Dyson, 2015; Younas et al., 2024).

HDAC6 has emerged as a key regulator of cellular functions related to protein quality control pathways (Boyault et al., 2007; Simões-Pires et al., 2013;

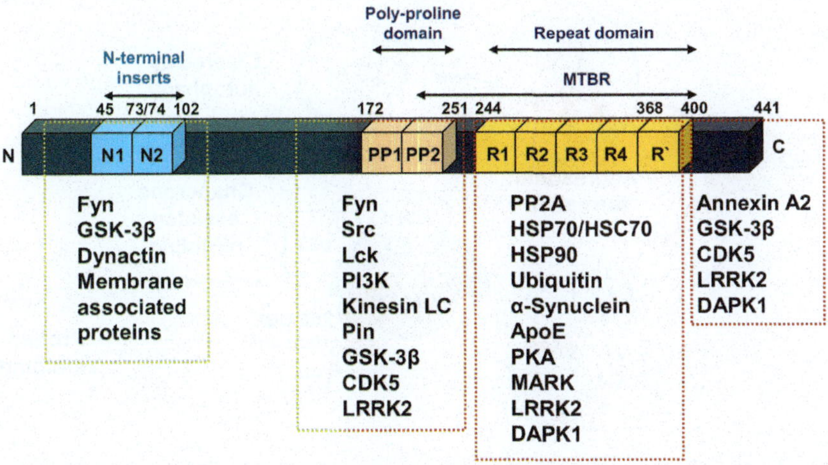

Fig. 1 Domain-specific interactions of tau with cellular proteins. Full-length human Tau comprises two N-terminal inserts (N1 and N2) and four C-terminal imperfect repeat regions of approximately 30 amino acids each (R1, R2, R3, and R4). The repeat domains are flanked by the proline-rich regions PP1 and PP2 on the N-terminal side and the pseudo-repeat region (R′) on the C-terminal side. Tau is an intrinsically disordered protein that interacts with numerous cellular proteins, which regulate its function through post-translational modifications such as phosphorylation and acetylation or through transient interactions that influence its local conformation. The figure illustrates the major proteins reported to interact with different regions of Tau.

Kwon et al., 2007). The unique domain organization of HDAC6 enables it to function as a deacetylase enzyme primarily acting on cytoplasmic proteins, including Tau, as well as an adaptor protein through its ZnF UBP domain (Valenzuela-Fernandez et al., 2008; Yan et al., 2013). HDAC6 is involved in multiple processes relevant to neurodegeneration such as modulation of microtubule dynamics, autophagy, clearance of aggregated proteins and aggresome formation (Hai & Christianson, 2016; Lee et al., 2010; Wang et al., 2024). The levels of HDAC6 have been found to be upregulated in AD pathology pertaining to its role in aggregate clearance whereas the deacetylase function of HDAC6 known to exacerbate neurodegenerative state (Fig. 2). This suggests that HDAC6 may have a dual role in AD in neurodegeneration and neuroprotection through its catalytic domains and ZnF UBP domains respectively. Several studies have implicated the role of HDAC6 in AD pathology affecting the axonal transport and synaptic plasticity. Furthermore, HDAC6 has been shown to influence actin cytoskeletal organization as well as Tau function via acetylation, suggesting its broader role in neuronal health and disease (Guo et al., 2017; Govindarajan et al., 2013; Min et al., 2010).

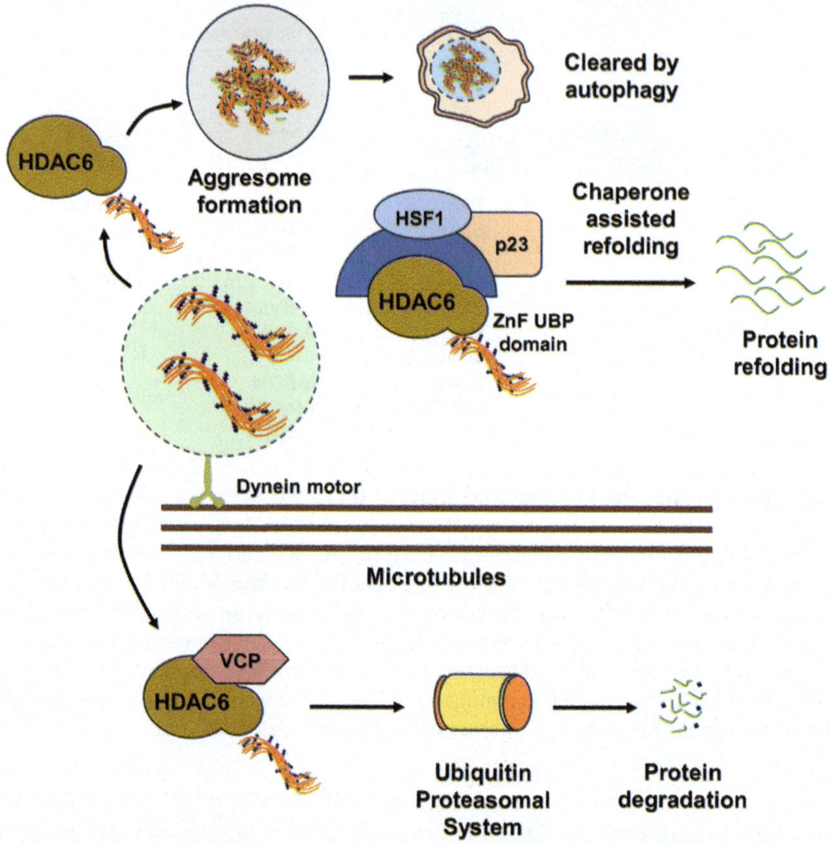

Fig. 2 Cellular mechanisms involved in protein aggregate clearance. Protein misfolding disrupts proteostasis and leads to the accumulation of toxic protein aggregates, including oligomeric and fibrillar species. Misfolded proteins activate molecular chaperones such as HSP90 and HSF1, which facilitate protein refolding. Proteins that cannot be efficiently refolded are targeted for degradation through the ubiquitin–proteasome system (UPS) following polyubiquitination. HDAC6 regulates both the chaperone-mediated refolding pathway and UPS and acts as a functional link between these processes. As aggregate burden increases, the degradative capacity of UPS becomes insufficient, leading cells to employ the aggresome–autophagy pathway. Aggresomes are perinuclear aggregate-rich structures that are subsequently degraded through autophagy. HDAC6 contributes to aggresome formation by binding polyubiquitinated aggregates through its ZnF UBP domain and recruiting dynein motor proteins for their transport and sequestration.

The role of HDAC6 as a direct modulator of Tau in a non-enzymatic manner through protein-protein interaction has been less explored. Tau-HDAC6 ZnF UBP interaction can provide important insight into Tau homeostasis with respect to its stability in functional native state and various pathological aggregation species. In this study, we utilized biochemical and biophysical approach complemented with computational methods to examine Tau-HDAC6 ZnF UBP association with an aim to determine the role of ZnF UBP domain in tau structure stability and function.

2. Rationale

Protein aggregation and impaired proteostasis are the central pathological feature of several neurodegenerative diseases (Zheng et al., 2024; Wright & Dyson, 2015). Tau aggregation results in neurotoxicity leading to disruption numerous neuronal functions such as axonal transport, disruptive energy transduction, synaptic loss (Farias et al., 2011). The function of Tau is dependent on its extensive PTM signature profile as well as its association with several interacting proteins which can modulate

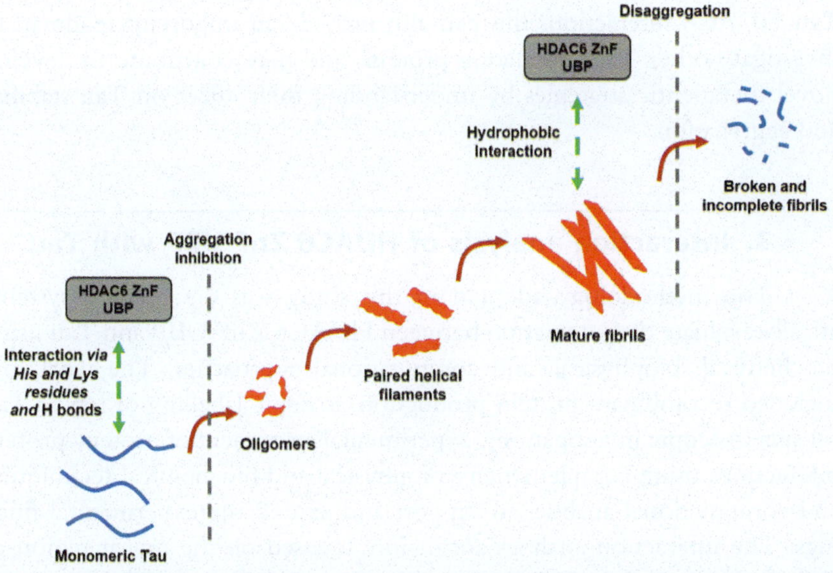

Fig. 3 Schematic representation of HDAC6 ZnF UBP-mediated regulation of Tau aggregation pathways. The ability of HDAC6 ZnF UBP to inhibit Tau aggregation and promote disaggregation of preformed Tau aggregates was evaluated using in vitro assays. HDAC6 ZnF UBP interacts with Tau primarily through hydrophobic contacts and interactions involving His and Lys residues within the repeat domain. These interactions weaken intermolecular Tau associations, resulting in aggregation inhibition and aggregate disassembly.

Tau's local conformation. Most of these proteins are involved in proteostatic mechanisms – molecular chaperones, ubiquitin proteasomal system (UPS) and autophagy, thus plays an important role in Tau turn over (Fig. 3) (Lee et al., 2013; Gorantla & Chinnathambi, 2018, 2021; Chinnathambi, 2025).

HDAC6 is a critical mediator of protein quality control as it binds to poly-ubiquitinated proteins through its ZnF UBP domain, facilitating their autophagic degradation via aggresome formation. HDAC6 association with Tau has been previously investigated but the mechanistic insights of their association and the effect of this association on Tau structure, stability and aggregation properties are yet to be explored (Balmik et al., 2020, 2021; Cook et al., 2012; Perez et al., 2009).

The present methodological approach offers a framework to investigate the Tau-HDAC6 ZnF UBP interactions by using NMR, ITC, pull-down assay and computational analyses providing useful insights on structural perturbation, binding energetics, binding affinity and intermolecular forces and validating their association. While providing important insights on Tau-HDAC6 interaction, this can also provide an important roadmap to investigate other Tau interacting proteins and may contribute to develop novel therapeutic strategies by understanding their effect on Tau stability and aggregation.

3. Interaction analysis of HDAC6 ZnF UBP with Tau

The methodology adopted in this study was designed to systematically evaluate the interaction between HDAC6 ZnF UBP and Tau using biochemical, biophysical, and computational approaches. The workflow included recombinant protein production, isotopic labelling of repeat Tau for spectroscopic investigations, experimental assessment of protein-protein interactions using complementary biophysical and biochemical techniques, and computational analyses to support and extend the experimental findings. The interaction analyses are mainly focused on the repeat region of Tau (amino acid 244–372, designated as R1, R2, R3 and R4), since it's role in Tau function in MT binding as well as its involvement in Tau aggregation (Goedert et al., 2017; Vaquer-Alicea & Diamond, 2019; Wright & Dyson, 2015). The details of each method and the corresponding experimental conditions are described in the following sections.

3.1 Purification of HDAC6 ZnF UBP

Recombinant protein expression was evaluated by culturing single isolated colonies of transformed Escherichia coli BL21* and BL21 CodonPlus RIL strains harboring the Tau (pT7C) and HDAC6 ZnF UBP (pET28a-LIC) expression plasmids, respectively. The plasmids confer antibiotic resistance, enabling the selective growth of transformed cells while suppressing non-transformed cells and potential contaminants. In both constructs, the gene of interest is under the control of the lac operator and T7 promoter system. Upon induction, T7 RNA polymerase initiates transcription of the target gene. Protein expression was induced using isopropyl β-D-1-thiogalactopyranoside (IPTG), a non-metabolizable analog of allolactose that activates the lac operon by relieving repression by the LacI protein. The expression levels of the recombinant proteins were assessed by comparing induced and uninduced bacterial cultures using SDS–PAGE analysis. Colonies exhibiting optimal protein expression were subsequently selected for large-scale culture and protein production.

3.1.1 Reagents

1. Lysis buffer (50 mM Tris-Cl pH 8.0, 500 mM NaCl, 20 mM imidazole, PMSF, protease inhibitors)
2. Wash buffer (50 mM Tris-Cl pH 8.0, 500 mM NaCl, 20 mM imidazole)
3. Elution buffer (50 mM Tris-Cl pH 8.0, 500 mM NaCl, 1 M imidazole, 5 % glycerol)
4. Dialysis buffer (50 mM Tris-Cl pH 8.0, 100 mM NaCl, 2 mM TCEP/DTT, 5 % glycerol)

3.1.2 Equipment's

1. High-pressure cell disruptor (Constant Systems Ltd.)
2. Ultracentrifuge (Optima XPN10 Ultracentrifuge - Beckman Coulter)
3. Ni-NTA Agarose (Qiagen)
4. Polypropylene chromatography columns (Bio-Spin - Bio-Rad)
5. Size-exclusion chromatography system (Superdex 75 HiLoad 16/600)
6. Chromatography system (AKTA Pure – GE)

3.1.3 Procedure

1. The HDAC6 ZnF-UBP construct cloned into the pET28a-LIC vector was transformed into Escherichia coli BL21 CodonPlus (RIL) cells and purified using a two-step procedure involving Ni-NTA affinity chromatography followed by size-exclusion chromatography *(Please see the notes 1 from 3.1.4 section)*.

2. A primary culture was prepared by inoculating transformed BL21 CodonPlus (RIL) cells from a glycerol stock into Terrific Broth (TB) supplemented with 50 μg/mL kanamycin and incubating overnight at 37 °C with shaking.
3. The overnight culture was used to inoculate a secondary culture in TB containing 50 μg/mL kanamycin at a 1:100 dilution. Cells were grown at 37 °C until an OD_{600} of approximately 0.6 was reached *(Please see the notes 1 from 3.1.4 section)*.
4. The culture was cooled to 16 °C and incubated for 1 h prior to induction.
5. Protein expression was induced with 0.2 mM IPTG, and the culture was further incubated at 16 °C for 16 h. Cells were then harvested by centrifugation and resuspended in 20–100 mL of lysis buffer, depending on the pellet size *(Please see the notes 2 from 3.1.4 section)*.
6. Cell pellets were homogenously resuspended by vortexing and lysed using a Constant Systems cell disruptor operated at 15,000 psi.
7. The lysate was clarified by centrifugation at 20,000 rpm for 30 min at 4 °C.
8. The supernatant was incubated with Roche Ni-NTA agarose resin at a resin-to-lysate ratio of 1:30 (v/v) for 1 h at 4 °C to facilitate protein binding.
9. Affinity purification was performed using 50 mM Tris-HCl (pH 8.0) containing 20 mM imidazole for washing, followed by elution with 50 mM Tris-HCl (pH 8.0) containing 1 M imidazole *(Please see the notes 2 from 3.1.4 section)*.
10. Eluted fractions were dialyzed against 50 mM Tris-HCl (pH 8.0), 100 mM NaCl, 2 mM TCEP (or DTT), and 5 % glycerol to remove excess imidazole.
11. Further purification was carried out by size-exclusion chromatography using a Superdex 75 HiLoad 16/600 column equilibrated with 50 mM Tris-HCl (pH 8.0), 100 mM NaCl, and 2 mM TCEP (or DTT). HDAC6 ZnF-UBP eluted at a retention volume of approximately 110 mL.
12. Fractions containing HDAC6 ZnF-UBP were pooled, concentrated, flash-frozen, and stored at −80 °C. Protein concentration was determined using the Bradford assay.

3.1.4 Notes

1. Maintain all purification steps under cold conditions (4 °C) to minimize protein degradation and preserve protein stability.

2. The imidazole concentration in the wash buffer is critical for efficient removal of non-specifically bound proteins while retaining the target protein on the Ni-NTA resin. Optimization of imidazole concentration may be required to achieve maximum purity.

3.2 Preparation of ^{15}N labelled repeat Tau

To facilitate NMR studies of Tau interactions with small molecules and protein partners, Tau was uniformly labeled with ^{15}N isotope. Briefly, *E. coli* cells harboring the Tau expression plasmid were cultivated in 6 L of LB medium at 37 °C until the culture reached an absorbance at 600 nm reaches 0.5–0.6. The cells were harvested by centrifugation at 5,000 rpm for 30 min and resuspended in unlabeled 1 × M9 salts containing 3 g/L of KH2PO4, 5 g of Na2HPO4 and 0.5 g of NaCl per liter media. After washing, the cells were pelleted again and the supernatant was discarded. The washed cells were then transferred to 1.5 L of minimal medium containing 1 × M9 salts, D-glucose, thiamine hydrochloride, MgSO4, CaCl2 and 15NH4Cl, along with ampicillin. The culture was allowed to recover at 37 °C with shaking for 1 h to facilitate adaptation to the minimal medium and incorporation of ^{15}N. Protein expression was induced with 0.5 mM IPTG, and the culture was incubated for an additional 4–5 h at 37 °C. Cells were harvested by centrifugation, and ^{15}N-labeled Tau protein was purified using cation-exchange chromatography followed by size-exclusion chromatography, as described above.

3.2.1 Reagents

1. 5X M9 salts [KH_2PO_4 (1.5 g/L), Na_2HPO_4 (3 g/L), NaCl (1.5 g/L), ^{15}N labelled NH_4Cl (1.5 g/L)]
2. 1X M9 salts [KH_2PO_4 (3 g/L), Na_2HPO_4 (5 g/L), NaCl (0.5 g/L)] pH 7.2
3. Minimal media [20% D-glucose, thiamine hydrochloride (5 mg/mL), 1 M $MgSO_4$, 1 M $CaCl_2$]
4. Ampicillin (100 μg/mL)

3.2.2 Equipment's

1. Incubator shaker (Infors HT)
2. Refrigerated centrifuge (Microcentrifuge 5418 R – Eppendorf)
3. Spectrophotometer (Jasco)
4. Autoclave
5. Filtration setup for sterilization (Millipore)

3.2.3 Preparation of minimal media

1. Add 300 mL of 5X M9 salts.
2. Add 30 mL of 20 % D-glucose solution.
3. Add 9 mL thiamine hydrochloride solution.
4. Filter sterilize glucose and thiamine solutions aseptically after autoclaving to prevent degradation caused by heat sterilization.
5. Add 3 mL of 1 M $MgSO_4$.
6. Add 150 μL of 1 M $CaCl_2$.
7. Make up the final volume to 1500 mL with Milli-Q water.
8. Autoclave and cool the minimal media prior to use.

3.2.4 Procedure

1. pT7C repeat Tau transformed BL21* cells were grown in 6 L LB medium containing 100 μg/mL ampicillin at 37 °C until OD_{600} reached ~0.6 *(Please see the notes 1 from 3.2.5 section).*
2. Cells were harvested by centrifugation and washed with sterile 1X M9 salts.
3. Thorough washing of cells with 1 × M9 salts is essential to minimize contamination from unlabelled nitrogen sources, thereby improving the efficiency of ^{15}N isotopic labeling *(Please see the notes 1 from 3.2.5 section).*
4. Washed cells were resuspended in minimal medium containing ^{15}N-labelled ammonium chloride as the sole nitrogen source *(Please see the notes 2 from 3.2.5 section).*
5. The culture was incubated at 37 °C for 1 h to deplete intracellular pools of unlabelled nitrogen-containing metabolites before induction *(Please see the notes 3 from 3.2.5 section).*
6. Protein expression was induced using 0.5 mM IPTG.
7. Cells were harvested after 4–5 h of induction.
8. The ^{15}N-labelled repeat Tau protein was purified using the same protocol employed for the purification of unlabelled Tau protein.

3.2.5 Notes

1. The pH of the M9 salts solution should be adjusted to 7.2 prior to autoclaving to ensure optimal growth conditions.
2. Glucose and thiamine solutions must be filter sterilized and added aseptically after autoclaving to prevent degradation caused by heat sterilization.
3. Thorough washing of cells with 1 × M9 salts is essential to minimize contamination from unlabelled nitrogen sources, thereby improving the efficiency of ^{15}N isotopic labeling.

3.3 Tau-HDAC6 interaction analysis

Ni-NTA pull-down assay is an affinity-based technique used to investigate protein–protein interactions through the co-elution of interacting proteins. The assay relies on the specific binding of a His-tagged protein to Ni^{2+} ions immobilized on nitrilotriacetic acid (NTA) resin via coordinate bonds. In a typical experiment, one protein is expressed with a His-tag and serves as the bait protein, while the potential interacting partner is used as the prey protein. The bait and prey proteins are incubated together with Ni-NTA resin under suitable binding conditions. Following incubation, the resin is washed several times with buffer containing a low concentration of imidazole to remove unbound and non-specifically bound proteins while retaining the His-tagged bait and any associated prey protein. The bound protein complex can then be recovered either by elution with a higher concentration of imidazole or by direct analysis of the resin-bound fraction. The presence of the prey protein in the bound fraction indicates a potential interaction with the bait protein. The pull-down assay was performed using Ni-NTA resin to immobilize HDAC6 ZnF UBP as the bait protein and Tau as the prey protein. HDAC6 ZnF UBP (20 μM) and Tau (20 μM) were mixed in a total volume of 200 μL of 1 × PBS (pH 7.4) and incubated to allow complex formation. Subsequently, 20 μL of Ni-NTA resin was added, and the mixture was incubated at 4 °C for 1 h with constant mixing. Following incubation, the resin was washed five times with 1 × PBS (pH 7.4) containing 30 mM imidazole to remove unbound and non-specifically bound proteins. The beads were collected by centrifugation after each wash. Samples collected from the input, flow-through, wash, and elution fractions were analyzed by SDS–PAGE to assess the interaction between HDAC6 ZnF UBP and Tau (Fig. 4).

3.3.1 Ni-NTA pull-down assay

3.3.1.1 Reagents

1. HDAC6 ZnF UBP protein
2. Full-length human Tau protein (hTau40wt)
3. Ni-NTA resin (Roche)
4. 1X PBS pH 7.4
5. Imidazole (MP Biomedicals)

3.3.1.2 Equipment's

1. Tube rotator mixer (Benchmark)
2. Refrigerated centrifuge (Eppendorf Microcentrifuge 5418 R)
3. SDS-PAGE apparatus (Bio-Rad)

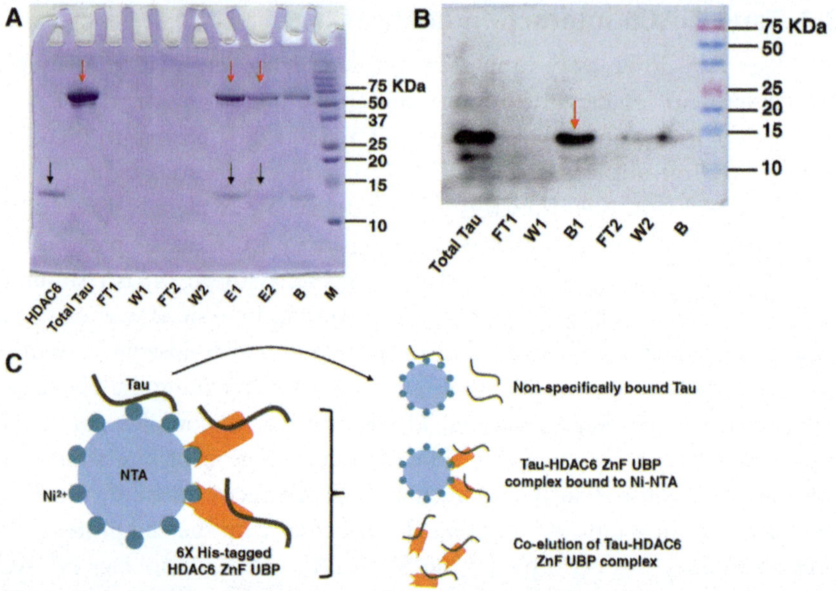

Fig. 4 In vitro interaction analysis of hTau40wt and HDAC6 ZnF UBP. (A) Pull-down analysis of Tau and HDAC6 ZnF UBP was performed using 6×His-tagged HDAC6 ZnF UBP as the bait protein and Tau as the prey protein. Elution with 50 mM Tris-Cl (pH 8.0) containing 1 M imidazole resulted in co-elution of Tau and HDAC6 ZnF UBP. A significant fraction of both proteins remained associated with the Ni-NTA beads. (B) Interaction analysis between K18wt and HDAC6 ZnF UBP was performed by co-incubating both proteins before incubation with Ni-NTA beads. A control experiment containing K18wt alone was included to assess non-specific binding. Flow-through, final wash, and bead fractions (FT1, W1, and B1 for K18wt with HDAC6 ZnF UBP; FT2, W2, and B2 for the control) were analyzed by SDS-PAGE and Western blotting using the total Tau antibody K9JA. Upon 0.5 s exposure, K18wt was detected in the bead fraction containing both proteins, indicating interaction with HDAC6 ZnF UBP, whereas only minimal signal was observed in the control bead fraction. (C) The Tau–HDAC6 ZnF UBP complex either co-eluted or was directly analyzed on SDS-PAGE in association with the Ni-NTA beads. In the control experiment, a small proportion of Tau exhibited non-specific binding to the beads.

3.3.1.3 Procedure

1. HDAC6 ZnF UBP (20 μM) and full-length Tau (10 μM) were mixed in PBS buffer and incubated with Ni-NTA resin to facilitate binding of the His-tagged HDAC6 ZnF UBP and associated protein complexes. All incubation and washing steps should be performed at 4 °C to preserve protein stability and minimize proteolytic degradation (Fig. 4) *(Please see the notes 1 from 3.3.1.4 section).*

2. The reaction mixture was incubated at 4 °C for 1 h with continuous gentle mixing to allow complex formation and binding to the resin.
3. The resin was washed five times with PBS containing 30 mM imidazole to remove unbound and non-specifically bound proteins. Low imidazole concentration helps in reducing non-specific protein binding while retaining His-tagged HDAC6 ZnF UBP on the resin *(Please see the notes 2 from 3.3.1.4 section).*
4. Multiple wash steps are necessary to remove unbound proteins and improve the specificity of the assay.
5. Bound protein complexes were recovered by centrifugation, and the final resin-bound fraction was collected for analysis *(Please see the notes 3 from 3.3.1.4 section).*
6. Appropriate controls (bait alone, prey alone, and resin alone) are included to confirm the specificity of the observed interaction *(Please see the notes 4 from 3.3.1.4 section).*
7. Wash and elution fractions were analyzed by SDS–PAGE followed by Western blotting to assess the interaction between HDAC6 ZnF UBP and Tau *(Please see the notes 5 from 3.3.1.4 section).*

3.3.1.4 Notes

1. All incubation and washing steps should be performed at 4 °C to preserve protein stability and minimize proteolytic degradation.
2. The concentration of imidazole in the wash buffer is important for reducing non-specific protein binding while retaining His-tagged HDAC6 ZnF UBP on the resin.
3. Adequate mixing during incubation is essential to facilitate efficient interaction between the bait and prey proteins.
4. Multiple wash steps are necessary to remove unbound proteins and improve the specificity of the assay.
5. Include appropriate controls (bait alone, prey alone, and resin alone) to confirm the specificity of the observed interaction.

3.3.2 1H-15N HSQC nuclear magnetic resonance spectroscopy

Heteronuclear Single Quantum Coherence (HSQC) Nuclear Magnetic Resonance (NMR) spectroscopy is a two-dimensional NMR technique widely used to investigate protein structure, dynamics, and molecular interactions. HSQC correlates the resonance frequencies of protons (^{1}H) with those of a heteronucleus, typically ^{15}N or ^{13}C, allowing the detection of individual nuclei within a protein. For protein interaction studies, proteins are commonly isotopically labelled with ^{15}N, which is incorporated into the backbone amide groups and certain side chains of amino acid residues during protein expression. In a ^{1}H–^{15}N HSQC spectrum, each non-proline amino acid residue generally gives rise to a distinct cross-peak corresponding to its backbone amide proton and attached ^{15}N nucleus. These peaks collectively form a characteristic spectral fingerprint of the protein, and individual resonances can be assigned to specific residues. Signals arising from many side-chain amide groups are often weak or absent due to rapid exchange with solvent protons. ^{1}H–^{15}N HSQC spectroscopy is extensively employed to study protein–ligand and protein–protein interactions. Spectra of the free protein are compared with those recorded in the presence of a binding partner. Interaction-induced changes in the local chemical environment of amino acid residues result in chemical shift perturbations, peak broadening, or intensity changes. Residues exhibiting significant spectral changes are typically located at or near the binding interface, providing residue-specific information about the interaction and facilitating the mapping of binding sites on the protein surface (Fig. 5).

3.3.2.1 Reagents

1. ^{15}N - labelled repeat Tau protein
2. HDAC6 ZnF UBP protein
3. Phosphate buffer (MP Biomedicals)

3.3.2.2 Equipment's

1. Optima MAX-XP Tabletop Ultracentrifuge (Beckman Coulter)
2. Avance III HD 700 MHz NMR spectrometer (Bruker)
3. TXI probe
4. NMR tubes
5. Bruker TopSpin software
6. Sparky software

3.3.2.3 Procedure

1. A 200 μM solution of ^{15}N - labelled repeat Tau was prepared in phosphate buffer.
2. Protein samples should be freshly prepared and maintained under reducing conditions (1 mM DTT) to prevent oxidation *(Please see the notes 1 from 3.3.2.4 section)*.
3. Consistent buffer composition is maintained for protein and ligand solutions to minimize spectral artifacts during titration. The percentage

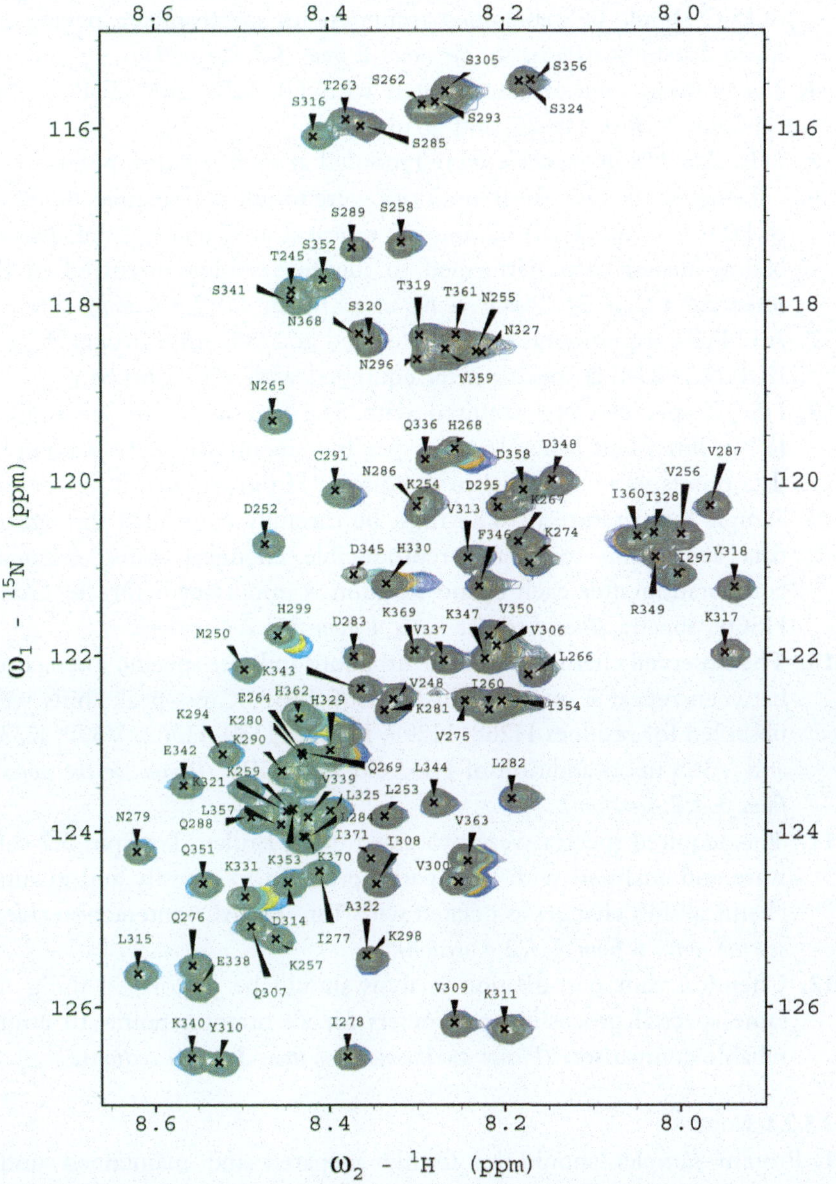

Fig. 5 ^{1}H-^{15}N HSQC NMR analysis of Tau and HDAC6 ZnF UBP interaction. Overlay of 1H–15N HSQC spectra acquired during titration of 200 μM ^{15}N -labelled repeat Tau with HDAC6 ZnF UBP at concentrations of 0 μM (gray), 10 μM (turquoise), 25 μM (orange), and 50 μM (blue). Peak assignments in the reference spectrum correspond to residues of repeat Tau [217]. Residues unaffected by HDAC6 ZnF UBP exhibit unchanged cross-peak positions, whereas interacting residues display chemical shift perturbations reflected by changes in cross-peak positions.

of D_2O should be maintained accurately for spectrometer locking and spectral stability *(Please see the notes 2 from 3.3.2.4 section).*

4. The protein solution was titrated with HDAC6 ZnF UBP at concentrations of 0, 10, 25, and 50 μM.
5. 1H−15N HSQC spectra were recorded at each titration point.
6. Changes in the cross-peak positions were monitored to assess the effect of HDAC6 ZnF UBP binding on repeat Tau. Chemical shift perturbation analysis was performed to identify residues involved in the interaction (Fig. 5) *(Please see the notes 3 from 3.3.2.4 section).*
7. All NMR experiments were performed at 278 K on a Bruker Avance III HD 700 MHz spectrometer equipped with a TXI probe.
8. HSQC spectra were acquired with 512 increments in the indirect (^{15}N) dimension using 16 scans per increment, while 2048 complex data points were collected in the direct ^{1}H dimension.
9. Sample temperature (278 K) must be carefully controlled throughout data acquisition to ensure reproducible chemical shifts. Adequate equilibration after each ligand addition is important before recording HSQC spectra *(Please see the notes 4 from 3.3.2.4 section).*
10. The observed chemical shift perturbations indicate specific interactions between repeat Tau and HDAC6 ZnF UBP. Cross-peak shifts were observed for residues H268, H299, H329, H330, H362, V300, K331, and V363 upon addition of HDAC6 ZnF UBP *(Please see the notes 5 from 3.3.2.4 section).*
11. The acquired spectra were processed using Bruker TopSpin 3.2 software and analyzed with the Sparky program to identify and quantify chemical shift changes associated with Tau–HDAC6 interaction *(Please see the notes 6 from 3.3.2.4 section).*
12. Chemical shift perturbation analysis should be performed using the same spectral processing parameters for all titration points to ensure reliable comparison *(Please see the notes 7 from 3.3.2.4 section).*

3.3.2.4 Notes

1. Protein samples should be freshly prepared and maintained under reducing conditions (1 mM DTT) to prevent oxidation.
2. Consistent buffer composition between protein and ligand solutions is essential to minimize spectral artifacts during titration.
3. The percentage of D_2O should be maintained accurately for spectrometer locking and spectral stability.

4. Sample temperature (278 K) must be carefully controlled throughout data acquisition to ensure reproducible chemical shifts.
5. Cross-peak shifts were observed for residues H268, H299, H329, H330, H362, V300, K331, and V363 upon addition of HDAC6 ZnF UBP.
6. Adequate equilibration after each ligand addition is important before recording HSQC spectra.
7. Chemical shift perturbation analysis should be performed using the same spectral processing parameters for all titration points to ensure reliable comparison.

3.3.3 Isothermal titration calorimetry

Isothermal titration calorimetry (ITC) is a biophysical technique used to characterize molecular interactions in solution by directly measuring the heat changes associated with binding events. It provides quantitative information on the thermodynamic parameters of an interaction, including the binding affinity (Kd), enthalpy change (ΔH), entropy change (ΔS), Gibbs free energy change (ΔG), and binding stoichiometry (n). An ITC instrument consists of a sample cell, a reference cell, an injection syringe, and a cleaning station. During an experiment, the protein solution is loaded into the sample cell, while the ligand solution is placed in the syringe for titration. The reference cell is filled with water or the same buffer used for the protein and ligand solutions. Upon each injection of ligand into the sample cell, binding between the interacting molecules results in either the release (exothermic) or absorption (endothermic) of heat. The instrument continuously measures these heat changes as differential power required to maintain both cells at the same temperature, and the data are recorded as heat flow (μcal/s) over time. The ITC data were analyzed using the MicroCal PEAQ-ITC Analysis Software and fitted using a one-set-of-sites binding model. Heat changes obtained from the buffer titration were used as controls and subtracted from the experimental data. Baseline correction and control fitting were performed using the line mode available in the analysis software. The resulting thermograms were used to determine the binding stoichiometry and thermodynamic parameters of the interaction.

3.3.3.1 Reagents

1. HDAC6 ZnF UBP protein
2. Repeat Tau (K18wt) protein
3. 20 mM BES buffer pH 7.4 (Sigma)

3.3.3.2 Equipment's

1. Malvern PEAQ-ITC instrument (MicroCal)
2. Syringes
3. 5 kDa molecular weight cut-off (MWCO) filters
4. Degassing unit

3.3.3.3 Procedure

1. HDAC6 ZnF UBP and Repeat Tau (K18wt) were exchanged into 20 mM BES buffer (pH 7.4) using 5 kDa molecular weight cut-off (MWCO) centrifugal filters to ensure buffer matching *(Please see the notes 1 from 3.3.3.4 section).*
2. Protein samples were prepared in the same buffer, with HDAC6 ZnF UBP adjusted to a concentration of 50 μM and Repeat Tau (K18wt) adjusted to 5 μM.
3. All solutions were filtered and degassed prior to the experiment to remove particulates and dissolved air bubbles *(Please see the notes 1 from 3.3.3.4 section).*
4. ITC measurements were performed at 25 °C using Repeat Tau (K18wt) loaded into the sample cell and HDAC6 ZnF UBP loaded into the injection syringe.
5. A preliminary injection of 0.4 μL was carried out to equilibrate the system and was excluded from data analysis.
6. The titration consisted of 19 successive injections of 2 μL each, with a spacing interval of 180 s between injections.
7. Raw thermograms were integrated and analyzed using MicroCal PEAQ-ITC Analysis Software. The binding isotherm was fitted using a one-set-of-sites binding model to determine the thermodynamic parameters and binding stoichiometry of the interaction.

3.3.3.4 Notes

1. Maintain identical buffer conditions for both protein and ligand samples to minimize heat artifacts arising from buffer mismatch.
2. Ensure that all solutions are thoroughly filtered, degassed, and free of particulates or air bubbles prior to loading into the ITC cell and syringe.

3.3.4 Molecular modelling and docking

The three-dimensional model of the HDAC6 Zinc Finger Ubiquitin-Binding Protein (ZnF UBP) domain was generated using the SWISS-MODEL server. The crystal structure of the human HDAC6 ZnF UBP domain (PDB ID: 3C5K) obtained from the RCSB Protein Data Bank was used as the template for model generation. The model exhibiting a Global Model Quality Estimate (GMQE) score of 0.99 was selected for subsequent analyses. The Repeat Tau protein model previously developed by our group was used for the interaction studies. Protein–protein docking between HDAC6 ZnF UBP and Repeat Tau was performed using the PatchDock server. Since the study involved protein–protein interactions, the clustering RMSD parameter was set to 4 Å. The top ten docking solutions generated by PatchDock were further refined using the FireDock server. Among the refined complexes, the model exhibiting the lowest binding energy was selected for subsequent simulation and interaction analyses. Structural visualization and analysis of the docked complexes were carried out using VMD (version 1.9.4a38, October 20, 2019) and PyMOL Molecular Graphics System (version 1.8.4.0) (Fig. 6).

Molecular dynamics (MD) simulations of the docked protein–protein complex were performed using GROMACS version 2019.3 with the AMBER03 force field. The TIP3P water model was employed for solvation. The docked complex was positioned at the center of a cubic simulation box with a minimum distance of 10 Å between the protein surface and the box boundaries. The initial dimensions of the simulation box were set to 88 × 88 × 88 Å. The system was solvated with water molecules, and chloride ions (Cl^-) were added to neutralize the overall charge of the system.Energy minimization was carried out using the steepest descent algorithm until the maximum force (Fmax) reached the desired convergence criterion. The system converged in 418 steps with an Fmax value below 1000 kJ/mol/nm. Following minimization, the system was subjected to temperature and pressure equilibration using the leap-frog integrator. Non-bonded interactions were calculated using the Verlet cut-off scheme with buffered neighbor searching, while long-range electrostatic interactions were treated using the Particle Mesh Ewald (PME) method. Temperature equilibration was performed at 310 K using the V-rescale thermostat (modified Berendsen thermostat), with initial velocities generated according to the Maxwell–Boltzmann distribution under periodic boundary conditions along the x-, y-, and z-axes. Pressure equilibration was subsequently carried out at 1.0 bar using Berendsen pressure coupling under periodic boundary conditions (Fig. 7).

The production MD simulation was performed for 50 ns, corresponding to 25,000,000 integration steps. The resulting trajectories were visualized using VMD software. Structural stability and dynamic behavior

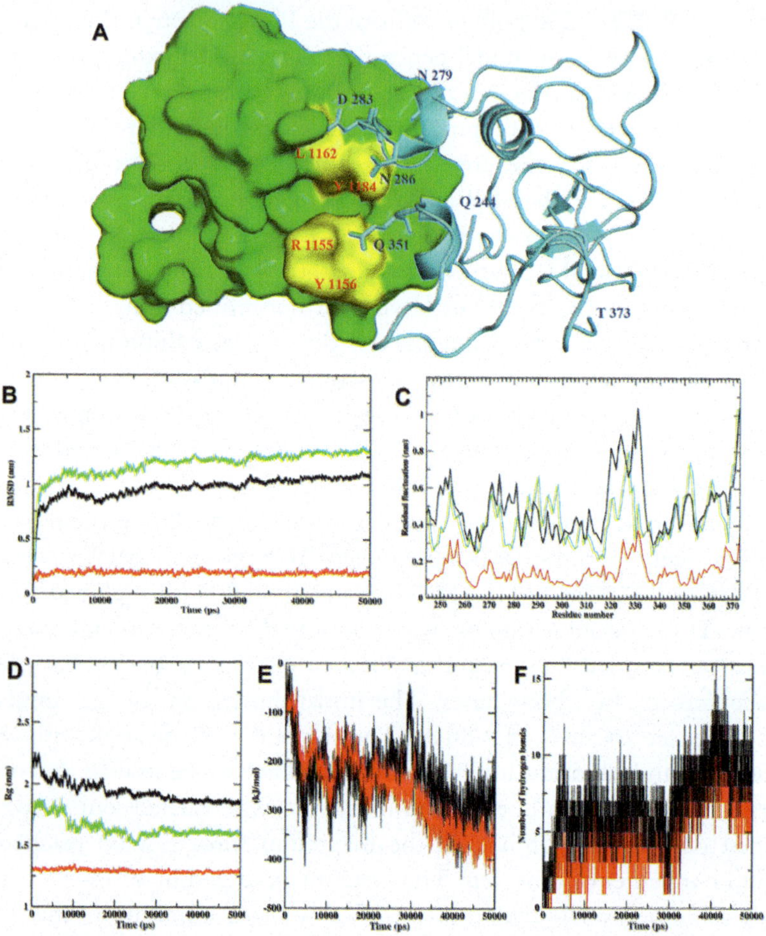

Fig. 6 Molecular docking and molecular dynamics simulation of HDAC6 ZnF UBP with repeat Tau (244–373). (A) Refined docked complex of repeat Tau (244–373) and HDAC6 ZnF UBP. HDAC6 ZnF UBP is shown in green, repeat Tau in cyan, and the active region of HDAC6 ZnF UBP in yellow. (B) RMSD profiles of HDAC6 ZnF UBP (red), repeat Tau (green), and the complex (black), plotted as RMSD (nm) versus simulation time (ps). (C) RMSF analysis of Tau residues. Residue number is plotted on the X-axis and fluctuation (nm) on the Y-axis. RMSF values for the initial 10 ns (red), final 10 ns (green), and entire 50 ns simulation (black) are shown. (D) Radius of gyration (Rg) profiles of HDAC6 ZnF UBP (red), repeat Tau (green), and the complex (black), plotted as Rg (nm) versus simulation time (ps). (E) Lennard–Jones (red) and Coulombic (black) interaction energy profiles representing hydrophobic and electrostatic interactions within the complex. Energy values are plotted against simulation time. (F) Hydrogen bond analysis showing the total number of intermolecular hydrogen bonds formed between HDAC6 ZnF UBP and repeat Tau (black) over the simulation period. Hydrogen bonds involving specific HDAC6 ZnF UBP residues (Q1131, R1155, R1155/Y1184, N1158, Q1163, and Q1187) and repeat Tau residues (N327/I328, D283, N286, E264/N265, I277/N279, and L282, respectively) are also shown (red).

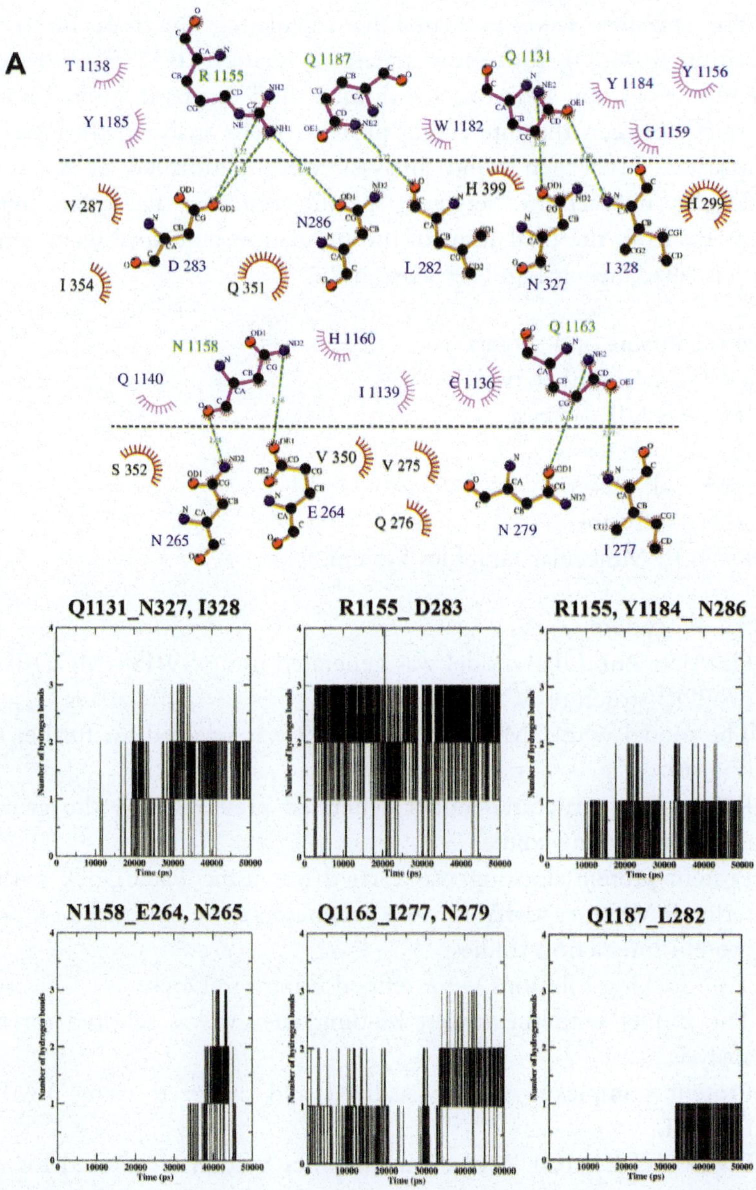

Fig. 7 Intermolecular interaction analysis of HDAC6 ZnF UBP with repeat Tau (244–373). (A) Residues participating in hydrogen-bonding and hydrophobic interactions were identified using LigPlot+analysis of structures obtained from the final 10 ns of the molecular dynamics simulation. (B) Individual hydrogen-bond profiles for interactions between HDAC6 ZnF UBP residues (Q1131, R1155, R1155/Y1184, N1158, Q1163, and Q1187) and repeat Tau residues (N327/I328, D283, N286, E264/N265, I277/N279, and L282, respectively).

of the complex were evaluated by calculating the root mean square deviation (RMSD), root mean square fluctuation (RMSF), and radius of gyration (Rg). In addition, Coulombic and Lennard–Jones interaction energies between the interacting proteins were analyzed throughout the simulation. Hydrogen bond analyses were performed to assess intermolecular interactions between specific residues. Residue interaction networks from the final 10 ns of the simulation trajectory were generated and visualized using LigPlot+ version 2.2.

3.3.4.1 Software and servers

1. SWISS-MODEL server
2. RCSB-PDB database
3. PatchDock server
4. FireDock server
5. VMD software
6. PyMOL Molecular Graphics System

3.3.4.2 Procedure

1. HDAC6 ZnF UBP model was generated using SWISS-MODEL based on PDB structure 3C5K.
2. The model with GMQE score of 0.99 was selected for further studies (Fig. 6).
3. Repeat Tau structural model generated previously by the group was used for docking studies.
4. Protein-protein docking was carried out using PatchDock with clustering RMSD set to 4Å. RMSD parameters were optimized for protein-protein interaction studies.
5. Top docking solutions were refined using FireDock.
6. The model with the lowest binding energy was selected for further analysis.
7. Protein complexes were visualized and analyzed using VMD and PyMOL.
8. VMD and PyMOL Molecular Graphics System were used for identifying interacting residues and structural orientation.

3.3.5 Molecular dynamics simulations

3.3.5.1 Software

1. GROMACS 2019.3
2. AMBER03 force field

3. TIP3P water model
4. VMD visualization software

3.3.5.2 Procedure

1. The docked protein complex was placed in the center of a cubic simulation box with a 10 Å distance from the edges (Fig. 7).
2. The box dimensions were set to 88 × 88 × 88 Å.
3. The system was solvated with TIP3P water molecules and neutralized using Cl− ions.
4. Energy minimization was performed using the steepest descent algorithm until Fmax was below 1000 kJ/mol/nm.
5. PME algorithm was used for electrostatic calculations while Verlet cut-off scheme was applied for non-bonded interactions.
6. Temperature equilibration was carried out at 310 K using the V-rescale thermostat.
7. Pressure equilibration was performed at 1 bar using Berendsen pressure coupling.
8. Final production MD simulation was run for 50 ns.
9. Final trajectories were analyzed for RMSD, RMSF, radius of gyration (Rg), hydrogen bonding, and interaction energies using VMD software.

4. Concluding remarks

The methodology provides molecular-level insights into interaction of Tau-HDAC6 ZnF UBP domain by integrating experimental and computational approaches. Although further validation in cellular and in vivo systems is required, the workflow established here offers a robust platform for future investigations of Tau-associated pathways and related neurodegenerative processes. The integrated methodological approach is particularly valuable for addressing the challenges associated with studying intrinsically disordered proteins, whose structural heterogeneity and dynamic conformations often limit characterization by individual techniques.

References

Arakhamia, T., Lee, C. E., Carlomagno, Y., Kumar, M., Duong, D. M., Wesseling, H., ... DeTure, M. (2020). Posttranslational modifications mediate the structural diversity of tauopathy strains. *Cell, 180*(4), 633–644.e612.

Balmik, A. A., Chidambaram, H., Dangi, A., Marelli, U. K., & Chinnathambi, S. (2020). HDAC6 ZnF UBP as the modifier of tau structure and function. *Biochemistry, 59*(48), 4546–4562.

Balmik, A. A., Sonawane, S. K., & Chinnathambi, S. (2021). The extracellular HDAC6 ZnF UBP domain modulates the actin network and post-translational modifications of Tau. *Cell Communication and Signaling, 19*(1), 49.

Boyault, C., Zhang, Y., Fritah, S., Caron, C., Gilquin, B., Kwon, S. H., ... Matthias, P. (2007). HDAC6 controls major cell response pathways to cytotoxic accumulation of protein aggregates. *Genes & Development, 21*(17), 2172–2181.

Chinnathambi, S. (2025). Tau proteotasis in Alzheimer's disease. *Advances in Protein Chemistry and Structural Biology, 147*, 333–353.

Cook, C., Gendron, T. F., Scheffel, K., Carlomagno, Y., Dunmore, J., DeTure, M., & Petrucelli, L. (2012). Loss of HDAC6, a novel CHIP substrate, alleviates abnormal tau accumulation. *Human Molecular Genetics, 21*(13), 2936–2945.

Di Lorenzo, D. (2024). Tau protein and tauopathies: Exploring tau protein–protein and microtubule interactions, cross-interactions and therapeutic strategies. *ChemMedChem, 19*(21), e202400180.

Farias, G., Cornejo, A., Jimenez, J., Guzman, L., & B. Maccioni, R. (2011). Mechanisms of tau self-aggregation and neurotoxicity. *Current Alzheimer Research, 8*(6), 608–614.

Goedert, M., Eisenberg, D. S., & Crowther, R. A. (2017). Propagation of tau aggregates and neurodegeneration. *Annual Review of Neuroscience, 40*, 189–210.

Gorantla, N. V., & Chinnathambi, S. (2021). Autophagic pathways to clear the tau aggregates in Alzheimer's disease. *Cellular and Molecular Neurobiology, 41*(6), 1175–1181.

Gorantla, N. V., & Chinnathambi, S. (2018). Tau protein squired by molecular chaperones during Alzheimer's disease. *Journal of Molecular Neuroscience, 66*(3), 356–368.

Govindarajan, N., Rao, P., Burkhardt, S., Sananbenesi, F., Schlüter, O. M., Bradke, F., ... Fischer, A. (2013). Reducing HDAC6 ameliorates cognitive deficits in a mouse model for Alzheimer's disease. *EMBO Molecular Medicine, 5*(1), 52–63.

Guo, W., Naujock, M., Fumagalli, L., Vandoorne, T., Baatsen, P., Boon, R., ... Vanwelden, T. (2017). HDAC6 inhibition reverses axonal transport defects in motor neurons derived from FUS-ALS patients. *Nature Communications, 8*(1), 861.

Hai, Y., & Christianson, D. W. (2016). Histone deacetylase 6 structure and molecular basis of catalysis and inhibition. *Nature Chemical Biology, 12*(9), 741–747.

Hanger, D. P., Byers, H. L., Wray, S., Leung, K.-Y., Saxton, M. J., Seereeram, A., ... Anderton, B. H. (2007). Novel phosphorylation sites in tau from Alzheimer brain support a role for casein kinase 1 in disease pathogenesis. *Journal of Biological Chemistry, 282*(32), 23645–23654.

Kwon, S., Zhang, Y., & Matthias, P. (2007). The deacetylase HDAC6 is a novel critical component of stress granules involved in the stress response. *Genes & Development, 21*(24), 3381–3394.

Lee, J. Y., Koga, H., Kawaguchi, Y., Tang, W., Wong, E., Gao, Y. S., ... Lu, J. (2010). HDAC6 controls autophagosome maturation essential for ubiquitin-selective quality-control autophagy. *The EMBO Journal, 29*(5), 969–980.

Lee, M. J., Lee, J. H., & Rubinsztein, D. C. (2013). Tau degradation: The ubiquitin–proteasome system versus the autophagy-lysosome system. *Progress in Neurobiology, 105*, 49–59.

Martin, L., Latypova, X., & Terro, F. (2011). Post-translational modifications of tau protein: Implications for Alzheimer's disease. *Neurochemistry International, 58*(4), 458–471.

Min, S.-W., Cho, S.-H., Zhou, Y., Schroeder, S., Haroutunian, V., Seeley, W. W., ... Mukherjee, C. (2010). Acetylation of tau inhibits its degradation and contributes to tauopathy. *Neuron, 67*(6), 953–966.

Perez, M., Santa-Maria, I., De Barreda, E. G., Zhu, X., Cuadros, R., Cabrero, J. R., ... Perry, G. (2009). Tau–an inhibitor of deacetylase HDAC6 function. *Journal of Neurochemistry, 109*(6), 1756–1766.

Rawat, P., Sehar, U., Bisht, J., Selman, A., Culberson, J., & Reddy, P. H. (2022). Phosphorylated tau in Alzheimer's disease and other tauopathies. *International Journal of Molecular Sciences, 23*(21), 12841.

Serrano-Pozo, A., Escott-Price, V., Grinberg, L. T., Pascoal, T., Suárez-Calvet, M., Dubois, B., & Sperling, R. A. (2026). Alzheimer's disease. *The Lancet, 407*(10544), 2241–2262.

Simões-Pires, C., Zwick, V., Nurisso, A., Schenker, E., Carrupt, P.-A., & Cuendet, M. (2013). HDAC6 as a target for neurodegenerative diseases: What makes it different from the other HDACs? *Molecular Neurodegeneration, 8*(1), 7.

Sonawane, S. K., Dubey, T., Balmik, A. A., Das, R., & Chinnathambi, S. (2022). Alzheimer's disease pathology: A tau perspective, 23(2), 2–65.

Stelzl, L. S., Pietrek, L. M., Holla, A., Oroz, J., Sikora, M., Köfinger, Jr, ... Hummer, G. (2022). Global structure of the intrinsically disordered protein tau emerges from its local structure. *Jacs Au, 2*(3), 673–686.

Valenzuela-Fernandez, A., Cabrero, J. R., Serrador, J. M., & Sanchez-Madrid, F. (2008). HDAC6: A key regulator of cytoskeleton, cell migration and cell–cell interactions. *Trends in Cell Biology, 18*(6), 291–297.

Vaquer-Alicea, J., & Diamond, M. I. (2019). Propagation of protein aggregation in neurodegenerative diseases. *Annual Review of Biochemistry, 88*(1), 785–810.

Wang, L., Shi, S., Unterreiner, A., Kapetanovic, R., Ghosh, S., Sanchez, J., ... Donovan, K. A. (2024). HDAC6/aggresome processing pathway importance for inflammasome formation is context-dependent. *Journal of Biological Chemistry, 300*(2), 105638.

Wang, Y., & Mandelkow, E. (2016). Tau in physiology and pathology. *Nature Reviews Neuroscience, 17*(1), 22–35.

Weingarten, M. D., Lockwood, A. H., Hwo, S.-Y., & Kirschner, M. W. (1975). A protein factor essential for microtubule assembly. *Proceedings of the National Academy of Sciences, 72*(5), 1858–1862.

Wright, P. E., & Dyson, H. J. (2015). Intrinsically disordered proteins in cellular signalling and regulation. *Nature Reviews Molecular Cell Biology, 16*(1), 18–29.

Yan, J., Seibenhener, M. L., Calderilla-Barbosa, L., Diaz-Meco, M.-T., Moscat, J., Jiang, J., ... Wooten, M. C. (2013). SQSTM1/p62 interacts with HDAC6 and regulates deacetylase activity. *PloS One, 8*(9), e76016.

Younas, A., Younas, N., Iqbal, M. J., Ferrer, I., & Zerr, I. (2024). Comparative interactome mapping of Tau-protein in classical and rapidly progressive Alzheimer's disease identifies subtype-specific pathways. *Neuropathology and Applied Neurobiology, 50*(1), e12964.

Zheng, H., Sun, H., Cai, Q., & Tai, H.-C. (2024). The enigma of tau protein aggregation: Mechanistic insights and future challenges. *International Journal of Molecular Sciences, 25*(9), 4969.

CHAPTER FIVE

Defining active and repressive chromatin states in neural crest cells using low-input CUT&RUN

Ayesha Nasreen[a,b,1], **Sribas Chowdhury**[a,1], **and Vivek T. Natarajan**[a,b,*]

[a]CSIR-Institute of Genomics and Integrative Biology, Mathura Road, New Delhi, India
[b]Academy of Scientific and Innovative Research (AcSIR), Ghaziabad, Uttar Pradesh, India
*Corresponding author. e-mail address: tnvivek@igib.in

Contents

Abstract

The transition of neural crest cells (NCCs) from a multipotent state to lineage-restricted derivatives, including melanocytes, is governed by tightly regulated epigenetic mechanisms that orchestrate cell type specific gene expression programs. Histone post-translational modifications (PTMs), in particular, play an important role in modulating chromatin accessibility, enhancer activation, and transcription factor occupancy, thereby facilitating dynamic chromatin and transcriptional reprogramming required during development. However, profiling such chromatin states in rare and transient Neural Crest Cell (NCC) populations in vivo remains technically

[1] Authors contributed equally

Methods in Enzymology, Volume 734
ISSN 0076-6879, https://doi.org/10.1016/bs.mie.2026.05.048

challenging. To address this, we present an optimized low-input Cleavage Under Targets and Release Using Nuclease (CUT&RUN) workflow tailored for fluorescence-activated cell sorting (FACS) isolated NCCs from zebrafish embryos. This approach enables high-resolution and low-background mapping of key histone modifications, including H3K27ac, H3K4me3, and H3K27me3, from limited cell numbers. Collectively, these methodologies provide a robust framework for dissecting chromatin state dynamics in developmental systems and can also offer insights into epigenetic dysregulation associated with disease.

1. Introduction

As a central pillar of epigenetic regulation, histone post-translational modifications (PTMs) form a highly dynamic and coordinated network that governs gene expression by modulating chromatin structure and accessibility (Allis & Jenuwein, 2016; Taverna et al., 2007). By altering the interaction between DNA and histone proteins, PTMs can either help to promote a more open, transcriptionally permissive chromatin state or enforce a compact, transcriptionally repressive configuration (Kikuchi et al., 2023; Park & Kim, 2020). Beyond simply acting as structural modifiers, histone PTMs serve as molecular cues that are interpreted by specialized "reader" proteins, which in turn recruit transcriptional machinery or chromatin remodelling complexes to specific genomic loci (Allfrey et al., 1964; Shahbazian & Grunstein, 2007). In this way, PTMs integrate intrinsic developmental signals and along with chromatin modifiers in turn fine-tune gene expression programs.

During development, these modifications end up playing a pivotal role in orchestrating cell fate decisions by establishing and maintaining lineage-specific transcriptional landscapes (Li et al., 2021). In highly dynamic and multipotent populations such as neural crest cells, histone PTMs contribute to the regulation of chromatin plasticity, enabling cells to transition between different states and adopt diverse lineage outcomes (Ignatius et al., 2008; Murko et al., 2013; Singh et al., 2011). Among these modifications, histone acetylation and methylation have emerged as key regulators of chromatin accessibility and enhancer activity, making it particularly relevant in the context of lineage specification and differentiation (Cedar & Bergman, 2009; Okano et al., 1999). Understanding how histone post-translational modifications (PTMs) orchestrate transcriptional programs provides a powerful framework to dissect the molecular mechanisms underlying neural crest cell (NCC) fate decisions.

Given the remarkable plasticity of NCCs and their ability to generate a wide spectrum of derivatives such as pigment cells or neurons and glia of PNS or the ectomesenchymal derivatives forming the head skeleton, transitions between multipotent and lineage-restricted states are accompanied by highly dynamic and spatially restricted changes in chromatin organization (Martik & Bronner, 2017; Simões-Costa & Bronner, 2015). Capturing these transient and cell state–specific epigenetic landscapes, however, presents a significant technical challenge, particularly when working with rare cell populations isolated directly from in vivo systems. Traditional chromatin profiling approaches such as chromatin immunoprecipitation followed by sequencing (ChIP-seq) have been instrumental in mapping histone modifications genome-wide (Ku et al., 2011). But their application is often limited by the requirement for large cell numbers, extensive crosslinking, and sonication-based chromatin fragmentation, all of which can introduce background noise and reduce sensitivity (Furey, 2012).

In this context, Cleavage Under Targets and Release Using Nuclease (CUT&RUN) has emerged as a powerful alternative that overcomes many of these limitations. CUT&RUN is performed under native conditions, eliminating the need for crosslinking and enabling more precise interrogation of protein–DNA interactions (Skene & Henikoff, 2017). By employing an antibody coupled micrococcal nuclease to selectively cleave DNA at sites bound by histone modifications or chromatin-associated proteins, the method generates highly specific fragments that are released into solution for downstream sequencing. This targeted cleavage strategy not only improves signal-to-noise ratios but also inherently enriches for mononucleosomal fragments, providing high-resolution mapping of histone modification landscapes. Importantly, CUT&RUN requires substantially fewer input cells when compared to ChIP-seq, making it particularly well suited for studies involving limited or difficult to isolate tissue/cellular populations (Abbasova et al., 2025; Hainer & Fazzio, 2019).

These advantages are especially hold true in the context of NCC biology, where cells are often obtained from in-vivo model systems like zebrafish in which isolating sufficient numbers of viable, lineage-specific cells remains a major bottleneck. The ability of CUT&RUN to generate robust and reproducible chromatin profiles from low-input samples enables the interrogation of histone acetylation and methylation dynamics within these rare populations. By applying this approach to sorted NCCs, it becomes possible to map epigenetic transitions with high sensitivity and resolution, thereby linking chromatin state changes to the gene regulatory programs that drive lineage commitment and differentiation within these NCCs. In the following sections, we outline a detailed workflow for implementing CUT& RUN in NCCs, from tissue dissociation and cell enrichment to nuclei isolation, chromatin profiling, and downstream data analysis.

2. Tissue dissociation and enrichment of neural crest cells

Zebrafish experiments were performed in strict accordance with the institutional animal ethics approval (IAEC) of the CSIR-Institute of Genomics and Integrative Biology (IGIB), India (Proposal No 45a). All efforts were made to minimize animal suffering.

2.1 Methods

1. Embryo Collection and Washing
2. Collect approximately 500−600 *sox10:GFP* embryos (stage 16−24 hpf) and wash thoroughly using a cell strainer with embryo water. Process stage-matched GFP-negative wild-type embryos in parallel as controls.
3. Dechorionation
4. Incubate embryos in 0.8–1 mg/mL pronase for 10 min to remove chorions. Transfer dechorionated embryos to fresh embryo water to eliminate residual pronase. Collect ~100 embryos and transfer to a 2 mL microcentrifuge tube.
5. Deyolking
6. Remove excess medium and add 200 μL ice-cold Ringer's solution. Keep the sample on ice and gently pipette (~20 times) until the yolks are fully disrupted. Centrifuge at 100 g for 1 min at 4 °C, repeat once, and carefully discard the supernatant.
7. Enzymatic Dissociation
8. Transfer deyolked embryos to Petri dishes containing 10 mL trypsin solution. Distribute embryos across multiple dishes to avoid overcrowding. Gently pipette once or twice to reduce clumping.
9. Incubation and Tissue Dissociation
10. Incubate at room temperature for:
    - 15 min (<24 hpf embryos)
    - 30 min (24–30 hpf embryos)
    - Gently triturate intermittently to facilitate dissociation into single cells.
11. Filtration

12. Pass the suspension through a 70 μm cell strainer into a 50 mL tube to obtain a single-cell suspension. Rinse dishes to recover residual cells.
13. Cell Pelleting and Washing
14. Centrifuge at 450 g for 5 min at 4 °C. Discard supernatant and resuspend the pellet in 1 mL ice-cold PBS. Repeat the wash once.
15. Final Resuspension
16. Resuspend cells in PBS supplemented with 5 % fetal bovine serum (FBS) and keep on ice until analysis.
17. Setup the Flow Cytometer:
    - FSC vs SSC plot for size and granularity
    - FITC fluorescence histogram for GFP detection
18. Gating Strategy
19. Load GFP-negative wild-type cells to:
    - Exclude debris and doublets using FSC/SSC
    - Set the FITC threshold for background fluorescence
20. Sample Acquisition
21. Load cells from Tg(*sox10:GFP*) embryos and acquire data using established parameters.
22. Quantification
23. Determine the number of GFP-positive cells within the FITC gate, corresponding to neural crest cells.

2.2 Materials

1. Pronase (0.8–1 mg/mL stock) (Roche)
2. TrypLE Express (Gibco)
3. PBS (calcium and magnesium free), sterile and cold
4. Fetal bovine serum (FBS)
5. FACS buffer (PBS with 2–5 % FBS or BSA)
6. 70 μm cell strainer
7. 15 mL and 1.5 mL sterile tubes
8. Tricaine or euthanasia solution for larva
9. Ringers solution

3. Isolation and enrichment of nuclei from sorted neural crest cells

Following fluorescence-activated cell sorting (FACS) of Sox10^{+} neural crest cells (NCCs) based on GFP fluorescence, it is critical to maintain the collected cells on ice to preserve viability and chromatin integrity for downstream applications. From Tg(*sox10:GFP*) embryos, approximately 2×10^5 to 4×10^5 cells can be recovered when starting with an initial pool of ~500–600 embryos, depending on the developmental stage (typically between 16–24 h post-fertilization). To enable efficient nuclei isolation from such low-input samples, we employed a modified protocol adapted from 10X single-cell ATAC-seq workflows, which are optimized for minimal cell numbers. This approach incorporates gentler lysis conditions and reduced washing steps to minimize nuclei loss during processing. Under these conditions, the protocol consistently yields intact nuclei with low debris levels, making them well suited for downstream CUT&RUN applications and other chromatin profiling assays requiring limited input material.

3.1 Methods

1. Centrifuge the cell suspension at 1000 g for 10 min at 4 °C. Carefully remove the supernatant and resuspend the pellet in PBS supplemented with FBS.
2. Transfer 1−3 × 10^5 cells into a 1.5 mL tube in a total volume of 50 μL PBS + FBS.
3. Centrifuge at 1000 g for 10 min at 4 °C.
4. Carefully remove 45 μL of supernatant without disturbing the cell pellet.
5. Add 45 μL of chilled lysis buffer and gently mix by pipetting three times.
6. Incubate the sample on ice for exactly 1.5 min.
7. Add 50 μL of chilled wash buffer without mixing.
8. Centrifuge at 500 g for 5 min at 4 °C.
9. Remove 95 μL of supernatant carefully, avoiding disruption of the nuclei pellet.
10. Add 100 μL of chilled PBS + FBS without mixing.
11. Carefully discard the supernatant without disturbing the nuclei pellet.
12. Resuspend the nuclei pellet in 50−100 μL of chilled PBS + FBS.
13. For counting, dilute 2 μL of nuclei suspension with 8 μL PBS, then add 10 μL Trypan Blue and determine nuclei concentration using an automated cell counter or hemocytometer.

3.2 Materials

1. Lysis Buffer (Tris-Cl pH 7.4 (10 mM), NaCl (10 mM), $MgCl_2$ (3 mM), Tween20 (0.1 %), IGEPAL (0.1 %), 5 % Digitonin (0.01 %), BSA (0.01 %), Protease inhibitor cocktail (1X))

2. Wash Buffer (Tris-Cl pH 7.4 (10 mM), NaCl (10 mM), $MgCl_2$ (3 mM), Tween20 (0.1 %), BSA (0.01 %), Protease inhibitor cocktail (1X))
3. 0.4 % Trypan Blue and DAPI for nuclei staining and to check nuclear integrity
4. Hemocytometer

4. CUT and RUN profiling of histone acetylation and methylation of the sorted NCCs

For CUT&RUN-based profiling of histone post-translational modifications (PTMs), we examined key marks associated with distinct chromatin states H3K27ac (active), H3K4me3 (poised/active promoters), and H3K27me3 (repressed) to characterize epigenetic dynamics during neural crest development. CUT&RUN assays were performed using a commercially available kit from EpiCypher to selectively enrich mononucleosomal DNA fragments, followed by library preparation using the NEB DNA library preparation kit for downstream sequencing. The protocol was optimized with minor modifications to improve yield and reproducibility, including antibody titration and adjustments tailored for zebrafish samples. Notably, these refinements were critical for profiling low-input, fluorescence-sorted neural crest cell populations, a context in which CUT&RUN applications remain limited. This optimized workflow enables robust and reproducible chromatin profiling from GFP-positive neural crest cells isolated in vivo, providing a valuable approach for studying epigenetic regulation in rare developmental cell populations (Fig. 1).

4.1 Methods

1. ConA Bead Preparation
2. Wash ConA beads (11 μL/reaction) twice with cold Bead Activation Buffer using a magnetic rack and resuspend in 11 μL buffer. Keep on ice.
3. Cell Binding to Beads
4. Pellet down the nuclei obtained from the previous section (600 g, 3 min) and resuspend in 100 μL Wash buffer. Add activated beads and incubate 10 min at room temperature for binding.
5. Antibody Binding
6. Separate on magnet, discard supernatant, and resuspend in 50 μL Antibody Buffer. Add primary antibody (H3K27ac 1:50 (abcam, ab4729), H3K4me3 1:50 (EpiCypher), H3K27me3 1:50 (EpiCypher)) and incubate overnight at 4 °C with gentle mixing.

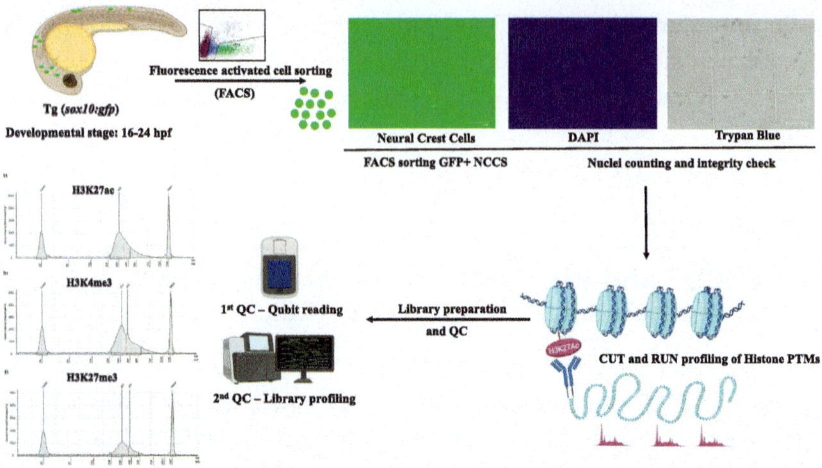

Fig. 1 Workflow for Low-Input CUT&RUN in Neural Crest Cells: Nuclei Preparation, Quality Assessment, and Library Profiling. Isolation and quality assessment of GFP$^+$ neural crest cell (NCC) nuclei for CUT&RUN. GFP$^+$ NCCs were isolated from zebrafish embryos and subjected to nuclei enrichment. Nuclei integrity was evaluated by DAPI and Trypan Blue staining. Preparations with optimal nuclei quality and input numbers were used for CUT&RUN. Library quality was assessed by measuring concentration and fragment size distribution using library profiling.

7. Washing and pAG-MNase Binding
8. Wash bead–cell complexes twice with Digitonin Buffer. Resuspend in 50 μL Digitonin Buffer, add pAG-MNase, and incubate 10 min at room temperature.
9. MNase Digestion
10. Wash twice, resuspend in 50 μL Digitonin Buffer, and add $CaCl_2$ to activate MNase. Incubate at 4 °C for 3 h to digest chromatin.
11. Reaction Termination and DNA Release
12. Add Stop Buffer, incubate at 37 °C for 10 min, then separate on magnet and collect supernatant containing released DNA.
13. DNA Purification
14. Purify DNA using magnetic beads (1.8× ratio), wash twice with 85% ethanol, air-dry briefly, and elute in 15–17 μL low-salt buffer.
15. Quantification
16. Quantify purified DNA using a fluorometric assay (e.g., Qubit).
    The concentrations obtained for the samples using qubit before library preparation were as follows: H3K27ac (0.52 ng/μL), H3K4me3 (0.41 ng/μL), H3K27me3 (0.34 ng/μL).

17. Proceed directly for library preparation using NEB DNA library preparation kit. Combine the eluted DNA (make up the volume to 25 μL using 0.1X TE or EB buffer) with 4.2 μL of End Prep Buffer and 1.8 μL of End Prep Enzyme Mix. Pipette mix 5X, vortex and spin.
18. Setup the thermocycler and incubate the samples at 20 °C for 20 min and then at 65 °C for 30 min
19. To setup the adaptor ligation reaction, combine 16.5 μL ligation mix with 0.55 μL ligation enhancer, vortex and spin and add 15.5 μL of this ligation master mix to the above End Prep reaction mixture and add 1.25 μL of 1.5 μM Adaptor. Pipette the entire volume up and down at least 10 times to mix thoroughly. Perform a quick spin to collect all liquid and incubate in a thermocycler at 20 °C for 15 min with the heated lid off.
20. Add 1 μL of USER Enzyme to the ligation mixture. Mix well and incubate at 37 °C for 15 min with the heated lid set to 50 °C.
21. Vortex SPRI selects beads to resuspend it properly. Add 47.75 μL (1X) of beads to the ligation reaction. Mix well by pipetting up and down for at least 10 times.
22. Incubate samples at room temperature for 5 min. Place the tubes on a magnetic stand to separate beads from the supernatant. After 5 min remove the supernatant without disturbing the beads.
23. While keeping tubes on the magnet, add 180 μL of freshly prepared 85% ethanol. Incubate at room temperature for 30 s and then discard it. Repeat this one more time.
24. Quick spin the tubes and remove any traces of ethanol if left by placing on magnet.
25. Air dry the beads for 2–3 min and then remove it from the magnet. Quickly resuspend the beads in a 17 μL EB buffer (Qiagen), quick spin and incubate for 2 min at room temperature. Place tubes on the magnet.
26. Transfer 15 μL of eluted DNA to new strip tubes.
27. Add 10 μL of Index primer mix along with 25 μL of Hot Start PCR mix to the DNA and mix it thoroughly.
28. Place the tubes in a thermocycler with the following PCR cycling conditions ((98 °C 45 s 1 cycle), (98 °C 15 s, 60 °C 30 s 16 cycles), (72 °C 60 s 1 cycle)). The cycle number has been optimised to ensure maximal yield and also to avoid PCR duplicates.
29. Vortex SPRI select beads to resuspend it properly. Add 47.75 μL (1X) of beads to the ligation reaction. Mix well by pipetting up and down for at least 10 times.

30. Incubate samples at room temperature for 5 min. Place the tubes on a magnetic stand to separate beads from the supernatant. After 5 min remove the supernatant without disturbing the beads.
31. While keeping tubes on magnet, add 180 μL of freshly prepared 85 % ethanol. Incubate at room temperature for 30 s and then discard it. Repeat this one more time.
32. Quick spin the tubes and remove any traces of ethanol if left by placing on magnet.
33. Air dry the beads for 2–3 min and then remove it from the magnet. Quickly resuspend the beads in 13 μL EB buffer (Qiagen), quick spin and incubate for 2 min at room temperature. Place tubes on magnet.
34. Transfer 11 μL of eluted DNA to a fresh tube. Quantify purified DNA using a fluorometric assay (e.g., Qubit). Also, check the size distribution on Agilent Tapestation.
35. All CUT&RUN libraries that were prepared using this protocol show predominant enrichment at ~300 bp DNA as expected, reflecting ~170 bp excised DNA and sequencing adapters.
36. The concentrations obtained for the samples using Qubit after library preparation were as follows: H3K27ac (79.4 ng/μL), H3K4me3 (55.4 ng/μL), H3K27me3 (68 ng/μL).

4.2 Materials

1. CUTANA EpiCypher Kit
2. NEB Next Ultra II DNA Library Prep Kit
3. Bead Activation Buffer (20 mM HEPES, pH 7.9, 10 mM KCl, 1 mM $CaCl_2$, 1 mM $MnCl_2$)
4. Wash Buffer (20 mM HEPES, pH 7.5, 150 mM NaCl, 0.5 mM Spermidine, 1X PIC)
5. Digitonin Buffer (Wash Buffer + 0.01 % digitonin)
6. Antibody Buffer (Digitonin Buffer + 2 mM EDTA)
7. 0.4 % Trypan Blue and DAPI for nuclei staining and to check nuclear integrity
8. Hemocytometer
9. Magnetic stand
10. PCR strip tubes
11. SPRI select beads
12. Qiagen EB buffer
13. Qubit and Tapestation reagents

Notes:

Neural crest cells represent a rare population; therefore, all steps from dissociation to sorting and nuclei isolation should be performed with minimal handling and gentle pipetting to preserve yield. Proper gating is essential to accurately isolate GFP-positive NCCs. Always include GFP-negative controls to define background fluorescence and exclude debris and doublets.

Given the limited number of sorted cells, use gentle lysis conditions and minimize washing steps to prevent nuclei loss. The strength of lysis buffer can be reduced if it affects nuclei morphology. We observed no nuclei clumping or loss to nuclei integrity with the lysis buffer used. But it was standardised for the developmental stages mentioned in the chapter. The nuclei pellet may not always be visible; proceed carefully to avoid accidental loss.

Antibody specificity and performance are critical for CUT&RUN success. Validate antibodies for CUT&RUN where possible, and optimize concentrations for low-input samples, particularly in zebrafish systems. Two or three sets of antibodies can be tested if looking for things other than histone PTMs.

5. Data acquisition and analysis

An appropriate Illumina sequencing platform should be chosen based on the number of CUT&RUN libraries and the level of coverage required. Paired-end sequencing (at least 2×50 cycles) is generally recommended for this application. In most cases, 3–8 million paired-end reads are sufficient to obtain reliable data; however, in this study, approximately 10 million paired-end reads were generated for each library to ensure adequate coverage. The CUT&RUN analysis workflow typically consists of the following steps: quality check, alignment to reference genome, normalizing for spike-in reads, duplicate removal, and peak calling (Fig. 2A).

5.1 Pre-processing sequencing files

Reproducibility and interpretability of outputs is crucial for genomic data analysis and standard established pipeline platforms such as Nextflow, NumPy etc. offer consistent workflows to ensure such standards are met (DI Tommaso et al., 2017; Harris et al., 2020). To streamline and standardize the steps of CUT&RUN analysis, Nextflow

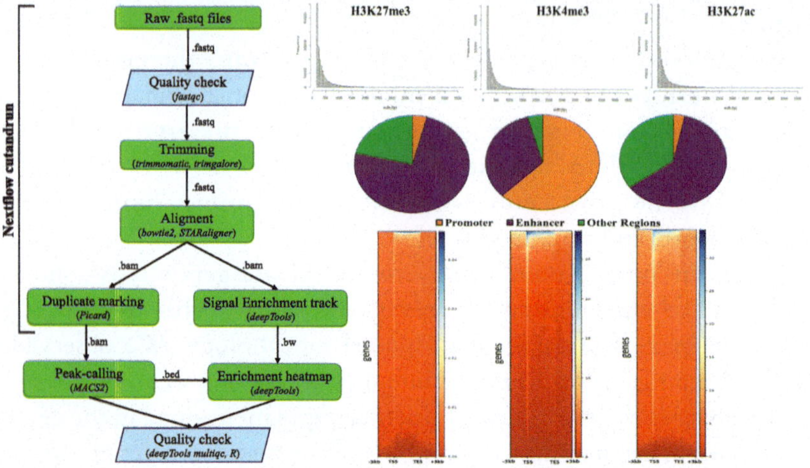

Fig. 2 Analysis of Cut&Run sequencing data: (A) Workflow of analysis; (B) Fragment length distribution of peaks called by MACS2; (C) Genomic distribution of called peaks across regulatory regions; (D) Enrichment of signal around Transcription Start Site (TSS) of genes.

nf-core/cutandrun pipeline was used for the initial processing of sequencing data, integrating all major analytical steps within a reproducible framework.

1. Install Nextflow in the Linux environment, either directly or in a conda virtual environment.
2. Prepare input samplesheet for the samples describing the location of sample fastq file, replicate number and control fastq file.
3. Launch Nextflow and prepare the script for analysis. For this experiment we used standard parameters, except in peak-calling where the –broad parameter was used for H3K27ac and H3K27me3 to account for their broad genomic distribution.
4. Run the Nextflow pipeline and prepare outputs for downstream processing.

Notes: Nextflow offers two peak-callers: MACS2 (Model-based Analysis of ChIP-Seq) and SEACR (Sparse Enrichment Analysis for CUT &RUN) for peakcalling. MACS2 models spike-in normalized reads in a bimodal distribution and calls peak based on quantile cutoff (default 0.05). SEACR computes genome-wide quantification of background signal and uses it at a stringent threshold for calling peaks. It is designed specifically for Cut&Run to account for its low background signal. In our analysis, we found MACS2 to be a more robust peakcaller in terms of peak numbers, peak length distribution and peak-associated regulatory regions (Fig. 2B, C, D). For our analysis, we used MACS2 peakcaller across all samples and replicates.

5.2 Normalizing library size across samples

Chromatin states during development do not exist in strictly binary configurations of activation or repression. A distinctive intermediate state that arises during development is a bivalent state characterized by the concurrent presence of repressive H3K27me3 mark and activating H3K4me3 mark (Bernstein et al., 2006). This bivalent state is typically enriched around the promoters of lineage-specifying genes during early developmental stages when cells retain a high degree of plasticity and subsequently resolves into either active or repressed chromatin depending on cell type as development proceeds. Capturing such regions requires the signal strength of Cut&Run samples to be normalized for relative library size, as technical variability can arise from differences in antibody affinity, input cell number, sequencing depth and overall genome alignment efficiency. The standard pipelines usually normalize for ratio of spike-in to genome aligned reads for each individual file, which can introduce inconsistencies in signal scaling across datasets. To address this, there are multiple methods to normalize for differences in library size across samples as described in a review (Patel et al., 2024). We have found that the ChIP Rx model of normalization which computes a scale factor based on relative ratio of spike-in aligned reads works the best for capturing bivalent regions across the stages of development. We modified the ChIP Rx model to compute the relative scale factor and called it the ChIP Rx model of relative normalization. The ChIP Rx model of relative normalization accounts for the relative difference in spike-in reads and calculates scale-factor based on the inverse ratio of net spike-in aligned reads. The model is defined as:

If Sq = Spike-in reads in query/condition and Sr = Spike-in reads in reference/control.

Then, Rescaling factor = rs = 1/Sr.

Scale Factor = (1/Sq)/rs.

1. Genome aligned bam files marked for duplicates were used to calculate scale-factor. Calculate the number of genome-aligned and spike-in aligned reads from logfiles of sample run.
2. Calculate rescaling factor and scale factor for H3K27me3 and H3K4me3 based on spike-in read counts using reciprocal normalization (i.e., use H3K4me3 as the reference for H3K27me3 and vice versa).

3. Generate signal intensity tracks of both marks using deepTools bamCoverage (parameters: –MNase –centerReads).
4. Compute average signal tracks across replicates using deepTools bigwigAverage.
5. Create an enrichment matrix of signal tracks over concatenated peak files of both PTMs using computeMatrix in deepTools (mode: reference-point; parameters: –referencePoint center –missingDataAsZero) to identify bivalent regions.
6. Generate an enrichment heatmap from the resulting matrix using plotHeatmap in deepTools (–sortRegions keep).

5.3 Tools required

1. Nextflow (version > 1.1.0)
2. deepTools
3. bowtie2
4. MACS2
5. Anaconda3
6. Miniconda
7. Python (version > 3.4.0)

6. Concluding remarks

Histone acetylation and methylation represent key chromatin-based regulatory mechanisms that control gene expression by modulating nucleosome architecture, chromatin accessibility, and the recruitment of DNA binding transcriptional machinery. In neural crest cells, which undergo rapid and dynamic transitions from a highly multipotent state to lineage-restricted states, these post-translational modifications provide a regulatory framework for the coordinated activation and repression of developmental gene networks. A major challenge in dissecting these processes has been the inability to profile chromatin states in small and transient cell populations in vivo. This chapter focuses on the application of low-input chromatin profiling approaches to interrogate the distribution of functionally distinct histone marks, including H3K27ac, H3K4me3, and H3K27me3, across the neural crest genome. The integration of low-input chromatin profiling approaches such as CUT&RUN now provides an opportunity to overcome these limitations, enabling high-resolution mapping of histone modifications in transitioning NCCs. By optimizing workflows for zebrafish-derived samples and minimizing cell loss, the protocol described here facilitates robust and reproducible interrogation of epigenetic landscapes in physiologically relevant contexts.

Future studies that integrate CUT&RUN with complementary multi-omics approaches at single-cell resolution hold significant promise for advancing our understanding of chromatin-mediated gene regulation at cellular level. While CUT&RUN provides high-resolution maps of histone acetylation and methylation, combining these data with single-cell chromatin accessibility profiles (e.g., scATAC-seq) and transcriptomic readouts (e.g., scRNA-seq) will enable the simultaneous interrogation of regulatory element activity, chromatin state, and gene expression within the same cellular framework, providing a comprehensive readout of underlying mechanisms mediating cell fate determination. Such integrative analyses can facilitate the reconstruction of lineage trajectories and reveal how dynamic changes in histone modification landscapes correspond to shifts in chromatin accessibility and transcriptional outputs during cell fate transitions.

References

Abbasova, L., Urbanaviciute, P., Hu, D., Ismail, J. N., Schilder, B. M., Nott, A., ... Marzi, S. J. (2025). CUT&Tag recovers up to half of ENCODE ChIP-seq histone acetylation peaks. *Nature Communications, 16*(1), https://doi.org/10.1038/S41467-025-58137-2.

Allfrey, V. G., Faulkner, R., & Mirsky, A. E. (1964). Acetylation and methylation of histones and their possible role in the regulation of RNA synthesis. *Proceedings of the National Academy of Sciences of the United States of America, 51*(5), 786. https://doi.org/10.1073/PNAS.51.5.786.

Allis, C. D., & Jenuwein, T. (2016). The molecular hallmarks of epigenetic control. *Nature Reviews. Genetics, 17*(8), 487–500. https://doi.org/10.1038/NRG.2016.59.

Bernstein, B. E., Mikkelsen, T. S., Xie, X., Kamal, M., Huebert, D. J., Cuff, J., ... Lander, E. S. (2006). A bivalent chromatin structure marks key developmental genes in embryonic stem cells. *Cell, 125*(2), 315–326. https://doi.org/10.1016/j.cell.2006.02.041.

Cedar, H., & Bergman, Y. (2009). Epigenetic silencing during early lineage commitment. *StemBook*. https://doi.org/10.3824/STEMBOOK.1.42.1.

DI Tommaso, P., Chatzou, M., Floden, E. W., Barja, P. P., Palumbo, E., & Notredame, C. (2017). Nextflow enables reproducible computational workflows. *Nature Biotechnology, 35*(4), 316–319. https://doi.org/10.1038/NBT.3820.

Furey, T. S. (2012). ChIP-seq and beyond: New and improved methodologies to detect and characterize protein-DNA interactions. *Nature Reviews. Genetics, 13*(12), 840–852. https://doi.org/10.1038/NRG3306.

Hainer, S. J., & Fazzio, T. G. (2019). High-resolution chromatin profiling using CUT&RUN. *Current Protocols in Molecular Biology, 126*(1), https://doi.org/10.1002/CPMB.85.

Harris, C. R., Millman, K. J., van der Walt, S. J., Gommers, R., Virtanen, P., Cournapeau, D., ... Oliphant, T. E. (2020). Array programming with NumPy. *Nature, 585*(7825), 357–362. https://doi.org/10.1038/S41586-020-2649-2.

Ignatius, M. S., Moose, H. E., El-Hodiri, H. M., & Henion, P. D. (2008). Colgate/hdac1 repression of foxd3 expression is required to permit mitfa-dependent melanogenesis. *Developmental Biology, 313*(2), 568–583. https://doi.org/10.1016/j.ydbio.2007.10.045.

Kikuchi, M., Morita, S., Wakamori, M., Sato, S., Uchikubo-Kamo, T., Suzuki, T., ... Umehara, T. (2023). Epigenetic mechanisms to propagate histone acetylation by p300/CBP. *Nature Communications, 14*(1), https://doi.org/10.1038/S41467-023-39735-4.

Ku, C. S., Naidoo, N., Wu, M., & Soong, R. (2011). Studying the epigenome using next generation sequencing. *Journal of Medical Genetics, 48*(11), 721–730. https://doi.org/10.1136/JMEDGENET-2011-100242.

Li, D., Shu, X., Zhu, P., & Pei, D. (2021). Chromatin accessibility dynamics during cell fate reprogramming. *EMBO Reports, 22*(2), e51644. https://doi.org/10.15252/EMBR.202051644.

Martik, M. L., & Bronner, M. E. (2017). Regulatory logic underlying diversification of the neural crest. *Trends in Genetics: TIG, 33*(10), 715–727. https://doi.org/10.1016/J.TIG.2017.07.015.

Murko, C., Lagger, S., Steiner, M., Seiser, C., Schoefer, C., & Pusch, O. (2013). Histone deacetylase inhibitor trichostatin A induces neural tube defects and promotes neural crest specification in the chicken neural tube. *Differentiation; Research in Biological Diversity, 85*(1–2), 55–66. https://doi.org/10.1016/J.DIFF.2012.12.001.

Okano, M., Bell, D. W., Haber, D. A., & Li, E. (1999). DNA methyltransferases Dnmt3a and Dnmt3b are essential for de novo methylation and mammalian development. *Cell, 99*(3), 247–257. https://doi.org/10.1016/S0092-8674(00)81656-6.

Park, S. Y., & Kim, J. S. (2020). A short guide to histone deacetylases including recent progress on class II enzymes. *Experimental & Molecular Medicine, 52*(2), 204–212. https://doi.org/10.1038/S12276-020-0382-4.

Patel, L. A., Cao, Y., Mendenhall, E. M., Benner, C., & Goren, A. (2024). The wild west of spike-in normalization. *Nature Biotechnology, 42*(9), 1343–1349. https://doi.org/10.1038/S41587-024-02377-Y.

Shahbazian, M. D., & Grunstein, M. (2007). Functions of site-specific histone acetylation and deacetylation. *Annual Review of Biochemistry, 76*, 75–100. https://doi.org/10.1146/ANNUREV.BIOCHEM.76.052705.162114.

Simões-Costa, M., & Bronner, M. E. (2015). Establishing neural crest identity: A gene regulatory recipe. *Development (Cambridge, England), 142*(2), 242. https://doi.org/10.1242/DEV.105445.

Singh, N., Trivedi, C. M., Lu, M., Mullican, S. E., Lazar, M. A., & Epstein, J. A. (2011). Histone deacetylase 3 regulates smooth muscle differentiation in neural crest cells and development of the cardiac outflow tract. *Circulation Research, 109*(11), 1240–1249. https://doi.org/10.1161/CIRCRESAHA.111.255067.

Skene, P. J., & Henikoff, S. (2017). An efficient targeted nuclease strategy for high-resolution mapping of DNA binding sites. *ELife, 6*. https://doi.org/10.7554/ELIFE.21856.

Taverna, S. D., Li, H., Ruthenburg, A. J., Allis, C. D., & Patel, D. J. (2007). How chromatin-binding modules interpret histone modifications: Lessons from professional pocket pickers. *Nature Structural & Molecular Biology, 14*(11), 1025–1040. https://doi.org/10.1038/NSMB1338.

CHAPTER SIX

Identification of 2,4-dihydroxy-azaflavanone as an allosteric activator of SIRT1 and its potential role in cellular models of Parkinson's disease

Chitrakshi Pant[a,b,*,1], Vidya Jyothi Alli[b,c], Genji Sukumar[c,d], Abhijit Mohapatra[a,b], Surender Singh Jadav[b,c], Srihari Pabbaraja[c,d], and Shasi V. Kalivendi[a,b,*]

[a]Department of Applied Biology, CSIR-Indian Institute of Chemical Technology, Hyderabad, India
[b]Academy of Scientific and Innovative Research (AcSIR), Ghaziabad, India
[c]Department of Natural Products and Medicinal Chemistry, CSIR-Indian Institute of Chemical Technology, Hyderabad, India
[d]Department of Organic Synthesis and Process Chemistry, CSIR-Indian Institute of Chemical Technology, Hyderabad, India
*Corresponding authors. e-mail address: chitrakshi428@gmail.com; kalivendi@iict.res.in

Contents

Abstract

Mitochondrial dysfunction is one of the significant aspects of Parkinson's disease (PD) pathophysiology, marked by a gradual decline in oxidative phosphorylation, an abnormal increase in free radical species, dysfunctional mitochondrial quality control, and faulty mitochondrial biogenesis. Sirtuin 1 (SIRT1), a NAD^+-dependent class-III deacetylase, acts as a crucial metabolic sensor that orchestrates transcriptional

[1] ***Present address:*** Department of Pediatrics - Neurology NRI, Baylor College of Medicine, Jan and Dan Ducan Neurological Research Institute (Moursund St., Houston, TX, 77030, United States

Methods in Enzymology, Volume 734
ISSN 0076-6879, https://doi.org/10.1016/bs.mie.2026.06.011

programs related to mitochondrial biogenesis, respiratory chain assembly, and stress resilience, mainly by way of deacetylation and activation of peroxisome proliferator-activated receptor gamma coactivator-1 alpha (PGC-1α). While screening for the small molecule activators, we have identified that 2,4-dihydroxy-Azaflavanone allosterically activates SIRT1. This chapter outlines a detailed, multi-layered methodological framework for assessing the allosteric activation of SIRT1 by 2,4-dihydroxy-azaflavanone and its downstream effects in a cellular models of PD. The validation process involves synthesis of small molecules, molecular docking studies utilizing crystallographic SIRT1 coordinates (PDB: 5BTR), in vitro fluorometric deacetylase assays with recombinant enzyme, and cellular thermal shift assays (CETSA) to confirm direct, isoform-selective target engagement. The activation of downstream pathways is evaluated by immunoblotting and quantitative PCR for PGC-1α, TFAM, and quantification of mitochondrial DNA (mtDNA) copy number. Functional restoration of mitochondria in cells is analyzed by assessing the overall mitochondrial bioenergetics parameters using Seahorse extracellular flux analyzer. Overall, this integrated approach offers robust, reproducible results for exploring SIRT1-activators in the mechanisms mediating neurodegenerative disease models.

1. Introduction

The convergence of genetic, biochemical, and toxicological data clearly suggests mitochondrial dysfunction as a key and unifying factor in the disease pathology of Parkinson's disease (PD) (Schapira, 2008; Park et al., 2018). Mutations associated with familial PD in genes, such as, PINK1, Parkin, and DJ-1 were shown to directly impair mitochondrial membrane integrity, disrupt the electron transport chain (ETC), and hinder mitophagy-dependent quality control. This leads to the accumulation of dysfunctional organelles and a gradual decline in bioenergetic function (Pickrell & Youle, 2015; Bonifati, 2014). Further, 1-methyl-4-phenylpyridinium (MPP^+), the active metabolite of 1-methyl-4-phenyl-1,2,3,6-tetrahydropyridine (MPTP), and rotenone are specific inhibitors of mitochondrial complex-I, are known to cause sporadic forms of PD. This accurately reflects the bioenergetic failure seen in neurons derived from patients (Przedborski et al., 2004). Among the regulatory pathways that oversee mitochondrial homeostasis, Sirtuin 1 (SIRT1) has surfaced as a notably intriguing therapeutic target. SIRT1 is a NAD^+-dependent class-III histone deacetylase, acting as a sensor that links cellular energy status to transcriptional programs that manage oxidative stress responses, mitochondrial biogenesis and DNA repair (Ramsay et al., 1986; Imai & Guarente, 2014; Haigis & Guarente, 2006; Nemoto et al., 2005). Its influence on mitochondria involves the deacetylation and transcriptional activation of peroxisome proliferator-activated receptor gamma coactivator-1 alpha (PGC-1α), which subsequently drives the expression of nuclear respiratory factor 1 (NRF1), NRF2, and mitochondrial transcription factor A (TFAM), the crucial factors involved in the mitochondrial DNA replication and respiratory chain assembly (Scarpulla et al., 2012; Zheng et al., 2010). Post-mortem brain tissues of substantia nigra from PD patients showed decrease in SIRT1 and PGC-1α expression, the restoration of which provides neuroprotection in various experimental models (Mudo et al., 2012). In addition to its role in mitochondrial biogenesis, SIRT1 influences cellular stress resilience by deacetylating FOXO transcription factors, NF-κB, and p53, which in-turn suppresses pro-apoptotic signaling and boosts antioxidant gene expression (Brunet et al., 2004; Vaziri et al., 2001) (Scheme 1).

The pharmacological activation of SIRT1 is an appealing therapeutic approach from a mechanistic standpoint, but it poses considerable methodological difficulties. Initial studies that reported robust SIRT1 activation by polyphenolic compounds, such as resveratrol, were later

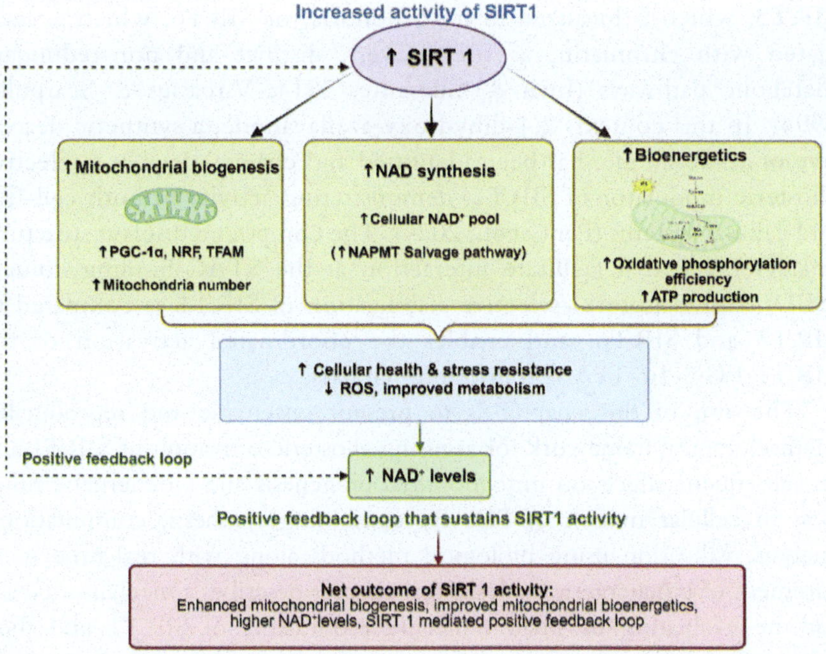

Scheme 1 Potential pathways modulated by SIRT1 that could be of significant relevance to PD mechanisms.

found to be mediated by indirect metabolic effects mediated by AMPK and substrate-dependent fluorescence artifacts in biochemical assays (Borra et al., 2005; Pacholec et al., 2010; Dai et al., 2018; Hubbard et al., 2013). These issues highlighted the critical need for orthogonal validation methods that can differentiate direct allosteric enzymatic modulation from secondary or assay-confounded effects. Structural investigations have revealed that SIRT1 contains an N-terminal regulatory domain known as the SIRT1-activating domain (STAC-binding domain, SBD), which facilitates allosteric activation. Small-molecule SIRT1 activating compounds (STACs) bind to this domain, stabilizing active conformations of the enzyme and improving catalytic efficiency without competing for the NAD^+ or substrate binding sites (Carrico et al., 2018; Bose & Beal, 2016). This allosteric mechanism is both structurally and mechanistically different from competitive substrate binding, providing the potential for selective and adjustable enzymatic enhancement without interfering broadly at the active site. However, due to the structural conservation within the sirtuin family, it is crucial to assess isoform selectivity to prevent unintended co-activation of SIRT3, which is localized in mitochondria, or SIRT6, which is associated with chromatin, as they govern distinct and non-redundant metabolic pathways (Imai & Guarente, 2014; Virbasius & Scarpulla, 1994). In this context, 2,4-dihydroxy-azaflavanone, a synthetic derivative of azaflavanone, has been identified and characterized as a selective allosteric modulator of SIRT1, demonstrating activity in both cell-free and cellular systems (Pant et al., 2024). The compound displays structural characteristics that facilitate interaction at the STAC-binding domain (SBD), shows isoform-selective stabilization of SIRT1 as compared to SIRT3 and SIRT6, and enables the coordinated activation of the SIRT1-PGC-1α-TFAM transcriptional axis.

The aim of this chapter is to present systematic and reproducible methodological framework for assessing allosteric activation of SIRT1 and its subsequent effects on mitochondrial biogenesis and bioenergetic function in cellular models of PD. By combining synthesis, computational analysis, validation using biological methods along with real-time measurement of bioenergetic analysis, this chapter describes a mechanistic basis for the evaluation of small molecule modulators of SIRT1 and their downstream effects.

2. Rationale

The convergence of impaired oxidative phosphorylation, excessive production of mitochondrial reactive oxygen species (ROS) and defective mitophagy leading to bioenergetic failure and proteotoxicity were surmised to be the contributing factors in the etiology of PD. SIRT1 is a NAD^+-dependent deacetylase and was shown to enhance mitochondrial biogenesis via its primary substrate PGC-1α. In this regard, SIRT1 activators could ameliorate biochemical manifestations of PD. The present protocol was based on three interconnected principles: (1) Synthesis and validation of SIRT1 activators by providing direct evidence of their allosteric interaction in a cell-free system and *in silico* analysis. (2) Assessment of the downstream transcriptional targets of SIRT1, viz., PGC-1α, TFAM in dopaminergic cells, which could lead to increased mitochondrial DNA copy number and biogenesis. (3) Assessing the role of SIRT1 activation in the restoration of MPP^+ mediated dysfunction in mitochondrial biogenesis and bioenergetics.

3. Methodology

3.1 Chemical synthesis of Azaflavanone derivative

Unless specified otherwise, all reagents were utilized directly as obtained from commercial sources. Solvents were appropriately dried prior to use according to established protocols. Reaction progress was monitored by thin-layer chromatography (TLC) on silica gel (SiO_2) plates, which were visualized under UV light (254 nm), by iodine staining, or with 2,4-dinitrophenylhydrazine reagent. Column chromatography was performed using silica gel (100–200 mesh) packed in glass columns, the reaction for synthesizing compounds is well described in the later Schemes 2 and 3.

NMR spectra were recorded at 500 MHz for proton (^1H) and 125 MHz for carbon (^{13}C). Chemical shifts (δ) are expressed in ppm and referenced to residual solvent signals in $CDCl_3$ (^1H: δ 7.26; ^{13}C: δ 77.16 ppm) and

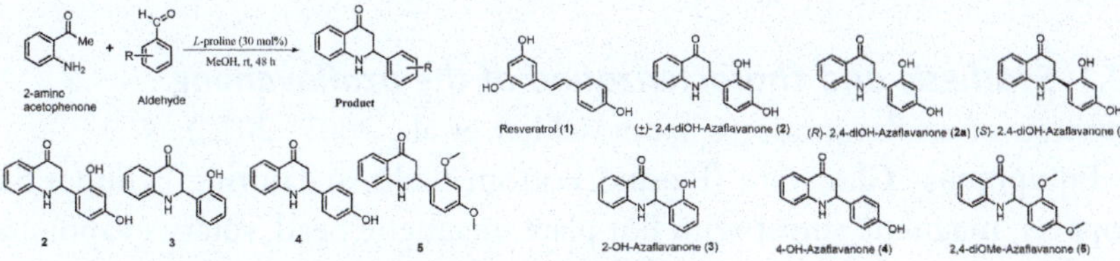

Scheme 2 Synthesis of aza-flavanone derivatives.

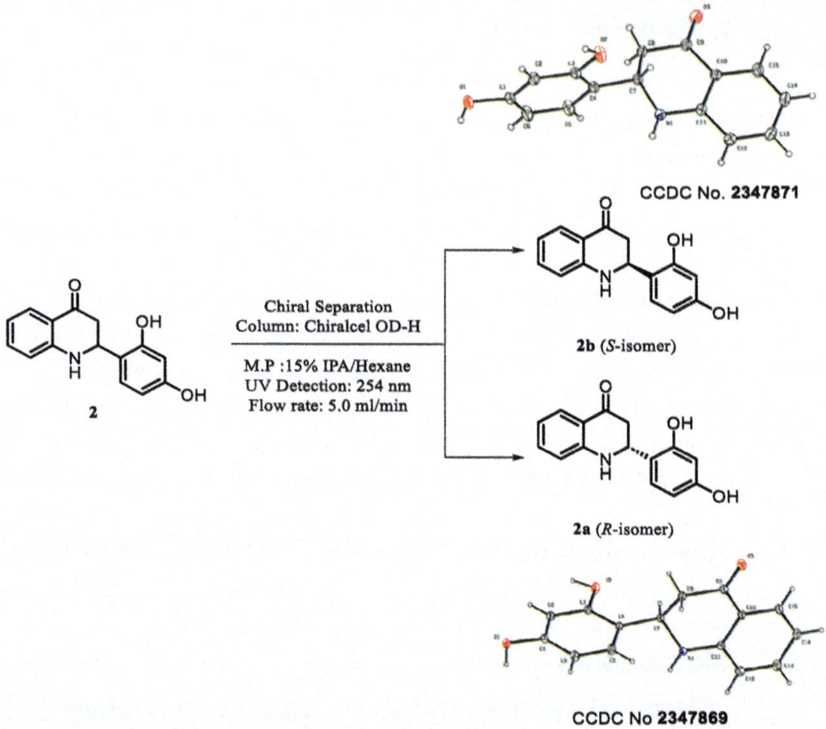

Scheme 3 Separation of aza-flavanone isomers.

DMSO-d_6 (^{1}H: δ 2.50; ^{13}C: δ 39.52 ppm), with coupling constants (J) reported in Hz. Chiral HPLC analyses were carried out using a Daicel Chiralcel OD-H column. Please refer to Figs. 1 and 2 below.

Infrared (IR) spectra were obtained using a Bruker Alpha Eco ATR FT-IR spectrometer with samples prepared in methanol and chloroform. Chemical structures and nomenclature were generated using ChemDraw Ultra 21.0. Melting points of compounds were determined using a POLMAN MP96 apparatus. Optical rotation measurements were recorded at 589 nm using an Anton Paar MCP 200 polarimeter.

3.1.1 Synthesis and characterization of the azaflavanone

Synthesis of azaflavanone (Chandrasekhar et al., 2007, 2012):

Equipment- Glassware (Round bottom flasks, measuring cylinder and dropper), magnetic stirrer with hot plate, magnetic bead, rotary evaporator, separating funnel, filter paper, filtration funnel, stopper column with stand.

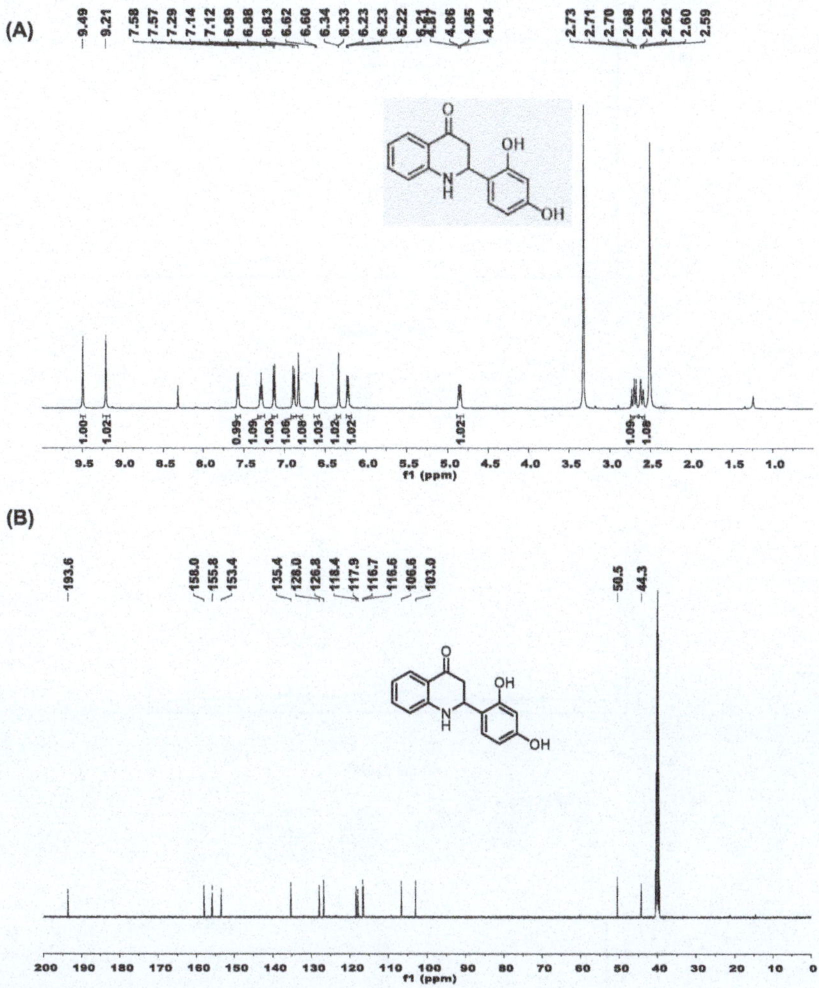

Fig. 1 (A) ^{1}H NMR Spectra of compound 2 in DMSO-d_6 at 500. (B) ^{13}C NMR Spectra of compound 2 in DMSO d_6 at 125 MHz.

Reagents- 2-aminoacetophenone, 2, 4-dihydroxy benzaldehyde, 2-hydroxy benzaldehyde, 4-hydroxy benzaldehyde, 2, 4-di methoxy benzaldehyde, L-Proline, Ammonium chloride, anhydrous $MgSO_4$.

Solvents- Methanol, hexane, Ethyl acetate and dichloromethane.

Procedure-

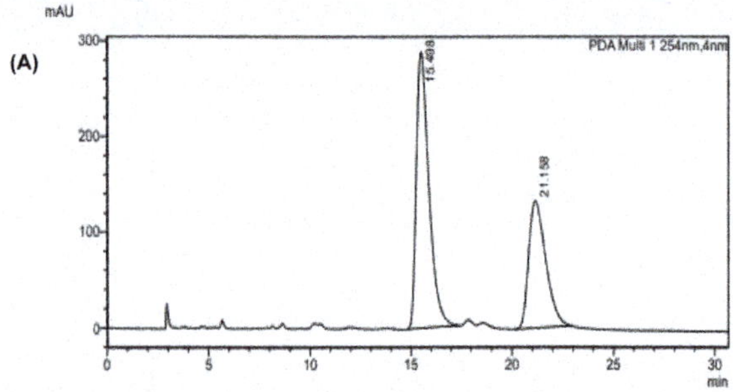

<Peak Table>

PDA Ch1 254nm

Peak#	Ret. Time	Area	Height	Area%	Height%
1	15.498	12563173	288323	61.994	68.419
2	21.158	7701934	133085	38.006	31.581
Total		20265107	421408	100.000	100.000

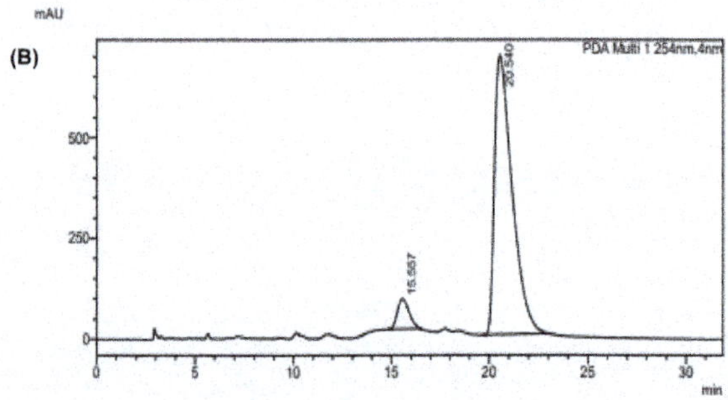

<Peak Table>

PDA Ch1 254nm

Peak#	Ret. Time	Area	Height	Area%	Height%
1	15.557	3093611	74090	6.690	9.684
2	20.540	43147320	690946	93.310	90.316
Total		46240931	765035	100.000	100.000

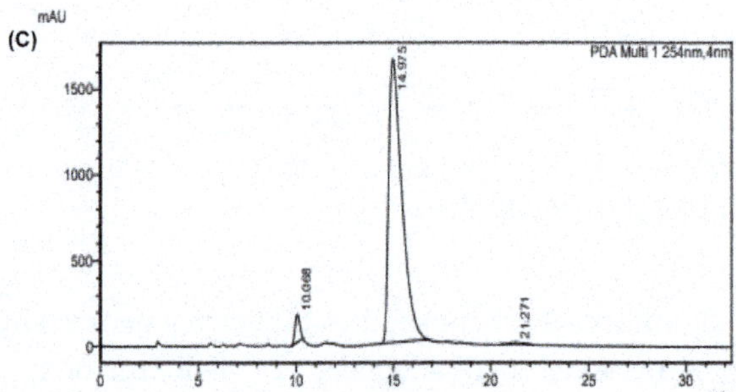

<Peak Table>

PDA Ch1 254nm

Peak#	Ret. Time	Area	Height	Area%	Height%
1	10.068	2699200	161194	3.289	8.836
2	14.975	78869904	1652555	96.105	90.588
3	21.271	497223	10514	0.606	0.576
Total		82066326	1824263	100.000	100.000

Fig. 2 (A) HPLC Chromatogram of racemic 2, (B) HPLC chromatogram of Product 2a, (C) HPLC chromatogram of the Product 2b.

3.1.2 *Racemic-2-(2,4-dihydroxyphenyl)-2,3-dihydroquinolin-4(1H)-one (2)*

1. In a round-bottom flask, add 2-amino acetophenone (1.48 mmol, 1 equiv.), L-Proline (0.44 mmol, 0.3 equiv.), and 2, 4 dihydroxy benzaldehyde (1.48 mmol, 1 equiv.) in Methanol (5 mL).
2. The reaction mixture was continuously stirred for 48 h, and the progression of the reaction was monitored using thin-layer chromatography (TLC).
3. Upon completion, aqueous ammonium chloride solution was added to the reaction mixture and extracted it with dichloromethane (3 × 10 mL).
4. All the organic layers were combined, dried over magnesium sulphate, and evaporated under vacuum to get the crude product.
5. The obtained crude product was purified by column chromatography using ethyl acetate and hexane (40:60) as eluting solvent to obtain pure product in 61.2 % yield.

The most active one **(2)** in the series was selected based on earlier studies. However, since this compound is racemic, both the enantiomers were separated by chiral HPLC preparative technique, and the structures of both enantiomers were confirmed by X-ray crystallography (Scheme 2). The biological properties of all these compounds, i.e., racemic compound **2**, *R*-enantiomer **2a** and *S*-enantiomer **2b** were compared with Resveratrol **1**.

3.2 Characterization of compounds

3.2.1 *Racemic-2-(2,4-dihydroxyphenyl)-2,3-dihydroquinolin-4(1H)-one (2)*

Compound **2** prepared following known procedure (Schapira, 2008).
Yield: 231 mg, 61.2 %, Yellow solid; m.p: 180–182 °C.
R_f: 0.2 (40 % EtOAc-Hexane).
^{1}H NMR (500 MHz, DMSO-d_6): δ 9.48 (s, 1 H), 9.20 (s, 1 H), 7.56 (d, J = 7.7 Hz, 1 H), 7.28 (t, J = 7.1 Hz, 1 H), 7.12 (d, J = 8.3 Hz, 1 H), 6.88 (d, J = 8.3 Hz, 1 H), 6.82 (s, 1 H), 6.59 (t, J = 7.4 Hz, 1 H), 6.32 (d, J = 1.9 Hz, 1 H), 6.21 (dd, J = 8.3, 1.9 Hz, 1 H), 4.84 (dd, J = 11.3, 4.0 Hz, 1 H), 2.70 (dd, J = 15.9, 11.7 Hz, 1 H), 2.60 (dd, J = 15.9, 4.1 Hz, 1 H).
^{13}C NMR (125 MHz, DMSO-d_6): δ 193.6, 158.0, 155.8, 153.4, 135.4, 128.0, 126.8, 118.4, 117.9, 116.7, 116.6, 106.6, 103.0, 50.5, 44.3.
IRυ_{max}: 3319, 1607 cm^{-1}.
HRMS (ESI) m/z: $[M + H]^+$ calculated for $C_{15}H_{14}NO_3$ 256.0974; found 256.0977.

3.3 (*R*)- 2-(2,4-dihydroxyphenyl)-2,3-dihydroquinolin-4(1 H)-one (2a)

Using semi-preparative chiral HPLC, the compound **2a** was separated from **2**. HPLC conditions are Daicel Chiralcel OD-H column (hexane/2-propanol = 85:15), $\lambda = 254$ nm, flow rate = 1 mL/min, $t_R = 16.85$ min. **Specific rotation:** $[\alpha]_D^{20} = -78.50$ (c = 0.2, MeOH). The absolute configuration of atom C7 of compound **2a** is confirmed as "***R***" using single X-ray crystallographic studies.

3.4 (*S*)- 2-(2,4-dihydroxyphenyl)-2,3-dihydroquinolin-4(1 H)-one (2b)

Using semi-preparative chiral HPLC, the compound **2b**was separated from **2**. HPLC conditions are Daicel Chiralcel OD-H column (hexane/2-propanol = 85:15), $\lambda = 254$ nm, flow rate = 1 mL/min, $t_R = 13.03$ min. **Specific rotation:** $[\alpha]_D^{20} = +56.00$ (c = 0.2, MeOH). The absolute configuration of atom C7 of compound **2b** is confirmed as "***S***" using single X-ray crystallographic studies.

Molecular Docking studies.

Equipment- Maestro v2021–1 (Schrodinger LLC) (Schrödinger, 2022-1, 2022) operating on a CentOS 7 platform on a Fusionstor Invento workstation with an Intel Xeon processor E5–2699 v4, 128 GB RAM, and a 5 TB HDD with NVIDIA Quadro P5000 and NVIDIA Quadro P600 was used for the current computational study (Scheme 4) (Schrödinger, 2022-1, 2022; Schrödinger Release, 2024-2, 2024).

Procedure-

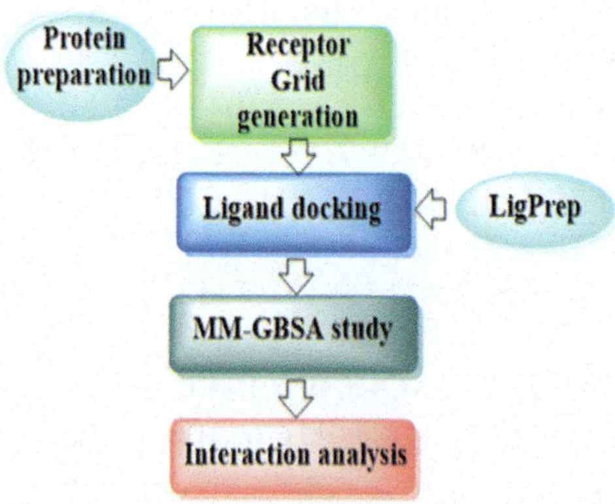

Scheme 4 Flow chart depicting the protocol for *in silico* study.

3.4.1 Protein and ligand preparation

1. From RCSB PDB (http://www.rcsb.org) the X-ray crystal structure of human SIRT1 containing p53-AMC peptide complexed with resveratrol (PDB ID: 5BTR) was retrieved.
2. The complex was then pre-processed and curated using "Protein Preparation" wizard from Schrodinger Suite using default parameters.
3. H-bond optimization and protein minimization were performed using Epik and OPLS_3e force fields, respectively.
4. All ligands were drawn directly using 2D sketcher of Schrodinger Suite.
5. The sketched ligands were subjected to "Ligprep" at pH = 7.0 using other default parameters to generate possible ionized and minimized isomers.

Notes

1. The selection of the target PDB was prioritized based on the chemical similarity between the co-crystal ligand and our library.
2. The target ligand was present in a single chain; all redundant chains were removed.
3. All duplicate chains, water molecules, and extraneous small molecules (e.g., buffers, salts) were deleted.
4. Missing loops and side chains were modelled and filled during pre-process to ensure complete protein architecture.
5. Ensure the ligand in the library is neutral to prevent electrostatic biases in the active site.

3.4.1.1 Receptor grid generation and ligand docking

SIRT1 accommodates three resveratrol molecules simultaneously in three distinct sub-sites; however, for the convenience we have randomly selected one sub-site to generate the receptor grid.

1. Using "Receptor Grid Generation" wizard, a grid with internal and external cube dimensions of 20 Å was constructed by taking default parameters and without any constraints.
2. The prepared ligands were subjected to molecular docking using GLIDE module against the generated grid.
3. For docking, XP (Extra Precision) protocol was used with flexible ligands and rigid protein.
4. The resultant docking scores and interaction patterns were compared with resveratrol for thorough analysis (Fig. 3).

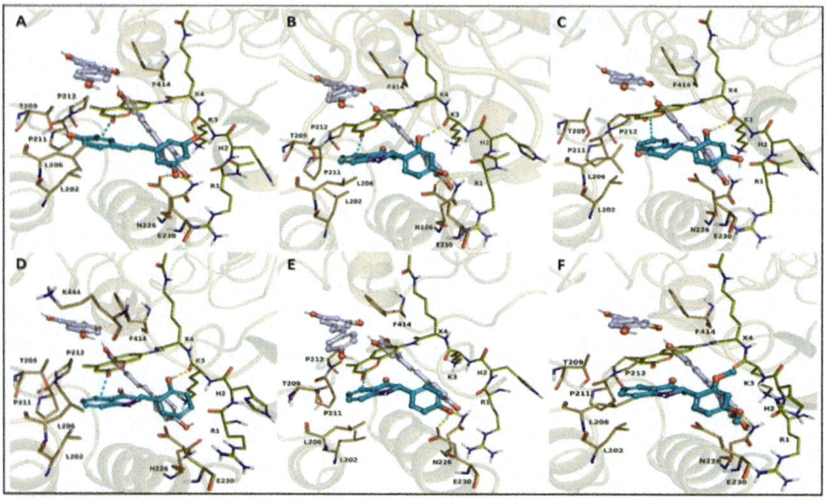

S.No.	Code	Dock score (kcal/mol)	MM/GBSA (kcal/mol)
A.	Resveratrol (**1**)	-8.933	-86.57
B.	(R)- 2,4-diOH-Azaflavanone (**2a**)	-9.098	-77.13
C.	(S)- 2,4-diOH-Azaflavanone (**2b**)	-8.263	-69.14
D.	2-OH-Azaflavanone (**3**)	-7.977	-28.62
E.	4-OH-Azaflavanone (**4**)	-7.462	-49.70
F.	2,4-diOMe-Azaflavanone (**5**)	-8.770	-50.45

Fig. 3 Molecular docking analysis illustrates the interaction of 2,4-dihydroxy-azaflavanone (B) within the allosteric STAC-binding domain of SIRT1. Comparative docking with resveratrol (A), (C) (S)-2,4-diOH-azaflavanone, (D) 2-OH-azaflavanone, (E) 4-OH-azaflavanone, (F) 2,4-diOMe-azaflavanone are shown for reference.

Notes-

1. Dock resveratrol alongside the ligands to validate and calibrate the docking protocol.
2. The interaction pattern of the library was compared with selected resveratrol to identify retained key interactions and distinct additional interactions.

3.4.1.2 MM/GBSA study

1. The binding free energy of each ligand- protein complex was calculated by performing Molecular Mechanic/Generalized Born Surface Area (MM/GBSA) study.
2. The docked complexes (pv.maez files) were directly subjected to MM/GBSA calculations.

3. The OPLS_3e force field was used along with VSGB solvation and other default parameters in Prime module.
4. The MM-GBSA calculates the difference between the energies of bound protein and unbound protein and ligand (Cao et al., 2015; Friesner et al., 2006; Wichapong et al., 2014).

$$\Delta G_{bind} = G_{complex} - (G_{protein} + G_{ligand})$$

Here, $G_{complex}$, $G_{protein}$, and G_{ligand} are the free energies of the complex, protein, and ligand, respectively.

Notes. For calculating MM/GBSA, ensure the docked complexes minimization sampling method.

Biochemical and structural validation of allosteric SIRT1 activation. Cell-free SIRT1 deacetylase activity assay.

To assess the effect of compounds on SIRT1 activation, recombinant human SIRT1 was used for the fluorometric deacetylase assays under defined conditions. This cell-free system provides direct evidence on the interaction of small molecules on the activation of SIRT1 is well depicted

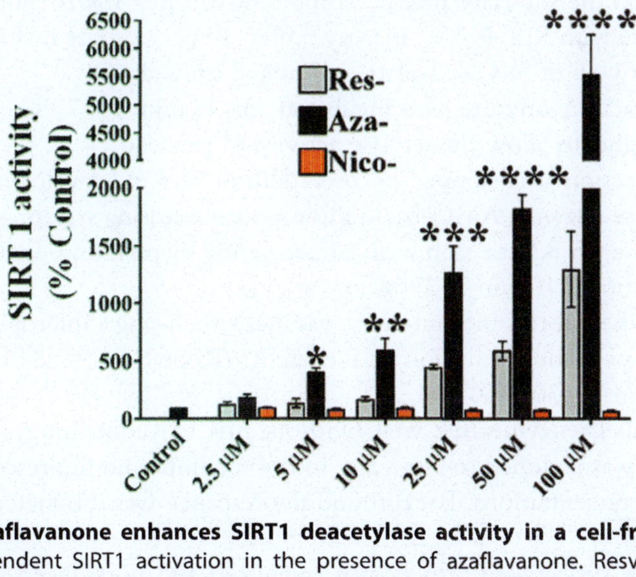

Fig. 4 Azaflavanone enhances SIRT1 deacetylase activity in a cell-free system. Dose-dependent SIRT1 activation in the presence of azaflavanone. Resveratrol was used as a positive control. Nicotinamide inclusion attenuated SIRT1 activation, confirming the specificity of the assay. Data represent mean ± SEM of three independent experiments.

in Fig. 4. On contrary, Analysis of SIRT1 activity in cells treated with compounds could be due to the various indirect confounding effects such as, cellular NAD^+ levels, AMPK signaling, or the transcriptional upregulation of SIRT1 expression etc.

Equipment- Fluorescence microplate reader (e.g., EnSpire, PerkinElmer) with excitation/emission filters, 384-well black flat-bottom microplates, microcentrifuge tubes (1.5 mL), incubator set to 37 °C, multichannel pipette, analytical balance.

Reagents- SIRT1 assay kit (e.g., Sigma-CS1040), 2,4-dihydroxy-azaflavanone dissolved in DMSO, resveratrol, nicotinamide.

Procedure-

1. Prepare the working stocks of 2,4-dihydroxy-azaflavanone, resveratrol and nicotinamide to achieve the desired final concentrations (2.5, 5, 10, 25, 50, and 100 μM) in 30 μL assay buffer (50 mM Tris-HCl- pH 8.0, 100 mM NaCl, 5 mM $MgCl_2$, 1 mM DTT)
2. Incubated the recombinant enzyme (SIRT1)-compound mixture in 50 μL final volume at 37 °C for 10 min to allow compound-enzyme interactions prior to the addition of substrate.
3. Prepared the substrate mix by combining 1.5 μG SIRT1 pure peptide substrate with 5 μL NAD^+ in assay buffer. 10 μL substrate mix was added to each well of 384 well-plate and mixed immediately.
4. The reaction mixture was incubated for 45 min at 37 °C (protected from light) to allow deacetylase activity to proceed.
5. The reaction was stopped by the addition of 5 μL developer solution (Protease enzyme, AMC-based fluorophore-releasing system, stabilizers, Tris-base pH 8.0) to each well. Mixed gently by pipetting. Incubated for an additional 10 min at 37 °C.
6. Lastly, the fluorescence intensity was measured using a microplate reader at an excitation/emission wavelength (Excitation = 340−380 nm; Emission = 430−460 nm).
7. In parallel, enzyme-free well (without SIRT1) containing only-compound was included to assess any intrinsic compound fluorescence at all tested concentrations. Background fluorescence was subtracted from all readings.
8. Relative deacetylase activity was calculated and normalized to vehicle control.

Notes.

1. Avoid freeze-thaw cycles of recombinant SIRT1 protein; prepare single-use aliquots and store at −80 °C.
2. Fluor-de-Lys substrate exhibits fluorescence at emission wavelengths overlapping with some aromatic compounds. Always include compound controls (no-enzyme) to account for background fluorescence.

Isoform selectivity and target engagement via cellular thermal shift assay (CETSA).

While the cell-free enzymatic assay establishes direct interaction with small molecules, it is pertinent to confirm whether the observed interaction also occurs in live cells to achieve the desired downstream effects. Towards this SIRT1 activity was assessed in rat dopaminergic N27 cells using endogenous SIRT1 as the enzyme source. This method combines the cellular metabolic status, including endogenous NAD^+ levels and other deacetylase enzymes, and demonstrates that the binding and selectivity of azaflavanone towards SIRT1 is maintained in the cellular milieu. The cellular thermal shift assay (CETSA) is based on the principle that ligand binding increases the thermodynamic stability of proteins, leading to shift in the protein melting curve, detectable through immunoblot analysis of soluble protein fractions after incremental thermal denaturation (Jafari et al., 2014). Consequently, CETSA offers direct evidence of compound-protein interaction in a physiologically relevant cellular environment and allows for simultaneous evaluation of isoform selectivity by examining the stability of other SIRTs under identical experimental conditions and has been effectively demonstrated in Fig. 5 below.

Equipment- Cell culture incubator (37 °C, 5 % CO_2), fluorescence microplate reader, refrigerated microcentrifuge (4 °C capability), 96-well black flat-bottom microplates, cell scraper, Bradford protein quantification kit, PCR thermocycler with ramp capability precise temperature control, refrigerated microcentrifuge, SDS-PAGE electrophoresis system and semi-dry transfer apparatus, chemiluminescence imaging system (e.g., Bio-Rad ChemiDoc), PCR strip tubes (0.2 mL) compatible with thermocycler, refrigerated storage unit at 4 °C for post-lysis processing.

Reagents- N27 rat dopaminergic cells (immortalized rat mesencephalic dopaminergic cell line), complete growth medium: RPMI 1640 supplemented with 10 % FBS and 1 % penicillin-streptomycin, 2,4-dihydroxy-azaflavanone, RIPA lysis buffer supplemented with protease inhibitor

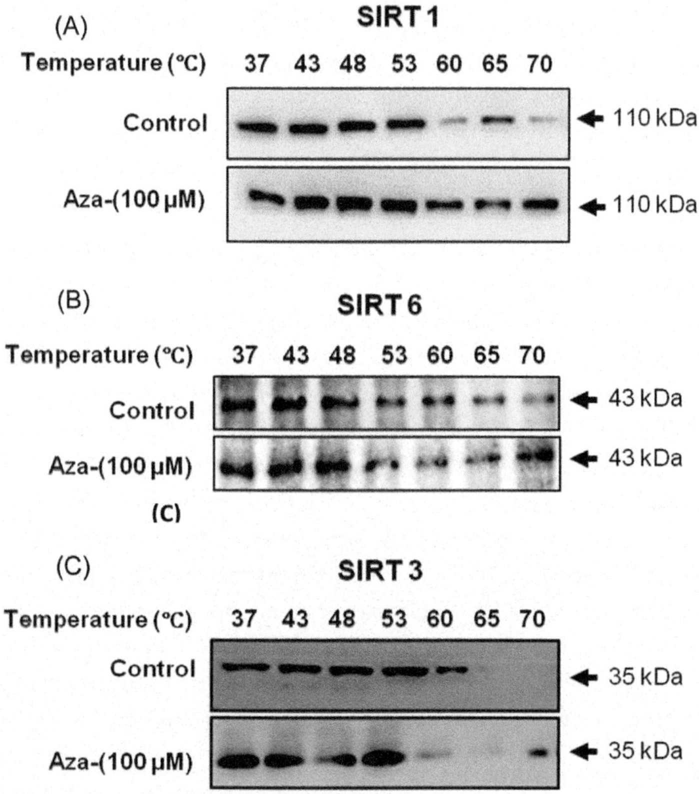

Fig. 5 Cellular thermal shift assay (CETSA) demonstrates selectivity of 2,4-dihydroxyflavanone towards SIRT1. N27 cells treated with azaflavanone (100 μM) demonstrated increased thermal stability of (A) SIRT1 across a temperature gradient. No comparable stabilization is observed in the representative immunoblots for (B) SIRT6 or (C) SIRT3.

cocktail and phosphatase inhibitors, Dulbecco's PBS (DPBS), SDS-PAGE reagents: Laemmli sample buffer (4×), SDS-PAGE gels (10 %), running buffer, nitrocellulose membranes, blocking buffer, primary antibodies: anti-SIRT1 (1:1000), anti-SIRT3 (1:1000), anti-SIRT6 (1:1000) validated for human/rat cross-reactivity, HRP-conjugated secondary antibodies (1:5000), enhanced chemiluminescence (ECL) detection reagents.

Procedure.

1. N27 cells were lysed in 1 mL of DPBS containing 0.2 % Triton-X-100, supplemented with a protease inhibitor cocktail and centrifuged at 12,000 rpm for 30 min at 4 °C.

2. The resulting clarified supernatants (1000 μL, 100 μg) were treated with azaflavanone (100 μM) for one hour at 37 °C. Subsequently, the samples were divided into seven separate tubes (120 μL each) and incubated at different temperatures (37, 43, 48, 53, 60, 65, and 70 °C) for an additional 10 min.
3. Lastly, the tubes were centrifuged at 12,000 rpm for 30 min to remove any denatured proteins, and 30 μL of the clarified supernatants from each aliquot were employed for Western blot analysis.
4. Equal volumes of 2x Laemmli sample buffer and clarified supernatant were mixed, boiled at 95 °C for 5 min, and the proteins were resolved on 10−12 % SDS-PAGE gels.
5. After resolving the proteins, they were transferred onto nitrocellulose membranes using a semi-dry transfer method. The membranes were then blocked with 5 % non-fat milk prepared in TBST for 1 h at room temperature to minimize non-specific binding.
6. The membranes were probed sequentially with anti-SIRT1, anti-SIRT3, and anti-SIRT6 antibodies (1:1000 in blocking buffer), incubating overnight at 4 °C with gentle agitation. Washed three times with TBST (5 min each).
7. Lastly, the membrane was incubated with appropriate HRP-conjugated secondary antibody (1:5000) for 1 h at room temperature. Three times wash with 1X TBST and the protein bands were visualized in a chemiluminescence imager following staining with ECL reagent.

Notes-

1. If concurrent NAD^+ measurements are required, collect a separate aliquot of lysate prior to the activity assay for NAD^+/NADH quantification to rule out indirect activation via elevated NAD^+.
2. The freeze-thaw lysis method is preferred over detergent lysis for CETSA, as detergents can alter protein aggregation kinetics and confound thermal stability measurements.
3. Antibody specificity for each sirtuin isoforms must be validated. Cross-reactive antibodies may generate misleading results.

Analysis of SIRT1-PGC-1α-TFAM Axis and mitochondrial biogenesis.

Activation of SIRT1 in dopaminergic neurons results in synchronized transcriptional activation of mitochondrial biogenesis pathways, along with observable alterations in mitochondrial number. This section assesses the downstream effects of azaflavanone-induced SIRT1 activation by assessing

SIRT1-PGC-1α-TFAM regulatory cascade leading to mitochondrial biogenesis and has been clearly exhibited by referring to Fig. 6.

Equipment- Cell culture incubator (37 °C, 5 % CO_2), fluorescence microplate reader, refrigerated microcentrifuge (4 °C capability), 96-well black flat-bottom microplates, cell scraper, Bradford protein quantification kit, PCR thermocycler with ramp capability precise temperature control, refrigerated microcentrifuge, SDS-PAGE electrophoresis system and semi-dry transfer apparatus, chemiluminescence imaging system (e.g., Bio-Rad ChemiDoc), refrigerated storage unit at 4 °C for post-lysis processing.

Reagents- Complete growth medium (RPMI 1640 supplemented with 10 % FBS and 1 % penicillin-streptomycin), 2,4-dihydroxy-azaflavanone, RIPA lysis buffer supplemented with protease inhibitor cocktail and phosphatase inhibitors, Dulbecco's PBS (DPBS), SDS-PAGE reagents: Laemmli sample buffer (4×), SDS-PAGE gels (10 %), running buffer, nitrocellulose membranes, blocking buffer, primary antibodies: anti- PGC-1α (1:1000),

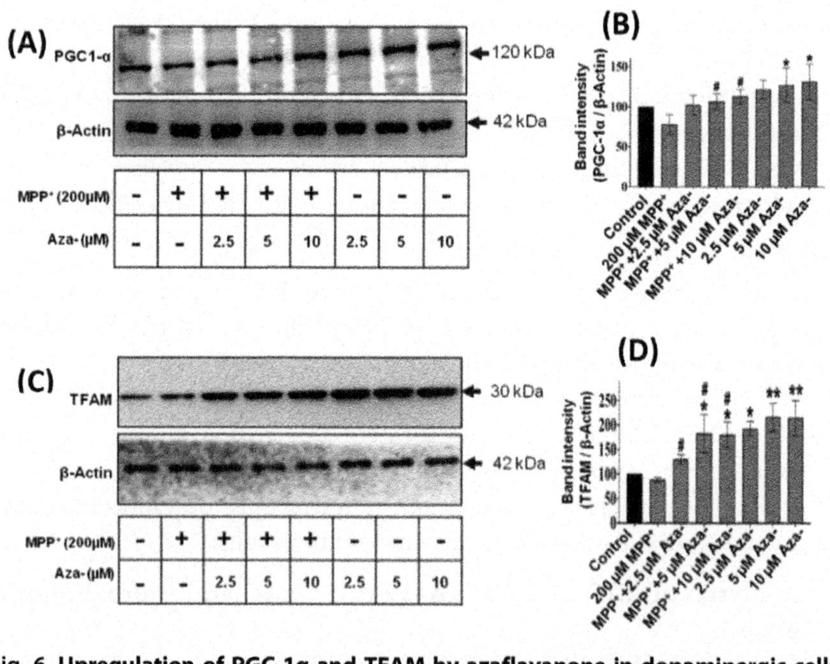

Fig. 6 Upregulation of PGC-1α and TFAM by azaflavanone in dopaminergic cells. Representative immunoblots of PGC-1α(A), TFAM (C) and (B and D) densitometric quantification normalized to housekeeping gene (β-actin). Data represent mean ± SEM from three independent experiments. Statistical significance indicated.

anti-TFAM (1:1000), anti- β-actin (1:1000) validated for human/rat cross-reactivity, HRP-conjugated secondary antibodies (1:5000), Enhanced chemiluminescence (ECL) detection reagents.

Procedure.

1. N27 cells in 6-well plates were pre-treated with azaflavanone (2.5, 5, and 10 μM, for 24 h), before the addition of MPP^+ treatment (150 μM, 24 h).
2. Cells were lysed in RIPA buffer supplemented with a protease inhibitor cocktail, followed by centrifugation at 12,000 rpm for 15 min at 4 °C.
3. The resulting clarified supernatants (calculated equal protein content using Bradford assay) with Laemmli sample buffer, boiled at 95 °C for 5 min, and loaded onto SDS-PAGE gels.
4. Transferred the resolved proteins on to nitrocellulose membranes using standard semi-dry transfer protocols. Blocked the membranes with 5 % non-fat milk in TBST for 1 h at room temperature.
5. The membranes were probed with anti- PGC-1α (1:1000), anti-TFAM (1:1000), anti- β-Actin (1:1000) in blocking buffer, incubated overnight at 4 °C with gentle agitation. Later, the membranes were washed with TBST (3 times × 5 min each) the next day.
6. Lastly, the membranes were incubated with appropriate HRP-conjugated secondary antibody (1:5000) for 1 h at room temperature. Washed three times with TBST and bands were visualized following ECL detection. The bands were imaged using a chemiluminescence gel documentation system. Densitometric analysis was performed using ImageJ software.

Quantification of mtDNA copy number.

The mtDNA copy number serves as a reliable indicator of mitochondrial biogenesis and is responsive to both transcriptional activation and mitochondrial degradation pathways (Clay Montier et al., 2009). Measuring the ratio of mitochondrial DNA (mtDNA): nuclear DNA offers a direct correlation of mitochondrial biogenesis as shown in Fig. 7.

Equipment- Real-time PCR system (e.g., Applied Biosystems), 96-well PCR plates with micro-seal adhesive film, microcentrifuge and benchtop vortex.

Reagents- Mitochondrial homogenization buffer, trypsin, TRIzol™, Rt-PCR primers. Primers (rat specific) sequences for assessing mt DNA (mtND1 forward: 5′-ATGGATTCGAGCATCCTACCC-3′ and mtND1 reverse: 5′-TCCTGCTAGGAAAATTGGCA-3′); primers sequences for

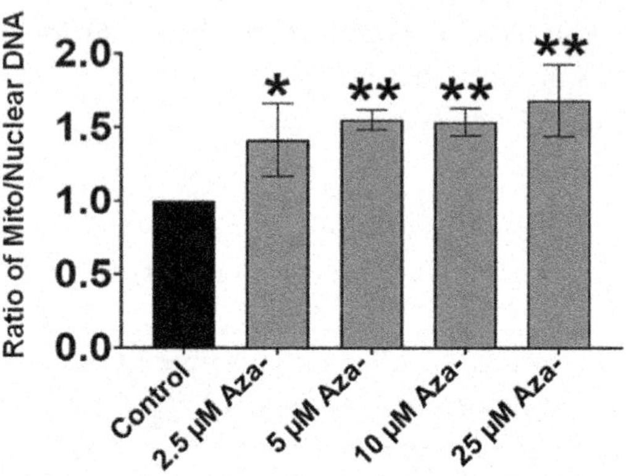

Fig. 7 Increased mitochondrial DNA copy number following SIRT1 activation by azaflavanone. Relative mtDNA content quantified by qPCR using mitochondria-encoded gene specific primers normalized to nuclear DNA. Data represent mean ± SEM of three independent experiments.

assessing nuclei DNA levels (β-actin exon forward: 5′-AGCCATGTAC GTAGCCATCC-3′ and β-actin exon reverse: 5′-CTCTCAGCTGTG GTGGTGAA-3).

Procedure.

1. N27 cells grown in a 90 mm culture dish were treated with different concentrations of azaflavanone (2.5–25 μM) for 24 h. At the end of incubation, cells were collected by trypsinization; half of the cells were used for nuclear DNA isolation, and the rest were used for mitochondrial DNA isolation.
2. Mitochondrial DNA was isolated by homogenizing the cells in 1 mL cold mitochondria homogenization buffer (250 mM sucrose, 10 mM Tris-HCl, 1 mM EDTA) at pH 7.4–7.5 and centrifuged the lysate at 4,000 rpm for 10 min at 4 °C to remove nuclear and cellular debris.
   - The supernatant was collected in a clean microfuge tube and centrifuged at 11,000 rpm for 15 min at 4 °C. After discarding the supernatant, the mitochondrial pellet was resuspended in 480 μL of high salt buffer (10 mM tris HCl, 10 mM KCl, 10 mM $MgCl_2$, 0.4 M NaCl, and 2 mM EDTA) and 75 μL of 10 % SDS.
   - The mixture was incubated at 55 °C for 10 min to completely denature and solubilize the mitochondrial proteins. Salting out of proteins was done by the addition of 200 μL of 6 M NaCl to this mixture and centrifuged at 11,500 rpm for 20 min at RT and the supernatant was transferred to a fresh tube.

- Two volumes of 100 % ethanol were added to the supernatant and centrifuged again at 14,000 rpm for 10 min at RT for the complete precipitation of the mitochondrial pellet.
- Further, the mitochondrial DNA pellet was washed twice with 70 % ethanol, and the obtained pellet was air-dried to remove any traces of ethanol and dissolved in 20−30 μL of nuclease free water.
- The mitochondrial DNA was quantified and used as a template for quantitative real-time PCR analysis of mtND1 gene which is exclusively encoded by mitochondrial DNA. The sample was stored in aliquots at −30 °C until use.

3. For nuclear DNA isolation, TRIzol™ method was employed.
- N27 cells were homogenized carefully in 1 mL of TRIzol™ reagent (it's the nuclei isolation buffer) and left at RT for 5 min 500 μL of chloroform was added to the homogenate, mixed vigorously and centrifuged at 12,000 rpm for 20 min at 4 °C.
- The cloudy interphase (the middle layer) was collected out of the three layers (which contain DNA) in a fresh tube. Equal volume of chilled 100 % iso-propanol was added to the interphase layer and mixed by inverting the tube four to five times.
- After 3 min of incubation, the mixture was centrifuged at 2000 *g*′ for 5 min at 4 °C. After discarding the supernatant, the DNA pellet was washed with 75 % ethanol and again centrifuged at 7000x for 7 min for pelleting down which was then air-dried to remove reminiscent ethanol.
- Lastly, the DNA pellet was resuspended in sterile nuclease-free water and used as the template for the amplification of β-actin gene, which is exclusively encoded by genomic DNA by real-time PCR.

To perform **quantitative real-time PCR** (qRtPCR), TB Green® Premix Ex Taq™ II (TIi RnaseH Plus) with ROX-dye II was utilized. PCR mix was put together following manufacturer's protocol. The ingredient list and amplification conditions are shown below Table 1 and 2 respectively:

- Real-time PCR was conducted using a 20 μL reaction volume containing 2x TB (SYBR) green and gene-specific primers (The qPCR primers are detailed in Table 3, utilizing a real-time PCR machine (7500-Applied Biosystems). β-Actin gene was used as reference for normalization of gene expression.

Table 1 Quantitative real-time PCR mix/ingredients list.

Ingredient	**Volume**
2X TB Green® buffer (+ ROX-II dye)	10 μL
cDNA	1 μL
Forward primer (10 pM)	1 μL
Reverse primer (10 pM)	1 μL
H_2O	7 μL
Total	**20 μL**

Table 2 Amplification conditions are described below.

Steps	**Temperature**	**Time**	
Initial denaturation	95 °C	5 minutes	
Denaturation	95 °C	30 seconds	}
Annealing	60 °C	30 seconds	} **X 40 cycles**
Extension	72 °C	1 minute	
Final extension	72 °C	10 minutes	
Hold	4 °C	----	

- The results are presented as mean ± SD, and statistical analyses were performed using GraphPad Prism 9.5.1. Parametric data were evaluated through one-way ANOVA, followed by Dunnett's post hoc multiple comparison test for significance.

Notes.

1. Ensure that there is no RNA contamination in the samples, which could affect the results. In such cases, the samples may be treated with RNase.

Table 3 Primers used.

Gene	Forward primer	Reverse primer
mtND1	ATGGATTCGAG CATCCTACCC	TCCTGCTAGGAAAATTGGCA
β- actin-exon'	AGCCATGTACGT AGCCATCC	CTCTCAGCTGTGGTGGTGAA

2. The designed primers on genomic DNA should be located on the same exon to get the desired product size.

Analysis of mitochondrial function using mitostress assay:

Analysis of mitochondrial oxygen consumption rate (OCR) in live cells offers a real-time, multi-parameter quantification of mitochondrial bioenergetics by sequentially measuring the oxygen consumption rate (OCR) after pharmacological perturbation of specific ETC components (Dranka et al., 2011). This platform allows for the resolution of unique respiratory parameters that together define mitochondrial bioenergetic capacity that cannot be deduced from single-endpoint analysis.

Equipment- Agilent Seahorse XFe24 Analyzer, Seahorse XF24 cell culture microplates, Seahorse XF24 sensor cartridges, calibration solution (Agilent Seahorse XF Calibrant Solution), non-CO2 incubator set to 37 °C for sensor cartridge hydration and pre-assay cell equilibration, microplate reader for protein quantification (post-assay normalization), automated cell counter or hemocytometer for seeding density determination.

Reagents- Seahorse XF Assay Medium: DMEM without sodium bicarbonate (Agilent; Cat. No. 103575–100), supplemented with 10 mM D-glucose, 2 mM L-glutamine, and 1 mM sodium pyruvate. Adjust pH to 7.4 at 37°C. Prepare fresh on day of assay inhibitors Oligomycin (ATP synthase inhibitor): 2.5 μM final concentration in assay medium; FCCP (carbonyl cyanide p-trifluoromethoxyphenylhydrazone; ETC uncoupler): 3 μM final concentration in assay medium; Rotenone (complex I inhibitor): 2 μM final concentration in assay medium; Antimycin A (complex III inhibitor): 2 μM final concentration in assay medium, Bradford protein assay kit, 0.1 N NaOH.

Procedure.

1. Preparation (Day 0):
   - The Agilent Seahorse XFe analyzer was switched on one day prior to the assay and allowed to warm up overnight (minimum of 16 h).

- Seed the cells at an optimized density (For 24-well plate; 30,000−40,000 cells per well are optimum) in the Seahorse XFe24 cell culture microplate using 200 μL of appropriate cell culture growth medium and incubate at 37 °C in 5% CO_2 condition (For pre-treatments, cells were seeded prior to this day and the treatment was performed on day zero).
- The sensor cartridge plate was hydrated with Calibrant solution (Seahorse) at 37 °C in a non-CO_2 incubator overnight. Meanwhile prepare fresh 5 mM stocks of the inhibitors to be used.

2. Experiment day (Day 1):
   - Replace the culture medium in the cell culture microplate with a warm XFe DMEM assay medium with fresh supplements (pH = 7.4) using a multichannel pipette and place the microplate into a 37 °C non-CO_2 incubator for 45 min to 1 h before the assay.
   - Meanwhile, the inhibitors and test compound were added in the respective ports of the sensor cartridge plate as described in the Table 4. The calibration plate with the loaded sensor cartridge was placed (with required commands according to the injection strategy and assay) on the instrument tray for calibration. Calibration will take approximately 15–30 min

Table 4 Modulators/inhibitors/test compounds loading sequence and concentrations are A→B→C→D and described below.

Compound	Injection port	Port concentration	Modulators concentration required (per port)	Final concentration (per well)
Oligomycin A/ Test compound	**A**	8X	20 μM/–	2.5 μM/–
Oligomycin A/ FCCP	**B**	9X	20 μM/27 μM	2.5 μM/3 μM
FCCP/ Rotenone + Antimycin A	**C**	10X	27 μM/20 μM	3 μM/2 μM (each)
Rotenone + Antimycin A/Empty	**D**	11X	20 μM/–	2 μM (each)/–

- Following calibration, cell culture microplate was placed after removing the calibration plate and then performed the assay using the Wave 2.6 software platform.
- After completion of the assay, cells were washed with 1X DPBS and proteins were solubilized using 20 μL of 0.1 N NaOH solution. Protein quantification was performed using the Bradford method to normalize the OCR values of assay results.

Each port volume is about 75 μL in total, and the volume of medium in each well was 525 μL for mitostress assay.

3. Mitostress assay-
   - Run the Mitochondrial Stress Test protocol with the following measurement cycle settings: 3 baseline OCR measurements (Mix: 3 min, Wait: 3 min, Measure: 3 min); Inject Port A (Oligomycin): 3 measurement cycles; Inject Port B (FCCP): 3 measurement cycles; Inject Port C (Rotenone/Antimycin A): 3 measurement cycles.
4. Post-Assay Protein Normalization-
   - Following the assay, carefully aspirate XF assay medium from each well. Add 20 μL of 0.1 N NaOH per well and incubate on ice for 10 min. Collect lysate and determine total protein concentration by Bradford assay.
   - Normalize all OCR values to total protein content per well (μg) to account for inter-well cell density variation.
5. Parameter Calculation-
   - Calculate the following bioenergetic parameters from the OCR trace using Seahorse Wave software or manual calculation (Dranka et al., 2011).
   - Basal respiration: Difference between the measurement before oligomycin injection and non-mitochondrial respiration
   - ATP-linked respiration: Difference between basal respiration and OCR after oligomycin injection
   - Maximal respiratory capacity: OCR after FCCP injection − non-mitochondrial respiration
   - Spare respiratory capacity: Difference between Maximal respiratory capacity and Basal respiration
   - Proton leak: Difference between the OCR after oligomycin injection and non-mitochondrial respiration
   - Non-mitochondrial respiration: Residual OCR after the addition of rotenone/antimycin A.

The mitostress assay was performed in N27 cells treated with and without MPP^+ (150 μM) in the presence of different concentrations (5 and 10 μM) of azaflavanone for 24 h. Cells treated with MPP^+ together with azaflavanone exhibited a significant improvement in bioenergetics parameters as compared with cells treated only with MPP^+ in terms of basal OCR and proton leak.

In fact, cells treated with azaflavanone per se exhibited significantly higher levels of basal OCR and proton leak as compared to control. Though MPP^+-mediated decline in the ATP-linked OCR was slightly improved, the values were not found to be significant. However, proton leaks significantly increased in the cells treated with azaflavanone + MPP^+ as compared with those treated with MPP^+ treatment alone (Fig. 8).

Notes-

1. Optimal seeding density must be empirically determined for N27 cells to avoid over-confluence or under-confluence (which gives low signal).
2. FCCP concentration must be optimized for the specific cell line. Sub-optimal FCCP leads to underestimation of maximal respiratory capacity, while excessive FCCP causes non-specific effects. Optimize using 0.125, 0.25, 0.5, 1.0, 1.5, 2.0, 2.5, and 3.0 μM FCCP in a titration experiment.
3. The non-CO_2 pre-incubation step is mandatory. Residual CO_2 in medium causes carbonate buffering artefacts that could produce erratic OCR baseline measurements.

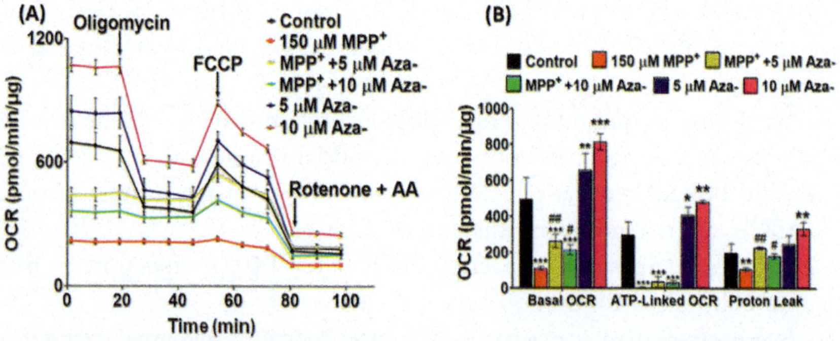

Fig. 8 Functional restoration of MPP^+ mediated mitochondrial dysfunction by azaflavanone. Seahorse XF analysis shows (A) OCR traces and (B) quantified respiratory parameters. MPP^+ reduced basal respiration, ATP-linked respiration, maximal respiratory capacity, and spare respiratory capacity; Pretreatment with azaflavanone ameliorated these parameters. Data was normalized to protein content; mean ± SEM from three independent experiments.

4. Concluding remarks

This chapter describes a series of experimental frameworks that demonstrates how 2,4-dihydroxy-azaflavanone selectively and directly activates SIRT1 at its allosteric STAC-binding domain. This has been established by multiple experiments involving molecular docking simulations, cell-free and cell based fluorometric deacetylase assays, CETSA-based isoform-selective target engagement. Subsequently, this biochemical activation corroborates with the downstream activation of SIRT1-PGC-1α-TFAM cascade, leading to a coordinated transcriptional upregulation of mitochondrial biogenic regulators, a measurable increase in mitochondrial genomic content (mtDNA copy number), and an improvement in the overall mitochondrial bioenergetic parameters in live cells. Most importantly, functional validation in the presence of the Complex-I inhibitor, MPP^+, demonstrates that activators of SIRT-1 could ameliorate PD mimetics mediated dysfunction in the pathways leading to mitochondrial biogenesis and bioenergetics. The methods outlined here integrating structural prediction, orthogonal enzymatic confirmation (both cell-free and cellular), cellular target engagement via CETSA, mtDNA quantification, and functional bioenergetic profiling could be considered as promising methods for assessing small-molecule modulators of SIRTs and their regulatory framework while studying the mechanisms mediating disease processes.

Acknowledgements

We acknowledge the financial support from CSIR, India (HCP 532401 and MLP-0066). We also express our gratitude to CSIR-IICT for performing plagiarism analysis of our book chapter (manuscript communication number: IICT/Pubs./2026/145).

Declaration

The authors declare that there are no conflicts of interest regarding the work presented in this chapter.

References

Bonifati, V. (2014). Genetics of Parkinson's disease—state of the art, 2013. *Parkinsonism & Related Disorders, 20*(1), S23–S28.

Borra, M. T., et al. (2005). Mechanism of human SIRT1 activation by resveratrol. *The Journal of Biological Chemistry, 280*(17), 17187–17195.

Bose, A., & Beal, M. F. (2016). Mitochondrial dysfunction in Parkinson's disease. *Journal of Neurochemistry, 139*(1), 216–231.

Brunet, A., et al. (2004). Stress-dependent regulation of FOXO transcription factors by SIRT1. *Science (New York, N. Y.), 303*(5666), 2011–2015.

Cao, D., Wang, M., Qiu, X., Liu, D., Jiang, H., Yang, N., & Xu, R. M. (2015). Structural basis for allosteric, substrate-dependent stimulation of SIRT1 activity by resveratrol. *Genes & Development, 29*(12), 1316–1325.

Carrico, C., et al. (2018). The mitochondrial acylome emerges proteomics, regulation by sirtuins. *Cell Metabolism, 27*(3), 497–512.

Chandrasekhar, S., Pushpavalli, S. N., Chatla, S., Mukhopadhyay, D., Ganganna, B., Vijeender, K., ... Bhadra, U. (2012). aza-Flavanones as potent crossspecies microRNA inhibitors that arrest cell cycle. *Bioorganic & Medicinal Chemistry Letters, 22*(1), 645–648.

Chandrasekhar, S., Vijeender, K., & Sridhar, C. (2007). L-Prolinecatalyzed one-pot synthesis of 2-aryl-2, 3-dihydroquinolin-4 (1H)-ones. *Tetrahedron Letters, 48*(28), 4935–4937.

Clay Montier, L. L., et al. (2009). Number matters: Control of mammalian mitochondrial DNA copy number. *Journal of Genetics and Genomics = Yi Chuan xue bao, 36*(3), 125–131.

Dai, H., et al. (2018). Sirtuin activators and inhibitors promises, achievements, and challenges. *Pharmacology & Therapeutics, 188*, 140–154.

Dranka, B. P., et al. (2011). Assessing bioenergetic function in response to oxidative stress by metabolic profiling. *Free Radical Biology & Medicine, 51*(9), 1621–1635.

Friesner, R. A., Murphy, R. B., Repasky, M. P., Frye, L. L., Greenwood, J. R., Halgren, T. A., ... Mainz, D. T. (2006). Extra precision glide: Docking and scoring incorporating a model of hydrophobic enclosure for protein− ligand complexes. *Journal of Medicinal Chemistry, 49*(21), 6177–6196.

Haigis, M. C., & Guarente, L. P. (2006). Mammalian sirtuins–emerging roles in physiology, aging, and calorie restriction. *Genes & Development, 20*(21), 2913–2921.

Hubbard, B. P., et al. (2013). Evidence for a common mechanism of SIRT1 regulation by allosteric activators. *Science (New York, N. Y.), 339*(6124), 1216–1219.

Imai, S., & Guarente, L. (2014). NAD+ and sirtuins in aging and disease. *Trends in Cell Biology, 24*(8), 464–471.

Jafari, R., et al. (2014). The cellular thermal shift assay for evaluating drug target interactions in cells. *Nature Protocols, 9*(9), 2100–2122.

Mudo, G., et al. (2012). Transgenic expression of PGC-1α protects dopaminergic neurons in the MPTP mouse model. *Cellular and Molecular Life Sciences: CMLS, 69*(7), 1153–1165.

Nemoto, S., et al. (2005). SIRT1 functionally interacts with PGC-1α. *The Journal of Biological Chemistry, 280*(16), 16456–16460.

Pacholec, M., et al. (2010). SRT1720, SRT2183, SRT1460, and resveratrol are not direct activators of SIRT1. *The Journal of Biological Chemistry, 285*(11), 8340–8351.

Pant, C., Sukumar, G., Alli, V. J., Jadav, S. S., Pabbaraja, S., & Kalivendi, S. V. (2024). Allosteric Activation of SIRT1 by 2, 4-Dihydroxy-azaflavanone Averts MPP+-mediated dysfunction in mitochondrial biogenesis and bioenergetics: Implications for Parkinson's disease. *ACS Chemical Neuroscience, 15*(15), 2870–2883.

Park, J. S., Davis, R. L., & Sue, C. M. (2018). Mitochondrial dysfunction in Parkinson's disease. *Current Neurology and Neuroscience Reports, 18*(5), 21.

Pickrell, A. M., & Youle, R. J. (2015). The roles of PINK1, parkin, and mitochondrial fidelity in Parkinson's disease. *Neuron, 85*(2), 257–273.

Przedborski, S., et al. (2004). MPTP as a mitochondrial neurotoxic model of Parkinson's disease. *Journal of Bioenergetics and Biomembranes, 36*(4), 375–379.

Ramsay, R. R., et al. (1986). Inhibition of mitochondrial NADH dehydrogenase by pyridine derivatives. *Biochemical and Biophysical Research Communications, 135*(1), 269–275.

Scarpulla, R. C., et al. (2012). Transcriptional integration of mitochondrial biogenesis. *Trends in Endocrinology and Metabolism: TEM, 23*(9), 459–466.

Schapira, A. H. (2008). Mitochondria in the aetiology and pathogenesis of Parkinson's disease. *Lancet Neurology*, 7(1), 97–109.

Schrödinger (2022−1). Schrödinger; LLC: New York, NY, 2022.

Schrödinger Release (2024−2). Desmond molecular dynamics system maestro-desmond interoperability tools; D. E. Shaw Research Schrödinger: New York, NY, 2024.

Vaziri, H., et al. (2001). hSIR2(SIRT1) functions as an NAD-dependent p53 deacetylase. *Cell, 107*(2), 149–159.

Virbasius, J. V., & Scarpulla, R. C. (1994). Activation of the human mitochondrial transcription factor A gene by nuclear respiratory factors. *Proceedings of the National Academy of Sciences of the United States of America, 91*(4), 1309–1313.

Wichapong, K., Rohe, A., Platzer, C., Slynko, I., Erdmann, F., Schmidt, M., & Sippl, W. (2014). Application of docking and QM/MM-GBSA rescoring to screen for novel Myt1 kinase inhibitors. *Journal of Chemical Information and Modeling, 54*(3), 881–893.

Zheng, B., et al. (2010). PGC-1α, a potential therapeutic target for early intervention in Parkinson's disease. *Science Translational Medicine, 2*(52), 52ra73.

CHAPTER SEVEN

Epigenetic regulation of fatty acid chain elongation in MAFLD and its implications in the liver-brain axis dysfunction

Atanu Mondal[a,b] and Chandrima Das[a,b,*]

[a]Biophysics and Structural Genomics Division, Saha Institute of Nuclear Physics, Kolkata, India
[b]Homi Bhabha National Institute, Mumbai, India
*Corresponding author. e-mail address: chandrima.das@saha.ac.in

Contents

Abstract

Lipid metabolism plays a crucial role in cellular health and physiology by acting as an energy storehouse, cell membrane component, brain development and signaling molecules. Crucial steps to metabolize dietary fat take place within the hepatic tissue. Any abnormalities in the hepatic fatty metabolic pathways cause abnormal accumulation of

Methods in Enzymology, Volume 734
ISSN 0076-6879, https://doi.org/10.1016/bs.mie.2026.05.042

lipid inside the liver, causing MAFLD, ranging from simple steatosis to more complex steatohepatitis and fibrosis. During high-fat-diet-induced hepatic inflammation, systemic proinflammatory cytokines disrupt the blood-brain barrier, resulting in neuroinflammation, cognitive impairment, brain damage and even neurodegeneration. Further, during this altered metabolic scenario, circulating metabolites pass through the impaired BBR and deregulate the epigenetic landscape of the central nervous system. Thus, it becomes crucial to understand the epi-metabolic crosstalk between two crucial organs of our body: the liver and the brain. Here in this chapter, we demonstrate the approach that we are using in our laboratory to study the epigenetic reprogramming in the context of metabolic gene expression in the liver, which is the causal for life style disorders like MAFLD. Remarkably, we intend to understand how liver dysfunction can have an implication in the brain function. Here, we discuss the concept of developing a diet-induced steatosis and steatohepatitis mouse model to understand the disease progression and its interconnection with brain physiology. Further, we also demonstrate 2D and 3D cell culture models to study the liver-brain cross-talk in greater molecular detail. Collectively, these approaches can provide a template for studying the role of epi-metabolic cross-talk in liver-guided brain dysfunction upon MAFLD.

1. Introduction

Fatty acids are indispensable structural constituents of cell membranes and serve as key precursors in numerous metabolic pathways, encompassing energy storage and lipid-derived signalling molecules (De Carvalho & Caramujo, 2018). They yield nearly twice the energy per gram compared to glucose (Robinson et al., 2022), which renders lipid storage a physiological adaptation advantageous during prolonged food shortages. However, contemporary dietary habits characterised by caloric surplus with a high-fat diet and excess sugar intake have led to abnormal lipid accumulation as di and triglycerides in hepatocytes (Im et al., 2021). This modern dietary regimen culminates in a pathophysiological state of the liver termed as metabolic dysfunction-associated fatty liver disease (MAFLD) (Im et al., 2021). MAFLD is a spectrum of conditions ranging from simple metabolic dysfunction-associated steatosis (MAS) to more severe forms like metabolic dysfunction-associated steatohepatitis (MASH), eventually advancing to fibrosis and hepatocellular carcinoma (HCC) (Eslam et al., 2021, 2022; Hagström et al., 2024; Younossi et al., 2021). Under conditions of excess lipid, hepatocytes initially attempt to oxidise surplus free fatty acids (FFAs) via mitochondrial β-oxidation (Chen et al., 2021). When this catabolic process is overburdened, short-chain saturated fatty acids (SFAs) are redirected towards the fatty acid elongation and desaturation pathway, yielding monounsaturated very long-chain fatty acids (VLCFAs), which are subsequently esterified into di- and tri-acylglycerols for storage (Matsuzaka et al., 2007; Wang et al., 2006).

Central to the regulation of hepatic lipid homeostasis is the coordinated activity of enzymes governing the fatty acid chain elongation cycle. This four-step cyclic process is catalysed sequentially by 3-ketoacyl-CoA synthases (ELOVLs), 3-ketoacyl-CoA reductases, 3-hydroxyacyl-CoA dehydratases (HACDs), and trans-2-enoyl-CoA reductases, with acyl-CoA synthetases facilitating substrate activation (Kihara, 2012; Wang et al., 2006). Among these, HACDs, comprising of four isoforms designated HACD1 (PTPLA), HACD2 (PTPLB), HACD3 (PTPLAD1), and HACD4 (PTPLAD2), catalyse the third, dehydration step of the elongation cycle (Mondal, Nandi, Singh, et al., 2025; Sawai et al., 2017). The ELOVL family of condensing enzymes, consisting of seven isoforms (ELOVL1–7) with distinct substrate specificities and tissue-specific expression profiles, governs the rate-limiting condensation step of fatty acid chain elongation (Ferrero et al., 2025; Nie et al., 2021) and is associated with human diseases, including metabolic and neurological disorders (Di Gregorio et al., 2014; Harkewicz et al., 2012; Matsuzaka et al., 2007; Moon et al., 2009; Zhang et al., 2001). It is plausible that any impairment in the fatty acid elongation pathway leads to the accumulation of short-chain SFAs, interfering with triglyceride assembly and inducing acute endoplasmic reticulum (ER) stress. Sustained ER stress, in turn, activates pro-inflammatory genes driving hepatic inflammation, fibrosis, and progressive cell death, a cascade of events that causes MAFL to MASH transition (Özcan et al., 2004).

The dysregulation of lipid metabolic enzyme expression in the context of hepatic steatosis is critically modulated by epigenetic mechanisms. Remarkably, a high-fat diet augments histone H3 lysine 4 trimethylation (H3K4me3) and altered promoter DNA methylation patterns of lipogenic genes in MASH mice relative to MAFL models (Lai et al., 2020; Melton et al., 2023; Nishida et al., 2016). Similarly, hyperacetylation of the fatty acid synthase (FASN) gene promoter has been demonstrated to drive de novo lipogenesis, amplifying triglyceride deposition in hepatic tissue (Gang et al., 2016; Gu et al., 2020; Lin et al., 2016). These findings collectively underscore the importance of epi-metabolic regulation in determining disease trajectory within the MAFLD spectrum.

Beyond hepatic dysfunction, MAFLD exerts systemic consequences with significant implications for central nervous system (CNS) function (De Cól et al., 2024; Huang et al., 2025). Aberrant hepatic metabolism in

MAFLD disrupts circulating metabolite pools and dysregulates cytokine secretion. Pro-inflammatory cytokines released from the diseased liver enter the systemic circulation and can compromise the integrity of the blood-brain barrier (BBB), increasing its permeability (Mondal et al., 2020) (Fig. 1). This permits the infiltration of circulating metabolites and inflammatory mediators into the CNS, where they have the capacity to alter gene expression patterns in neural and glial cells, potentially contributing to neuroinflammation and cognitive impairment. This inter-organ cross-talk, encompassing both metabolic and immunological perspectives, represents a critical but underexplored dimension of MAFLD pathophysiology (Cao et al., 2025; Hadjihambi et al., 2023). Understanding this axis from an epi-metabolic perspective is essential to delineate the pathophysiological aspect of hepatic lipid dysregulation.

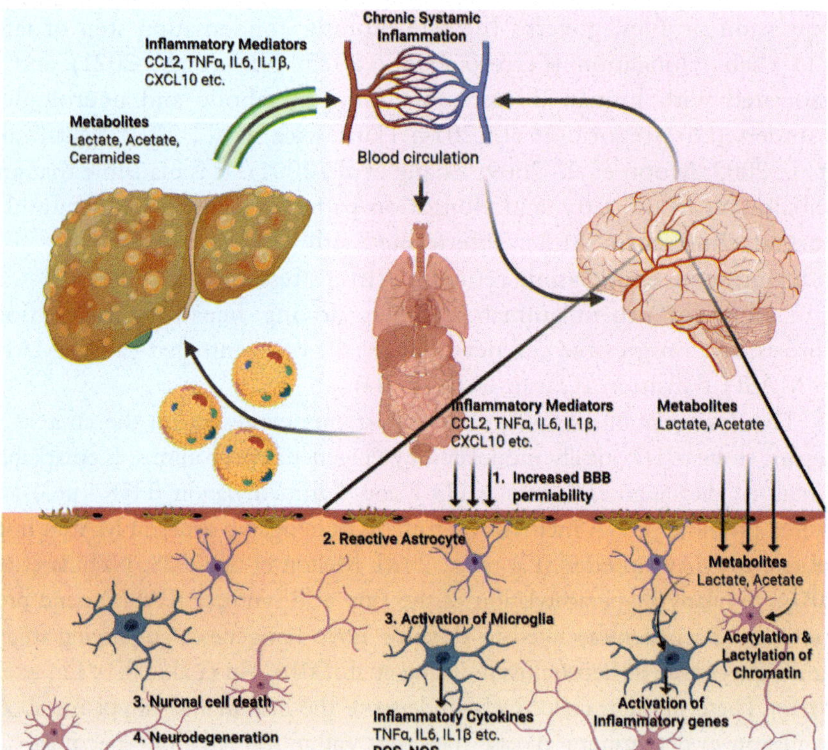

Fig. 1 The liver-brain circuit during metabolic dysfunction associated fatty liver diseases (MAFLD), causing neuroinflammation and brain dysfunction.

Thus, understanding the liver-brain cross-talk in complex, multifactorial diseases like MAFLD requires the development of a robust and integrative cellular and animal model system. This system would capture the molecular, metabolic, and epigenetic alterations across the MAFLD spectrum and would faithfully delineate the inter-organ communication for identifying novel therapeutic targets and mechanistic insights. The present study is therefore conceived to address this critical gap by establishing a comprehensive model system to elucidate the liver-brain cross-talk. The study focuses on delineating the function of hepatic lipid chain elongation programs and their implications in brain function at the organismal level.

2. Establishing the cell culture model to study the liver-brain axis

To study the cross-talk between liver and brain, we demonstrate a 2D cell culture method to mimic a high-fat diet scenario by treating the cells with free fatty acids. After treating the cells with free fatty acids, it is crucial to identify an epigenetic factor that is responsible for the deregulation of the lipid metabolic pathways. To identify the deregulated gene set, we perform RNA sequencing experiments and revalidate the data by performing lipidomic or metabolomic experiments. Once the master epigenetic regulator is identified, we knock down the factor and perform similar multi-omics experiments. Further, we study the mechanistic details of the epigenetic regulatory axis by performing ChIP sequencing and ATAC sequencing experiments.

This section provides a detailed outline of a 2D cell culture method to establish an MAFLD-like scenario, in the simplest way, for studying the separate molecular metabo-epigenetic events happening in the liver and the brain.

2.1 Equipment and reagents

1. HepG2 cell line
2. SHSY5Y cell line
3. Complete DMEM and DMEM-F12 media
4. Incomplete DMEM media
5. CO_2 incubator
6. Palmitic acid
7. Oleic acid
8. Stearic acid

9. 0.25 % Trypsin
10. Cell scraper
11. Centrifuge
12. Cell counter
13. Thermomixer

2.2 Procedure

1. Culture HepG2 cells in DMEM media supplemented with 10 % FBS, 1 % antibiotics, 1 % Glutamax and 1 % non-essential amino acids, and maintain at 37 °C incubator in 5 % CO_2 conditions.
2. Culture SHSY5Y cells in DMEM-F12 media supplemented with 10 % FBS, 1 % antibiotics, and 1 % Glutamax, and maintain at 37 °C incubator in 5 % CO2 conditions.
3. When HepG2 cells reach 60–70 % confluence, transfect the cell with shRNA to knock down a factor and control shRNA with Lipofectamine 3000. After 24 h of transfection free fatty acid treatment was done.
4. For free fatty acid treatment, mix Palmitic acid, Oleic acid and Stearic acid in 50 % ethanol in separate tubes and heat at 65 °C for 15 min to get a final concentration of 150 mM. After heating, add fat-free BSA (pre-heated at 37 °C) to a final concentration of 20 mM and incubate at 37 °C for 60 min in a thermos mixer to form a lipo-protein complex. Then, add PA-BSA (final concentration 300 μM), OA-BSA (final concentration 100 μM) and SA-BSA (final concentration 100 μM) complexes to an incomplete DMEM media (DMEM media without phenol red, without glucose, supplemented with 100 mM Sodium Pyruvate and 1 % Glutamax). Replace the HepG2 cells' media with fatty acid complex media and incubate the culture at a 37 °C incubator in 5 % CO_2 conditions for 16 h.
5. After the free fatty acid treatment, harvest the HepG2 cells and store for further analysis. Collect the used media and centrifuge at $2000 \times g$ for 3 min to remove cell debris. Then add the used media to the 60–70 % confluent SHSY5Y cells and incubate at 37 °C, 5 % CO_2 incubator for 16 h.
6. After 16 h of treatment, harvest the SHSY5Y cells and store for further analysis.

2.3 Note

1. Selecting a hepatic cell line and a brain cell line is crucial and depends upon the research question.

3. Establishing the multicellular spheroid model to study the liver-brain axis

To study the detailed molecular mechanism in the liver-brain communication, a simplified in-vitro cellular model is essential. We try to validate and study, in greater detail, the physiological importance of the previously identified epigenetic and metabolic factors during liver-brain interaction upon high-fat diet-induced MAFLD conditions. For that purpose, we use a 3D cell culture system to develop liver in a dish (Multicellular Liver spheroid) and brain in a dish (Multicellular Brain Spheroid) (Adhikari et al., 2024; Mondal, Nandi, Mandal, et al., 2025). By altering the expression of the previously identified (identified by a 2D cell culture system) epigenetic factor in the liver spheroid, we try to score the epi-metabolic changes in the brain spheroid during free fatty acids-treated MAFLD-like scenario.

Here, we provide a step-by-step approach to generate multicellular liver and brain spheroids and a protocol outline to mimic MAFLD to study hepatic epigenetic alteration-mediated brain dysfunction (Fig. 2).

3.1 Equipment and reagents

1. Huh7 cell line
2. HUVEC cell line
3. NIH-3T3 cell line
4. SHSY5Y cell line
5. U87MG cell line
6. HOG cell line
7. Complete DMEM, RPMI and DMEM-F12 media (Gibco)
8. Endothelial Cell Growth Medium (Sigma)
9. Incomplete DMEM media (Gibco)
10. CO_2 incubator
11. Palmitic acid (Sigma)
12. Oleic acid (Sigma)
13. Stearic acid (Sigma)
14. 0.25 % Trypsin (Gibco)
15. Cell scraper
16. Centrifuge (Eppendorf)
17. Cell counter (Invitrogen)
18. Thermomixer (Eppendorf)
19. Collagen Coated plate (Gibco)
20. 96-well spheroid forming plate (Invitrogen)

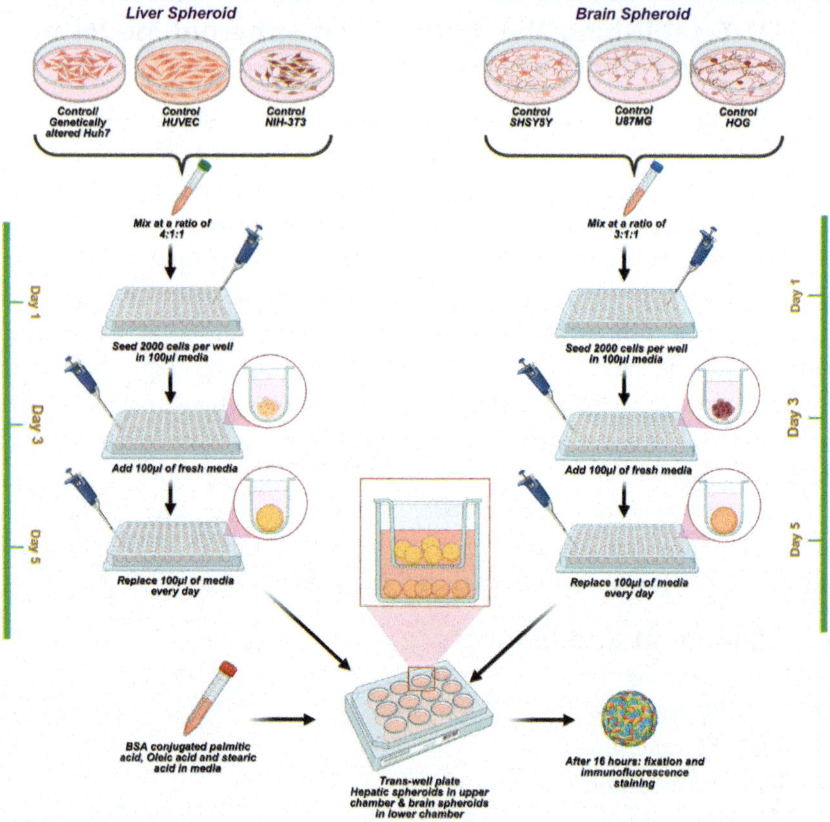

Fig. 2 Schematic representation for generating liver and brain spheroids and study of the liver-brain axis.

21. 12-well dish
22. Trans-well plate with 8–10 µm pore size (Invitrogen)

3.2 Procedure

1. Culture Huh7 cells in DMEM media supplemented with 10 % FBS, 1 % antibiotics, 1 % Glutamax and 1 % non-essential amino acids, and maintain at a 37 °C incubator in 5 % CO_2 conditions. To study the role of a protein, it is better to make a stable knockdown cell or knockout cell before forming spheroids.
2. Culture HUVEC in a collagen-coated plate in Endothelial Cell Growth Medium at 37 °C incubator in 5 % CO_2 incubator.
3. Culture NIH-3T3 cells in RPMI media supplemented with 12 % FBS, 1 % antibiotics and maintain at a 37 °C incubator in 5 % CO2 conditions.

4. Culture SHSY5Y cells in DMEM-F12 media supplemented with 10 % FBS, 1 % antibiotics, and 1 % Glutamax, and maintain at 37 °C incubator in 5 % CO2 conditions.
5. Culture U87MG cells in DMEM media supplemented with 10 % FBS, 1 % antibiotics, 1 % Glutamax and 1 % non-essential amino acids, and maintain at 37 °C incubator in 5 % CO_2 conditions.
6. Culture HOG cells in RPMI media supplemented with 12 % FBS, 1 % antibiotics, and maintain at 37 °C incubator in 5 % CO2 conditions.
7. To form a hepatic multicellular spheroid (Liver in a dish), mix Huh7, HUVEC and NIH-3T3 cells at 4:1:1 ratio in "spheroid-forming media" (DMEM F12 media supplemented with 20 ng/ml FGF, 20 ng/ml EGF, 5 μg/ml Insulin, 1X B27 supplement and 0.4 % fat free BSA). Put a total of 2000 cells in 100 μl of "spheroid-forming media" in one well of a "96-well spheroid forming plate". Spin the plate at 500 × *g* for 2 min and keep it in a 37 °C incubator in 5 % CO2 conditions. Next day, add 100 μl of fresh "spheroid-forming media" in each well. After two days of seeding, replace 100 μl of old "spheroid-forming media" with fresh "spheroid-forming media" every day.
8. To form a neuronal multicellular spheroid (Brain in a dish), mix SHSY5Y, U87MG and HOG cells at 3:1:1 ratio in "spheroid-forming media" (DMEM F12 media supplemented with 20 ng/ml FGF, 20 ng/ml EGF, 5 μg/ml Insulin, 1X B27 supplement and 0.4 % fat free BSA). Put a total of 2000 cells in 100 μl of "spheroid-forming media" in one well of a "96-well spheroid forming plate". Spin the plate at 500 × *g* for 2 min and keep it in a 37 °C incubator in 5 % CO2 conditions. Next day, add 100 μl of fresh "spheroid-forming media" in each well. After two days of seeding, replace 100 μl of old "spheroid-forming media" with fresh "spheroid-forming media" every day.
9. For free fatty acid treatment, mix Palmitic acid, Oleic acid and Stearic acid in 50 % ethanol in separate tubes and heat at 65 °C for 15 min to get a final concentration of 150 mM. After heating, add fat free BSA (pre-heated at 37 °C) to a final concentration of 20 mM and incubate at 37 °C for 60 min in a thermos mixer to form a lipo-protein complex. Then, add PA-BSA (final concentration 300 μM), OA-BSA (final concentration 100 μM) and SA-BSA (final concentration 100 μM) complexes to an incomplete DMEM media (DMEM media without phenol red, without glucose, supplemented with 100 mM Sodium Pyruvate and 1 % Glutamax). Place the Brain spheroids

(25−30 in number) and Liver spheroids (45−50 in number) in the lower and upper chambers of a 12-well trans-well plate, respectively and add 1 ml of media with 500 μM fatty acid complex.

10. Incubate the culture in an incubated at 37 °C 5 % CO2 incubator for 16 h.
11. Harvest the spheroids for further analysis.
12. and incubated at 37 °C, 5 % CO_2 incubator for 16 h.

3.3 Note

1. Cell health is always crucial for intact spheroid formation.
2. Always cut the front part of the tip to avoid damaging the spheroids while pipetting the spheroids out.
3. The spheroid-forming ability varies from cell line to cell line. For example, Huh7 cells form better spheroids than HepG2.

4. Development of the MAFLD mice model and genetic depletion

To study the pathophysiology, it is important to develop a disease model and intervene in the expression of the epigenetic factor in the hepatic tissue to score the effect of liver secretome on neuronal dysfunction. Hence, we have developed a diet-fed mouse model that mimics the MAFLD condition into two separate MAS and MASH mouse models. This section provides step-by-step details to develop a diet-induced Metabolic Dysfunction Associated Fatty Liver (MAFL, MAS, Steatosis) and Metabolic Dysfunction Associated Steatohepatitis (MASH) mouse model. We also discuss an approach for genetically knocking down a factor to study its role in MAFLD development and progression (Fig. 3).

4.1 Equipment and reagents

1. Three to four weeks old male BALB/c mice
2. Liver Targeted Liposomal Delivery System
3. Epigenetic factor-specific shRNA plasmid
4. Non-targated
5. 1 ml syringe
6. PBS
7. Chow diet (D12450K)
8. High-fat diet (D12492)
9. High-fat, high-fructose and high-cholesterol diet (D09100310)

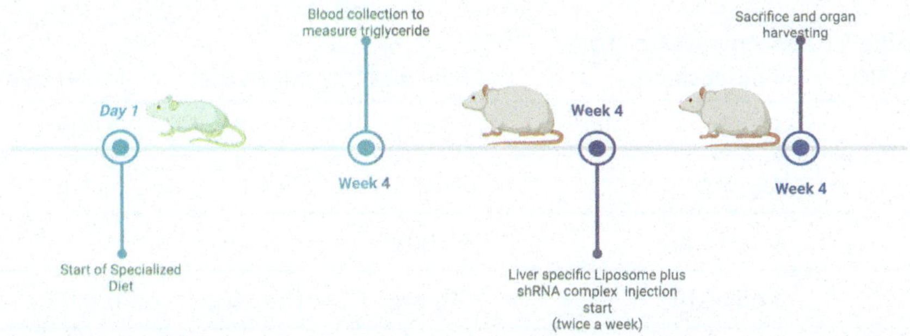

Fig. 3 Time course for the development of the MAFLD mice model to study the liver-brain axis.

10. Autoclaved double-distilled water
11. Capillary tube
12. Triglyceride measurement kit
13. Weighing Balance
14. Ventilated cage (IVC system)

4.2 Procedure

1. Subdivide the three to four weeks BALB/c mice into four groups (8–10 mice per group) in an individually ventilated cage (IVC system).
2. **The group 1** (Control): Provide chow diet (D12450K) containing 10 % fat for 12 weeks (Table 1).
3. **The group 2** (MAFL): Provide a high-fat diet (D12492) containing 60 % fat as palmitic acid for 12 weeks (Table 1).
4. **The group 3** (MAFL + ASO): Provide a high-fat diet (D12492) containing 60 % fat as palmitic acid for 12 weeks (Table 1).
5. **The group 4** (MASH): Provide a high-fat, high-fructose and high-cholesterol diet (D09100310) containing 40 % fat as palmitic acid, 20 % fructose and 2 % cholesterol for 12 weeks (Table 1).
6. Monitor the animal health, behavior, and food consumption rate every day.
7. Measure body weight weekly.
8. After eight weeks of feeding, collect 50 μl of blood by the retro-orbital transfusion method with capillary tubes and measure the blood triglyceride level by using a triglyceride measurement kit.
9. When the blood triglyceride level is heightened significantly in high-fat-diet-fed mice (groups 2 & 3), initiate the liposome plus shRNA complex injection in group 3.

Table 1 Diet composition chart.

SL NO.	Composition	Chaw diet	MAFL diet	MASH diet
1	Casein	200.00 g	200.00 g	200.00 g
2	Cystine	3.00 g	3.00 g	3.00 g
3	Corn Starch	550.00 g	—	—
4	Lodex 10	150.00 g	125.00 g	100.00 g
5	Sucrose	4.00 g	72.80 g	100.00 g
6	Solka Floc	50.00 g	50.00 g	50.00 g
7	Soybean Oil	25.00 g	25.00 g	25.00 g
8	Lard	20.00 g	245.00 g	20.00 g
9	Minerals (S10026B)	50.00 g	50.00 g	50.00 g
10	Choline Bitartrate	2.00 g	2.00 g	2.00 g
11	Vitamin (V10001C)	1.00 g	1.00 g	1.00 g
12	Blue Dye	0.03 g	0.05 g	0.03 g
13	Red Dye	0.03 g	—	0.03 g
14	Fructose	—	—	200.00 g
15	Palm Oil	—	—	135.00 g
16	Cholesterol	—	—	18.00 g

10. Inject 100 μl of liposome plus shRNA complex twice a week for the next four weeks through the tail vein at a dose of 10 mg per Kg body weight. Inject a liposome plus non-targeted shRNA complex into all the other three groups of mice.
11. After 12 weeks, sacrifice all the mice. Collect liver, blood serum and brain, and cryopreserve for further analysis.

4.3 Note

1. It is crucial to monitor the mouse's health regularly.
2. It is essential to maintain extra caution during collecting blood by the retro-orbital transfusion method.

5. Verifying and analysis of the liver-brain axis

After successfully conducting experiments on cell and animal models, we scrutinize the epi-metabolic circuits by multi-omics approaches. Beyond multi-omics, we use a few fundamental techniques to revalidate the MAFLD disease progression and its correlation with the liver-brain axis by monitoring mRNA and protein level expression of key epigenetic and metabolic factors. Here, in this section, we provide step-by-step protocols of a few crucial fundamental techniques for the analysis of the liver-brain circuit upon MAFLD.

5.1 Equipment

1. Microtome
2. Hot air oven
3. Water bath
4. Microwave oven
5. Orbital shaker
6. 4 °C incubator
7. Light Microscope
8. Confocal microscope
9. 96-well plate reader
10. Fixed-angle refrigerated Centrifuge
11. Chemidoc
12. Dounce Tissue homogenizer
13. Sonicator
14. Qubit
15. Nanodrop

5.2 Reagents

1. Paraffin
2. DAB chromogen
3. DAB substrate
4. Hematoxylin
5. Primary antibody
6. HRP-conjugated secondary antibody
7. Fluorophore-conjugated secondary antibody
8. Ethanol
9. Xylene
10. 10 % neutral buffered formalin
11. 100 mM Phosphate buffer

12. 1X PBST
13. RIPA
14. PBS
15. 5 % BSA
16. Tris-EDTA buffer (pH 9.0)
17. Protease inhibitor cocktail
18. Phosphatase inhibitor
19. TriZol
20. Chloroform
21. Isopropanol
22. NaCl
23. RNase-free water

5.3 Procedure

5.3.1 Immunohistochemistry and Immunofluorescence study to check the protein expression in the mouse tissue

1. Fix the brain and liver tissues 10 % neutral buffered formalin for two days.
2. Prepare a paraffin block with the formalin-fixed tissue, and cut a 4 μm section using a microtome.
3. Place all the tissue sections on poly-L-lysine-coated glass slides and heat them at 65 °C temperature for 5 min.
4. For immunohistochemical (IHC) and immunofluorescence (IF) analysis, deparaffinize the slides by heating them at 65 °C for 60 min and dip immediately into xylene for 10 min. Then wash the slide in 100 %, 90 %, 80 % and 70 % ethanol for 10 min each and finally hydrate it in distilled water for 20 min.
5. Boil the slides in Tris-EDTA buffer (pH 9.0) for 10−15 min for antigen retrieval and leave them at room temperature to cool down.
6. Wash the slides once with 100 mM phosphate buffer and incubate them with peroxidase for 10 min to block all the peroxides.
7. Wash the slides once with 100 mM phosphate buffer and incubate them in 5 % BSA (dissolved in water) for 45−60 min to block non-specific antigen binding.
8. Then, dilute the primary antibody at a 1:200 ratio in 5 % BSA dissolved in PBST and incubate the slides at 4 °C for 12 h.
9. Wash the slides in 100 mM phosphate buffer three times.
10. After that, dilute the secondary antibody (HRP conjugated for IHC and fluorophore conjugated for IF) at a 1:5000 ratio in PBST and incubate the slides at 25 °C for 45−60 min.

11. Wash the slides in 100 mM phosphate buffer three times.
12. For IHC, add DAB chromogen and DAB-substrate and incubate at 25 °C for 2 min in the dark and wash three times in phosphate buffer. Wash the slide with running water and counterstain with hematoxylin for 20 s, and wash in running tap water. Air-dry the slides and mount with cover slips using DPX. After drying, grab images using a bright-field light microscope.
13. For IF, counterstain the slides with DAPI for 20 min and wash with 100 mM phosphate buffer three times. Mount the slides with cover slips using glycerol-containing DABCO mounting media and seal the cover slips with transparent nail polish. After drying, grab images using a fluorescent microscope.

5.3.2 Immunofluorescence study to check the protein expression in the liver and brain spheroid

1. Put the spheroids in a 24-well dish (20–30 spheroids in one well).
2. Fix the brain and liver spheroids with 4 % paraformaldehyde for 20–30 min.
3. Wash the paraformaldehyde-fixed spheroids with PBS three times for 5 min each.
4. Permeabilize the spheroids with 1 % triton X–100 (diluted in PBS) for 20 min.
5. Wash the spheroids with PBS three times for 5 min each.
6. Incubate them in 5 % BSA (dissolved in PBST) for 45–60 min at 25 °C to block nonspecific antigen binding.
7. Then, dilute the primary antibody at a 1:200 ratio in 5 % BSA dissolved in PBST and incubate the spheroids at 4 °C for 12 h on an orbital shaker at 60 rpm.
8. Wash them in PBST five times for 5 min each.
9. After that, dilute the fluorophore-conjugated secondary antibody at a 1:2000 ratio in PBST and incubate the spheroids at 25 °C for 45–60 min.
10. Wash them in PBST five times for 5 min each.
11. Counterstain the spheroids with DAPI for 20 min and wash them in PBST five times for 5 min each.
12. Mount the slides with cover slips using glycerol-containing DABCO mounting media and seal the cover slips with transparent nail polish. After drying, grab images using a fluorescent microscope in z-stack mode.

5.3.3 Enzyme-linked immunosorbent assay to check the protein expression from the mouse tissue lysate

1. Cut the desired part of the tissue (100–150 mg) and put it in 500 µl RIPA buffer. Add 1X protease inhibitor cocktail and 1X phosphatase inhibitor.
2. Homogenized with a Dounce homogenizer.
3. Sonicate to break the extracellular matrix and to lyse the cell.
4. Centrifuge the lysate at 10,000 × g for 10 min at 4 °C temperature. Discard the debris and collect the supernatant.
5. Measure the protein concentration by Brad ford assay.
6. For coating the plate with primary antibody, dilute the primary antibody at a 1:200 ratio in PBS. Add 100 µl of dilatated primary antibody in each well of a 96 well plate and incubate at 4 °C for 12 h on an orbital shaker at 60 rpm. Wash the plate twice with 200 µl of 100 µM phosphate buffer.
7. After coating block the plat with 5 % BSA (dissolved in PBS) for 45–60 min at 25 °C.
8. Add 100 µl of dilatated primary antibody (polyclonal or primary antibody that recognize different region of the antigen) in each well of the 96 well plate and incubate at 4 °C for 12 h on an orbital shaker at 60 rpm. Wash the plate twice with 200 µl of 100 µM phosphate buffer.
9. Block the plat with 5 % BSA (dissolved in PBS) for 45–60 min at 25 °C.
10. Dilute the fluorophore-conjugated secondary antibody at a 1:2000 ratio in PBS. Add 100 µl of dilatated fluorophore-conjugated secondary antibody in each well of the 96 well plate and incubate at 25 °C for 45–60 min on an orbital shaker at 60 rpm.
11. Wash the plate twice with 200 µl of 100 µM phosphate buffer. Add 100 µl of 100 µM phosphate buffer in each well and measure the fluorescence intensity.

5.3.4 RNA extraction from tissue to check mRNA level expression

1. Cut the desired part of the tissue (10–20 mg) and put it in 1 ml TriZol.
2. Homogenized with a Dounce homogenizer.
3. Add 200 µl of chloroform and vortex vigorously for 30–40 s, and incubate it at 25 °C temperature for 5 min.
4. Centrifuge it at 10,000 × g for 10 min at 4 °C temperature.
5. Collect the upper layer in a separate tube. Add 500 µl of isopropanol and 5 M NaCl to the final concentration of 500 mM.

6. Mix and incubate at 25 °C for 20–30 min.
7. Centrifuge at 10,000 × g for 15 min at 4 °C. Discard the supernatant.
8. Wash the pellet with 80 % ethanol and centrifuge it at 1000 × *g* for 15 min at 4 °C temperature.
9. Discard the supernatant and air-dry the pellet.
10. Dissolve the pellet in 25–30 μl of RNase-free water and incubate it at 65 °C temperature for 15 min.
11. Put it in ice for 20 min and centrifuge it at 1000 × *g* for 3 min at 4 °C temperature.
12. Measure the concentration and quality of the RNA by using a Qubit or Nanodrop.

5.4 Note

1. The antigen retrieval is very crucial for IHC and IF of tissue samples. Tris-EDTA (pH 9.0) buffer works well for both nuclear (chromatin-bound) and cytosolic proteins, whereas sodium citrate buffer is useful for cytosolic proteins.
2. The duration of color development in IHC depends upon the availability of the protein in the cell. Overexposure may generate unnecessary background.
3. Sandwich ELISA works only with polyclonal primary antibodies. For monoclonal antibodies, it is better to use two different monoclonal antibodies that recognize distinct regions of the same antigen; otherwise, one may perform direct ELISA.
4. During RNA purification from tissue, collect the upper aqueous layer carefully. The in-between white protein layer is very loose and thick. Try not to take even a little part of the layer. After the ethanol wash, the RNA pellet becomes very loose; be extra careful while discarding the supernatant. After air-drying, the RNA pellet becomes transparent; not a single drop of ethanol should remain in the tube.

6. Concluding remarks

Extensive research is ongoing to understand the contribution of the bioenergetic pathways in human pathophysiological states. Our metabolic hub, the liver, is involved in sensing the nutritional demand of the body and programming the metabolic cascades in accordance. While the liver health is extensively impacted by the dietary regimen and a fat-enriched diet is the cause for a myriad of liver dysfunctions, like MAFLD, the indirect role of this vital organ in the homeostatic function of other human organs has remained elusive. Here, we intend to establish a detailed protocol for an approach to understand inter-organ cross-talk with a special emphasis on the liver-brain axis. Starting from a 2D cell culture model, where epi-metabolic screens can be set, further complexities can be explored in 3D-spheroid models and finally in the specialized diet-fed mice models. Remarkably, the epigenetic regulators in the liver fine-tune the expression of the liver-specific genes, whose deregulation is the cause of an altered metabolic niche, both in the liver and at the systemic level. These metabolites are the causal for eliciting their effect on distant organs like the brain. The passage of these metabolite pools through the blood-brain barrier, ultimately reaching the brain, could be implicated in the immune inflammation causing the neurodegenerative diseases, including Alzheimer's or Parkinson's disease, schizophrenia, etc. The present protocol is a step-by-step guide to address these complex problems involving inter-organ cross-talk. Further, the regulatory network guiding the metabolic niche in the liver is fine-tuned by the epigenetic factors, which play a seminal role in the disease manifestation and hence could be playing an important role in modelling the neurodegenerative disease states in an individual. Hence, comprehensively, this protocol projects a new paradigm of epi-metabolic signaling network that directly precludes the individual disease state.

Acknowledgement

This study is funded by Basic and Applied Research in Biophysics and Material Science (RSI 4008) by the Department of Atomic Energy Govt. of India, Core Research Grant (CRG/2022/005052) by the Department of Science and Technology and BT/PR52628/MED/30/2553/2024 by the Department of Biotechnology.

References

Adhikari, S., Singh, V., Nandi, S., Ghosal, M., Raj, N. S., Khanna, J., ... Das, C. (2024). UBR7 in concert with EZH2 inhibits the TGF-β signaling leading to extracellular matrix remodeling. *Cell Reports, 43*(7), https://doi.org/10.1016/j.celrep.2024.114394.

Cao, Y., Zhao, Y., Deng, T., Zhou, Q., Hu, G., Hu, Z. L., ... Chen, J. G. (2025). Hepatic acetyl-CoA metabolism modulates neuroinflammation and depression susceptibility via acetate. *Cell Metabolism, 37*(11), 2185–2201.e8. https://doi.org/10.1016/j.cmet.2025.08.010.

Chen, Q., Du, J., Cui, K., Fang, W., Zhao, Z., Chen, Q., ... Ai, Q. (2021). Acetyl-CoA derived from hepatic mitochondrial fatty acid β-oxidation aggravates inflammation by enhancing p65 acetylation. *IScience, 24*(11), 103244. https://doi.org/10.1016/j.isci.2021.103244.

De Carvalho, C. C. C. R., & Caramujo, M. J. (2018). The various roles of fatty acids. *Molecules: A Journal of Synthetic Chemistry and Natural Product Chemistry, 23*(10), 2583. https://doi.org/10.3390/molecules23102583.

De Cól, J. P., de Lima, E. P., Pompeu, F. M., Cressoni Araújo, A., de Alvares Goulart, R., Bechara, M. D., ... Barbalho, S. M. (2024). Underlying mechanisms behind the brain–gut–liver axis and metabolic-associated fatty liver disease (MAFLD): An update. *International Journal of Molecular Sciences 2024, 25*(7), *25*. https://doi.org/10.3390/ijms25073694.

Di Gregorio, E., Borroni, B., Giorgio, E., Lacerenza, D., Ferrero, M., Lo Buono, N., ... Brusco, A. (2014). ELOVL5 mutations cause spinocerebellar ataxia 38. *American Journal of Human Genetics, 95*(2), 209–217. https://doi.org/10.1016/j.ajhg.2014.07.001.

Eslam, M., Alkhouri, N., Vajro, P., Baumann, U., Weiss, R., Socha, P., ... George, J. (2021). Defining paediatric metabolic (dysfunction)-associated fatty liver disease: An international expert consensus statement. *Lancet Gastroenterol. Hepatol, 6*(10), 864–873. https://doi.org/10.1016/S2468-1253(21)00183-7.

Eslam, M., El-Serag, H. B., Francque, S., Sarin, S. K., Wei, L., Bugianesi, E., & George, J. (2022). Metabolic (dysfunction)-associated fatty liver disease in individuals of normal weight. *Nature Reviews Gastroenterology & Hepatology, 19*(10), 638–651. https://doi.org/10.1038/s41575-022-00635-5.

Ferrero, E., Vaz, F. M., Cheillan, D., Brusco, A., & Marelli, C. (2025). The ELOVL proteins: Very and ultra long-chain fatty acids at the crossroads between metabolic and neurodegenerative disorders. *Molecular Genetics and Metabolism, 144*(3), 109050. https://doi.org/10.1016/j.ymgme.2025.109050.

Gang, X., Yang, Y., Zhong, J., Jiang, K., Pan, Y., Jeffrey Karnes, R., ... Huang, H. (2016). P300 acetyltransferase regulates fatty acid synthase expression, lipid metabolism and prostate cancer growth. *Oncotarget,* 7(12), 15135. https://doi.org/10.18632/oncotarget.7715.

Gu, L., Zhu, Y., Lin, X., Tan, X., Lu, B., & Li, Y. (2020). Stabilization of FASN by ACAT1-mediated GNPAT acetylation promotes lipid metabolism and hepatocarcinogenesis. *Oncogene, 39*(11), 2437–2449. https://doi.org/10.1038/s41388-020-1156-0.

Hadjihambi, A., Konstantinou, C., Klohs, J., Monsorno, K., Le Guennec, A., Donnelly, C., ... Pellerin, L. (2023). Partial MCT1 invalidation protects against diet-induced non-alcoholic fatty liver disease and the associated brain dysfunction. *Journal of Hepatology, 78*(1), 180–190. https://doi.org/10.1016/j.jhep.2022.08.008.

Hagström, H., Vessby, J., Ekstedt, M., & Shang, Y. (2024). 99% of patients with NAFLD meet MASLD criteria and natural history is therefore identical. *Journal of Hepatology, 80*(2), e76–e77. https://doi.org/10.1016/j.jhep.2023.08.026.

Harkewicz, R., Du, H., Tong, Z., Alkuraya, H., Bedell, M., Sun, W., ... Zhang, K. (2012). Essential role of ELOVL4 protein in very long chain fatty acid synthesis and retinal function. *The Journal of Biological Chemistry, 287*(14), 11469–11480. https://doi.org/10.1074/jbc.M111.256073.

Huang, X., Sang, Z., Liu, H., Yang, J., Wang, K., Qu, Y., & Deng, H. (2025). From nonalcoholic fatty liver disease to neuroinflammation: The role of chronic systemic inflammation. *Metabolic Brain Disease, 40*(6), 240. https://doi.org/10.1007/s11011-025-01669-9.

Im, Y. R., Hunter, H., de Gracia Hahn, D., Duret, A., Cheah, Q., Dong, J., ... Mann, J. P. (2021). A systematic review of animal models of NAFLD finds high-fat, high-fructose diets most closely resemble human NAFLD. *Hepatology (Baltimore, Md.), 74*(4), 1884–1901. https://doi.org/10.1002/hep.31897.

Kihara, A. (2012). Very long-chain fatty acids: Elongation, physiology and related disorders. *Journal of Biochemistry, 152*(5), 387–395. https://doi.org/10.1093/jb/mvs105.

Lai, Z., Chen, J., Ding, C., Wong, K., Chen, X., Pu, L., ... Wang, L. (2020). Association of hepatic global DNA methylation and serum one-carbon metabolites with histological severity in patients with NAFLD. *Obesity, 28*(1), 197–205. https://doi.org/10.1002/oby.22667.

Lin, H. P., Cheng, Z. L., He, R. Y., Song, L., Tian, M. X., Zhou, L. S., ... Xiong, Y. (2016). Destabilization of fatty acid synthase by acetylation inhibits de novo lipogenesis and tumor cell growth. *Cancer Research, 76*(23), 6924–6936. https://doi.org/10.1158/0008-5472.CAN-16-1597.

Matsuzaka, T., Shimano, H., Yahagi, N., Kato, T., Atsumi, A., Yamamoto, T., ... Yamada, N. (2007). Crucial role of a long-chain fatty acid elongase, Elovl6, in obesity-induced insulin resistance. *Nature Medicine, 13*(10), 1193–1202. https://doi.org/10.1038/nm1662.

Melton, P. E., Burton, M. A., Lillycrop, K. A., Godfrey, K. M., Rauschert, S., Anderson, D., ... Huang, R. C. (2023). Differential DNA methylation of steatosis and non-alcoholic fatty liver disease in adolescence. *Hepatology International, 17*(3), 584–594. https://doi.org/10.1007/s12072-022-10469-7.

Mondal, A., Bose, D., Saha, P., Sarkar, S., Seth, R., Kimono, D., ... Chatterjee, S. (2020). Lipocalin 2 induces neuroinflammation and blood-brain barrier dysfunction through liver-brain axis in murine model of nonalcoholic steatohepatitis. *Journal of Neuroinflammation, 17*(1), 201. https://doi.org/10.1186/s12974-020-01876-4.

Mondal, A., Nandi, S., Mandal, B., Banerjee, I., Ddoza Hazari, M. W. A., & Das, C. (2025). Protocol for 3D tumor spheroid generation, immunostaining, and imaging through comparative approaches. *STAR Protocols, 6*(4), 104168. https://doi.org/10.1016/j.xpro.2025.104168.

Mondal, A., Nandi, S., Singh, V., Chakraborty, A., Banerjee, I., Sen, S., ... Das, C. (2025). TCF7l2 regulates fatty acid chain elongase HACD3 during lipid-induced stress. *Biochemistry*. https://doi.org/10.1021/acs.biochem.4c00491.

Moon, Y. A., Hammer, R. E., & Horton, J. D. (2009). Deletion of ELOVL5 leads to fatty liver through activation of SREBP-1c in mice. *Journal of Lipid Research, 50*(3), 412–423. https://doi.org/10.1194/jlr.M800383-JLR200.

Nie, L., Pascoa, T. C., Pike, A. C. W., Bushell, S. R., Quigley, A., Ruda, G. F., ... Carpenter, E. P. (2021). The structural basis of fatty acid elongation by the ELOVL elongases. *Nature Structural & Molecular Biology, 28*(6), 512–520. https://doi.org/10.1038/s41594-021-00605-6.

Nishida, N., Yada, N., Hagiwara, S., Sakurai, T., Kitano, M., & Kudo, M. (2016). Unique features associated with hepatic oxidative DNA damage and DNA methylation in non-alcoholic fatty liver disease. *Journal of Gastroenterology and Hepatology (Australia), 31*(9), 1646–1653. https://doi.org/10.1111/jgh.13318.

Özcan, U., Cao, Q., Yilmaz, E., Lee, A. H., Iwakoshi, N. N., Özdelen, E., ... Hotamisligil, G. S. (2004). Endoplasmic reticulum stress links obesity, insulin action, and type 2 diabetes. *Science (New York, N. Y.), 306*(5695), 457–461. https://doi.org/10.1126/science.1103160.

Robinson, E., Khuttan, M., McFarland-Lesser, I., Patel, Z., & Jones, A. (2022). Calorie reformulation: A systematic review and meta-analysis examining the effect of manipulating food energy density on daily energy intake. *International Journal of Behavioral Nutrition and Physical Activity, 19*(1), 1–19. https://doi.org/10.1186/S12966-022-01287-Z/FIGURES/5.

Sawai, M., Uchida, Y., Ohno, Y., Miyamoto, M., Nishioka, C., Itohara, S., ... Kihara, A. (2017). The 3-hydroxyacyl-CoA dehydratases HACD1 and HACD2 exhibit functional redundancy and are active in a wide range of fatty acid elongation pathways. *Journal of Biological Chemistry, 292*(37), 15538–15551. https://doi.org/10.1074/jbc.M117.803171.

Wang, Y., Botolin, D., Xu, J., Christian, B., Mitchell, E., Jayaprakasam, B., ... Jump, D. B. (2006). Regulation of hepatic fatty acid elongase and desaturase expression in diabetes and obesity. *Journal of Lipid Research, 47*(9), 2028–2041. https://doi.org/10.1194/jlr.M600177-JLR200.

Younossi, Z. M., Rinella, M. E., Sanyal, A. J., Harrison, S. A., Brunt, E. M., Goodman, Z., et al. (2021). From NAFLD to MAFLD: Implications of a premature change in terminology. *Hepatology (Baltimore, Md.), 73*, 1194–1198.

Zhang, K., Kniazeva, M., Han, M., Li, W., Yu, Z., Yang, Z., ... Petrukhin, K. (2001). A 5-bp deletion in ELOVL4 is associated with two related forms of autosomal dominant macular dystrophy. *Nature Genetics, 27*(1), 89–93. https://doi.org/10.1038/83817.

CHAPTER EIGHT

Critical structural features of HDAC11 inhibitors

Indrasis Dasgupta, Srijoni Sinha, and Shovanlal Gayen*
Laboratory of Drug Design and Discovery, Department of Pharmaceutical Technology, Jadavpur University, Kolkata, India
*Corresponding author. e-mail address: shovanlal.gayen@gmail.com; sgayen.pharmacy@jadavpuruniversity.in

Contents

Abstract

Histone deacetylase 11 (HDAC11), the sole member of the class IV HDACs, has gained significance as a key regulator of a wide range of physiological and pathological processes. The structural characterization of HDAC11 remains challenging due to the absence of an experimentally determined crystal structure. Proper analysis of reported HDAC11 inhibitors can yield valuable insights into the key structural fingerprints essential for inhibition. Using a diverse range of descriptor sets, including various 2D descriptors and fingerprint-based descriptors, it becomes possible to effectively identify critical structural features associated with HDAC11 inhibitory activity. These fingerprints enable the recognition of substructural features that play an important role in binding within the HDAC11 catalytic site and in achieving selective inhibition. In the present chapter, various computational methodologies, including Bayesian classification, Recursive partitioning approaches and machine learning-based classification models, have been discussed in a user-friendly manner to identify the critical structural features required for effective

Methods in Enzymology, Volume 734
ISSN 0076-6879, https://doi.org/10.1016/bs.mie.2026.05.045

HDAC11 inhibition. The insights derived from these approaches are expected to assist researchers in the rational design of selective HDAC11 inhibitors, which may be further validated through experimental studies.

1. Introduction

Histone deacetylases (HDACs) are a family of enzymes that are responsible for catalysing the removal of acetyl groups from lysine residues present in proteins (Yoshida et al., 2017). Through this deacetylation mechanism, HDACs serve as key modulators of various cellular signalling pathways (Xu et al., 2007). Alterations in their activity can influence the expression of proteins implicated in a wide spectrum of human disorders, including cancers, cardiovascular diseases, inflammatory conditions, and neurological complications (Falkenberg & Johnstone, 2014). Based on the sequence homology, the HDAC family is categorised into four major classes. Class I includes HDAC1, HDAC2, HDAC3, and HDAC8, whereas Class II is divided into subclass IIa (HDAC4, HDAC5, HDAC7, HDAC9) and subclass IIb (HDAC6 and HDAC10). Class III HDACs, called sirtuins, comprise seven mammalian enzymes (SIRT1-SIRT7) and are unique among HDAC families in requiring NAD^+ for their catalytic activity (Khatun et al., 2024a). HDAC11 represents the only enzyme within Class IV. Although structurally related to both Class I and Class II enzymes, HDAC11 is distinct, with its predicted polypeptide sequence consisting of 347 amino acids, corresponding to a molecular weight of approximately 39 kDa (Gao et al., 2002).

HDAC11 plays an important role in neurodegenerative diseases, being highly expressed in the brain and involved in regulating histone acetylation, neuronal differentiation, and oligodendrocyte maturation. Dysregulation of HDAC11 has been associated with several neurodegenerative and neuropsychiatric disorders, including depression, Parkinson's disease, schizophrenia, and CNS demyelinating diseases. Pharmacological inhibition of HDAC11 has shown promising therapeutic effects, including enhancing histone H3 acetylation, modulating neuroinflammation, regulating autophagy in microglia, and influencing disease-related gene expression and microRNA expression (Khatun et al., 2024b; Bhagat et al., 2025). HDAC11 plays a key role in the regulation of gene expression and protein function. Elevated HDAC11 expression has been reported to promote tumour progression in several malignancies, including non-small cell lung cancer (NSCLC) and endocrine tumours (Bora-Singhal et al., 2020). In contrast, reduced HDAC11 expression has been observed in tumour tissues compared to adjacent normal tissues in certain cancers, such as colorectal cancer (CRC), where it appears to suppress tumour progression. These contrasting findings indicate that the functional role of HDAC11 in cancer is highly dependent on the specific cancer type (Wen et al., 2022).

Considering the growing interest in HDAC11 as a drug target for neurodegenerative diseases, it is crucial to identify the key structural features responsible for HDAC11 inhibition. Currently, no experimentally resolved three-dimensional structure of HDAC11 is available in the Protein Data Bank (RCSB PDB) (Rose et al., 2012), which poses a significant challenge for structure-based inhibitor design. Under these circumstances, ligand-based drug design approaches, particularly classification-oriented modelling strategies, offer a valuable alternative to elucidate the key structural determinants necessary for effective HDAC11 inhibition. Understanding these critical molecular features will support the rational design and optimization of potent and selective HDAC11 inhibitors.

Various classification-based modelling approaches, including Bayesian classification, Recursive partitioning, and other advanced machine learning (ML)-based approaches, can be utilized for the identification of critical structural features of HDAC11 inhibitors. To effectively discriminate between potent inhibitors and non-inhibitors, a variety of 2D descriptors capturing distinct structural information, alongside fingerprint-based molecular representations such as Extended-Connectivity Fingerprints (*ECFPs*), Functional-Class Fingerprints (*FCFPs*), *MACCS* keys, and Klekota-Roth fingerprints can be employed. Based on the comprehensive evaluation metrics, it is possible to determine which algorithm most effectively discriminates the key structural features responsible for HDAC11 inhibition. Furthermore, detailed interpretation and SHAP (SHapley Additive exPlanations)-based analysis should be used to elucidate the underlying feature importance and decode the decision-making processes of the predictive models.

2. Rational

HDAC11 has emerged as a significant target due to its involvement in various physiological and pathological processes, including neurodegenerative disorders, cancer, and immune regulation. The primary limitations are the absence of an experimentally resolved three-dimensional crystal structure of HDAC11, which restricts the application of structure-based drug design strategies. In the absence of reliable structural information, ligand-based approaches provide a practical and effective alternative for understanding the important structural features of HDAC11 inhibition.

However, the available data on HDAC11 inhibitors are scattered across multiple sources and often lack systematic analysis to identify the key molecular determinants responsible for HDAC11 inhibitory activity. Furthermore, the structural diversity of reported inhibitors necessitates the use of various computational strategies capable of extracting meaningful patterns from complex chemical data. With the rapid advancement of machine learning methodologies and the availability of diverse molecular descriptor sets (e.g., *ECFP*, *FCFP* and *MACCS*), there is a growing opportunity to integrate these approaches for improved predictive performance and interpretability. The incorporation of interpretability techniques such as SHAP analysis further enhances the ability to decode model predictions, thereby bridging the gap between predictive accuracy and mechanistic understanding. The present work is designed to identify critical structural fingerprints associated with HDAC11 inhibitory activity.

3. Methodology

3.1 Preparation of the dataset

3.1.1 Software/packages

- Discovery Studio (DS) 3.0
- RDKit

3.1.2 Procedure

1. Data Collection and Quality Control
   (a) For reliable identification of critical structural fingerprints associated with HDAC11 inhibition, the preparation of a well-curated, carefully standardized dataset is an essential prerequisite.
   (b) Experimentally validated HDAC11 inhibitors must be compiled from publicly available bioactivity databases, including ChEMBL, BindingDB, and ZINC, as well as from peer-reviewed literature, with only those compounds possessing well-defined chemical structures and unambiguous biological annotations (Wassermann & Bajorath, 2011; Sterling & Irwin, 2015).

   (c) To reduce experimental variability, activity data generated using comparable biochemical assay conditions are preferentially selected, and compounds with conflicting or inconsistent activity reports should be excluded.

2. Structure Standardization and Removal of Duplicates

   (a) Chemical structures should be initially standardized by removing salts, solvents, and counterions, followed by normalization of charges and valence states.

   (b) Subsequently, duplicate entries must be identified based on canonical SMILES and InChI keys (de Sousa, 2017), and only a single representative structure should be retained, with preference given to the most reliable experimental activity value.

3. Treatment of Stereochemistry

   (a) In the case of 2D QSAR modelling, stereoisomers were not treated as distinct entities, as 2D molecular representations do not explicitly encode three-dimensional stereochemical information and therefore assign identical descriptor values to stereoisomeric structures.

   (b) Consequently, the inclusion of multiple stereoisomers would result in duplicate entries and could introduce bias into the predicted model. Therefore, such redundant stereoisomeric representations should be removed from the dataset.

   (c) In contrast, when the explicit retention and evaluation of stereochemical differences are of interest, a 3D QSAR modelling strategy is more appropriate, as it can capture spatial and conformational variations that influence biological activity.

   (d) However, for 3D QSAR approaches, molecular structures must undergo thorough and consistent geometry optimization using appropriate force fields to generate energetically reasonable conformations, ensuring that the calculated 3D descriptors accurately reflect the true physicochemical and steric features relevant to the biological activity (Verma et al., 2010).

4. Drug-Likeness Filtering

   (a) Drug-likeness filtering is an important step in dataset preparation to ensure that selected compounds possess physicochemical properties consistent with known orally active drugs.

   (b) This process typically involves applying established rules such as Lipinski's rule of 5 and Veber's criteria, which evaluate parameters including molecular weight, lipophilicity, hydrogen bond donors and acceptors, polar surface area, and molecular flexibility.

(c) Compounds that violate these guidelines are more likely to exhibit poor absorption, permeability, or bioavailability and are therefore excluded.
(d) By retaining only drug-like molecules, this filtering step improves the reliability and relevance of subsequent modelling and enhances the likelihood of identifying biologically viable candidates.

5. Choice of suitable activity cutoff
   (a) The selection of an appropriate activity cutoff is an important step in developing classification-based models for identifying critical structural features associated with the HDAC11 inhibitory activity.
   (b) Continuous bioactivity data, such as IC_{50}, EC_{50} and K_i values, are typically transformed into binary classes (active and inactive) using a predefined threshold, which directly influences model performance and interpretability.
   (c) An optimal cutoff should be chosen based on biological relevance and data distribution. Care must be taken to avoid arbitrary cutoff selection, as overly strict thresholds can lead to class imbalance or misclassification of compounds.
   (d) A well-justified cutoff ensures that the resulting model reliably captures meaningful structure-activity relationships and accurately identifies key structural determinants.

3.2 Division of the dataset

3.2.1 Software

- Discovery Studio (DS) 3.0
- scikit-learn package

3.2.2 Procedure

(a) In the absence of an independent external set, the entire dataset is partitioned into two subsets: a training set and a test set. The training set is utilized exclusively for model development, including feature selection and optimization of model hyperparameters. In contrast, the test set is reserved for validation to evaluate the predictive performance of the constructed models.
(b) Several dataset splitting strategies are considered to ensure robust and unbiased model evaluation, including random division, random per-cluster division, Kennard-Stone-based division, Euclidean distance-based division, and activity-based division (Martin et al., 2012).
(c) Generally, the dataset is separated into training and test sets using commonly adopted ratios such as 70:30, 75:25, or 80:20. These splitting

proportions are widely used to ensure that a sufficiently large subset is available for model development while retaining an adequate number of compounds in the test set for reliable validation of the predictive performance.

3.3 Bayesian classification

3.3.1 *Software*

- Discovery Studio (DS) 3.0
- scikit-learn package

3.3.2 *Concept*

Bayesian classification is a statistical approach governed by probability theory and based on Bayes' theorem, as represented in the equation below.

$$P(A/B) = \frac{P\left(\frac{B}{A}\right)P(A)}{P(B)}$$

It estimates the posterior probability of a class given the observed data by combining prior knowledge with evidence derived from the dataset. When sufficient and representative data are available, Bayesian classifiers perform probabilistic predictions, enabling not only the assignment of a sample to a specific class but also the estimation of the likelihood (probability) of that classification. Here, A denotes the hypothesis or model, and B represents the observed data. P(A) is the probability of the hypothesis before observing any data. P(B) is the marginal probability of the data. P(A/B) is the posterior probability (probability of hypothesis A given observed data B). P(B/A) is the likelihood (probability of data B if hypothesis A is true) (Khatun et al., 2025; Liu et al., 2014).

3.3.3 *Procedure*

1. Construction of Bayesian models
   (a) Various Bayesian classification models can be generated using the 'Create Bayesian Model' module of the Discovery Studio (Studio, 2008).
   (b) The combination of various molecular properties (*MPs*), which include lipophilicity (*AlogP*), molecular weight (*MW*), number of aromatic rings (*nAR*), number of hydrogen bond acceptors (*nHBA*), number of hydrogen bond donors (*nHBD*), number of rings (*nR*), number of rotatable bonds (*nRB*) and molecular fractional polar surface area (*M_FPSA*), *ECFP_6* and *FCFP_6* were served as input features for this classification study (Liu et al., 2014; Khatun et al., 2024c).

- **(c)** *ECFP_6*, or Extended connectivity fingerprints with a maximum diameter of 6 bonds, are circular topological fingerprints that encode substructural features by iteratively expanding circular neighbourhoods around each atom. *ECFP_6* captures detailed structural information, converting molecules into fixed-length bit vectors suitable for activity/property prediction (Rogers & Hahn, 2010).
- **(d)** On the other side, *FCFP_6* generates circular fingerprints by encoding functional-class-based substructural fragments up to a diameter of six bonds using an iterative search starting from each heavy atom (Sun, 2005).
- **(e)** Various combinations of *MPs*, *ECFP_6* and *FCFP_6* can be used for capturing structural features of molecules, thereby improving the predictive performance of constructed models.

2. Evaluation of the constructed models

- **(a)** In the case of classification-based modelling, compounds are assigned to one of four outcome categories by comparing model-predicted class labels with their corresponding experimental annotations.
- **(b)** Compounds that are experimentally active and correctly predicted as active are defined as true positives (TP).
- **(c)** Conversely, false positives (FP) correspond to inactive compounds that are incorrectly classified as active by the model.
- **(d)** False negatives (FN) denote active compounds that are misclassified as inactive, whereas true negatives (TN) represent inactive compounds that are correctly identified as inactive.
- **(e)** These four outcomes are organized into a two-by-two confusion matrix, which provides a concise summary of the classification results.
- **(f)** The confusion matrix forms the foundation for computing various statistical performance measures, which are commonly used to evaluate the predictive quality and reliability of classification models.
- **(g)** Various classification-based validation metrics can also be employed to evaluate the performance of the developed Bayesian models. These metrics include accuracy, precision, recall, specificity and F1 score (Khatun et al., 2025; Roy et al., 2015). The mathematical expressions for the classification performance metrics employed are presented below:

$$\text{Accuracy} = \frac{(TP + TN)}{(TP + TN + FP + FN)}$$

$$\text{Precision} = \frac{TP}{(TP + FP)}$$

$$\text{Recall} = \frac{TP}{(TP + FN)}$$

$$\text{Specificity} = \frac{TN}{(TN + FP)}$$

$$\text{F1Score} = \frac{2 * (Precision * Recall)}{(Precision + Recall)}$$

(h) In addition, receiver operating characteristic (ROC) curves are also generated for both the training and test sets to evaluate the discriminative ability of the constructed models.

(i) To further ensure model robustness and minimize overfitting, cross-validation strategies can be implemented using k-fold cross-validation with k values of 5, 10, 15, and 20. These cross-validation schemes enabled a comprehensive assessment of model stability by repeatedly partitioning the dataset into training and validation subsets and averaging the performance metrics across multiple folds (Amin et al., 2020).

(j) Based on these evaluation criteria, the most suitable combination of input features can be identified to achieve optimal discrimination between HDAC11 inhibitors and non-inhibitors.

(k) After selecting the best model, the identified favourable and unfavourable fingerprints for HDAC11 inhibition can be systematically analyzed. These fingerprints provide valuable insights into the structural features that enhance or reduce HDAC11 inhibitory activity and can be effectively used to guide the rational design of selective HDAC11 inhibitors.

3.4 Recursive partitioning

3.4.1 Software

Discovery Studio (DS) 3.0.

3.4.2 Concept

Recursive partitioning (RP) is a multivariate statistical technique for classification that operates by systematically dividing a dataset into homogeneous subsets through the construction of one or more decision trees. In this approach, the classification outcome is represented in the form of hierarchical decision rules, which are interpretable. RP facilitates the elucidation of relationships between a binary dependent variable (e.g., inhibitor versus non-inhibitor class) and multiple independent variables, including molecular properties and fingerprint-based descriptors (*ECFP_6* and *FCFP_6*). During the tree-building process, all descriptors are iteratively evaluated at each node to identify the most informative variable for partitioning the dataset. Once the optimal splitting criterion is selected, the same procedure is recursively applied to each resulting subset (Khatun et al., 2025; Zeileis et al., 2008; Amin et al., 2021). The outcome of this iterative process is a decision tree generated through successive data partitioning, providing a transparent model for the classification of HDAC11 inhibitors and non-inhibitors.

3.4.3 Procedure

1. Construction of Recursive Partitioning (RP) models
   (a) Various RP models can be generated using the 'Create RP Model' module of the DS 3.0 (Accelrys Inc., San Diego, USA, 2015).
   (b) As discussed previously, here also the combination of various molecular properties, which include *AlogP*, *MW*, *nAR*, *nHBA*, *nHBD*, *nR*, *nRB* and *M_FPSA*, *ECFP_6* and *FCFP_6* served as input features. Different combinations of *MPs*, *ECFP_6* and *FCFP_6* can be used for capturing structural features of molecules, thereby improving the predictive performance of constructed RP models.
   (c) Additionally, the maximum tree depth can be systematically varied from 1 to 20 across different feature combinations to evaluate the influence of tree complexity on predictive performance and to identify the optimal depth that yielded the best classification results (Amin et al., 2021).
2. Evaluation of constructed RP models
   (a) Here, also various classification-based validation metrics like accuracy, precision, sensitivity, specificity and F1 score can be employed to evaluate the performance of the developed RP models.
   (b) Receiver operating characteristic (ROC) curves are also generated for both the training and test sets of various RP models to evaluate the discriminative ability of the constructed models (Khatun et al., 2025; Amin et al., 2021).

(c) To further ensure model robustness, cross-validation strategies can be implemented using k-fold cross-validation with k values of 5, 10, 15, and 20.

(d) Based on these assessment criteria, the most appropriate combination of input features (*MPs*, *ECFP_6*, and *FCFP_6*) can be identified to achieve the best discrimination between HDAC11 inhibitors and non-inhibitors.

(e) The best RP model enables the identification of both favourable and unfavourable molecular fragments, providing insights into critical structural fingerprints that can be exploited for the rational design of HDAC11 inhibitors.

3.5 Other machine learning-based classification

3.5.1 Concept

Machine learning (ML) refers to a branch of artificial intelligence in which algorithms are designed to learn from data, uncover patterns, and make informed predictions without being manually programmed (Sarker, 2021). Based on the availability of labelled data, ML techniques are broadly classified into supervised and unsupervised learning. In supervised learning, models are trained using labelled datasets, where the input features are associated with known output values. This approach is commonly used for prediction and classification tasks (Saravanan & Sujatha, 2018). In contrast, unsupervised learning deals with unlabelled data and aims to discover hidden patterns, structures, or relationships within the data, such as clustering or dimensionality reduction (Chander & Gopalakrishnan, 2023). Commonly used supervised learning algorithms include linear and logistic regression, which capture linear relationships for continuous and categorical outcomes, respectively; k-nearest neighbors (kNN), which classifies instances based on similarity measures; support vector machines (SVMs), which identify optimal decision boundaries by maximizing class separation; and decision tree-based methods, such as decision trees, random forests, and gradient boosting, which capture complex non-linear patterns through hierarchical feature splitting. Additionally, artificial neural networks and deep learning models learn high-level representations from data through layered architectures (Shetty et al., 2022; LeCun et al., 2015).

3.5.2 Software/packages

- The scikit-learn library was primarily employed for constructing ML-based classification models.

- Hyperparameter optimization was performed using Grid Search CV (Belete & Huchaiah, 2022) and Randomized Search CV (Nugroho & Suhartanto, 2020), which are built-in utilities provided by scikit-learn.
- In addition to these approaches, Bayesian optimization was carried out using the Optuna framework (Akiba et al., 2019).
- SHAP analysis was performed using the Python library SHAP (SHapley Additive exPlanations)

3.5.3 Procedure

1. Calculation of different descriptors
   - **(a)** For the computation of molecular descriptors, several platforms and software can be employed, including the OCHEM web platform (Sushko et al., 2011), PaDEL-Descriptor calculating software (Yap, 2011), alvaDesc software (Mauri, 2020), and the Mordred descriptor package (Moriwaki et al., 2018), which provide a wide range of physicochemical, topological, and structural descriptors.
   - **(b)** In addition, fingerprint-based representations, such as *ECFPs*, *FCFPs*, and *MACCS* keys (Rogers & Hahn, 2010; Fernández-de Gortari et al., 2017), are also widely used to encode molecular substructures.
   - **(c)** Different 2D molecular descriptors can be utilised for the development of various ML-based classification models, as they reduce the computational burden associated with conformational analysis and energy minimization.
   - **(d)** In case of using 3D descriptors, the molecular structures must be properly energy-minimized before descriptor calculation to ensure a reliable and meaningful representation of their conformations.
2. Identification of the most contributing descriptors
   - **(a)** After the computation of different descriptors, data pre-treatment should be performed to refine the descriptor matrix by eliminating highly intercorrelated descriptors and those exhibiting negligible variance (Ballabio et al., 2014).
   - **(b)** After pre-treatment, proper feature selection strategies should be employed for the selection of the most informative descriptors.
   - **(c)** These feature selection strategies include filter methods (select features based on statistical criteria independent of any specific learning algorithm, such as Chi-square test, Information Gain, and correlation coefficients) (Cherrington et al., 2019), Wrapper methods (evaluate different subsets of features by training a specific

classifier and assessing its predictive performance; examples include Recursive Feature Elimination (RFE) and forward selection) (Awad & Fraihat, 2023).

(d) Although wrapper methods often produce higher predictive accuracy, they are computationally expensive, especially for high-dimensional datasets.

(e) Embedded methods incorporate feature selection directly into the model training process, as observed in the case of techniques such as LASSO regularization (Muthukrishnan & Rohini, 2016) and feature importance measures obtained from Random Forest models.

3. Optimization of hyperparameters and construction of ML models

(a) The descriptors identified from the feature selection strategies should be used to develop various ML-based classification models aimed at effectively distinguishing between HDAC11 inhibitors and non-inhibitors.

(b) The successful implementation of any ML-based classification model depends on the fine-tuning of necessary hyperparameters, as they have a significant impact on model accuracy and overall performance.

(c) Different techniques can be used for hyperparameter optimization, such as Grid Search (Belete & Huchaiah, 2022), Randomized Search (Nugroho & Suhartanto, 2020), and Bayesian Optimization (Akiba et al., 2019), which systematically explore the hyperparameter space to identify optimal configurations.

(d) These approaches, when combined with cross-validation, ensure that the selected hyperparameters yield consistent and reliable model performance, minimizing the risk of overfitting to a specific train-test split.

(e) Various important hyperparameters used in the development of different ML-based classification models are presented in Table 1 (Bisong, 2019).

(f) After proper optimization of the relevant hyperparameters, ML-based classification models can be developed using the scikit-learn library in Python (Bisong, 2019), enabling effective and reliable discrimination between HDAC11 inhibitors and non-inhibitors.

4. Evaluation of ML-based classification models

As discussed, various validation metrics like accuracy, precision, recall and F1 score were employed for the identification of the best ML model that most effectively discriminates HDAC11 inhibitors from non-inhibitors (Bhagat et al., 2025).

5. Mechanistic interpretation of the best model

Table 1 Necessary hyperparameters used for the development and optimization of different ML-based classification models.

ML-model	Hyperparameters
Random Forest Classifier (RFC)	*n_estimators, criterion, max_depth, min_samples_split, min_samples_leaf*
Support Vector Classifier (SVC)	*C, kernel, degree, gamma*
Linear Discriminant Analysis (LDA)	*solver*
Logistic regression (LR)	*penalty, C, l1_ratio, solver, max_iter*
Gradient Boosting Classifier (GBC)	*loss, learning_rate, n_estimators, subsample, criterion, min_samples_split, min_samples_leaf, max_depth,*
Extreme Gradient Boosting classifier (XGBC)	*n_estimators, max_depth, learning_rate, subsample, gamma, reg_lambda*
Multi-Layer Perceptron (MLP)	*hidden_layer_sizes, activation, solver, alpha, learning_rate*
k-nearest neighbors (kNN)	*n_neighbors, weights, algorithm, leaf_size, p, metric*

(a) After identifying the best ML-based classification model capable of discriminating HDAC11 inhibitors from non-inhibitors, a mechanistic interpretation of the model becomes essential.

(b) Such interpretation enables the identification of key structural fingerprints and molecular features that are most relevant to HDAC11 inhibition, thereby providing meaningful insights into the structure-activity relationships governing inhibitory activity.

(c) For the interpretation of the predictions of black-box ML models, SHAP analysis was employed.

(d) SHAP analysis is an interpretability technique rooted in cooperative game theory that explains the output of machine learning models by assigning an "importance value" (Shapley value) to each input feature.

(e) It bridges the gap between complex "black box" algorithms and human understanding by decomposing a prediction into the additive contributions of its parts, showing how much each feature pushes the result away from the baseline average toward the final output.

(f) SHAP analysis is critical not just for identifying global drivers of model behaviour (e.g., which molecular descriptors matter most across the dataset), but also for providing local explanations, revealing exactly why a specific individual instance (e.g., a single inhibitory molecule) received its particular prediction (Ponce-Bobadilla et al., 2024; Yang et al., 2022).

4. Data analysis and interpretation

An example study compiled 1191 compounds with experimentally reported HDAC11 inhibitory activities, expressed as IC_{50} values in nanomolar units, retrieved from the BindingDB database. The dataset was curated using the "Prepare Ligands for QSAR" module in Discovery Studio (DS) 3.0 (Accelrys Inc., San Diego, USA, 2015), which involved removing salts and solvent molecules, standardizing structures, neutralizing charges where appropriate, and eliminating redundant entries. This process removed 382 duplicate compounds, yielding 809 unique molecules. The curated dataset was then subjected to drug-likeness filtering using the "Filter using Lipinski and Veber's rules" protocol in DS 3.0 (Accelrys Inc., San Diego, USA, 2015) to exclude compounds with unfavourable physicochemical properties. After filtering, 712 drug-like compounds satisfying both Lipinski's rule of five and Veber's criteria were retained and used for all subsequent analyses. As a classification-based modelling strategy was adopted, the continuous activity data were converted into binary classes, with compounds having $IC_{50} \leq 1000$ nM labelled as active (1) and those with $IC_{50} > 1000$ nM labelled as inactive (0). The dataset division was carried out using the random per-cluster approach using DS 3.0 (Accelrys Inc., San Diego, USA, 2015) in the current example. Using this strategy, 554 compounds were allocated to the training set, while the remaining 158 compounds constituted the test set.

Various Bayesian classification models were generated using the 'Create Bayesian Model' module of DS 3.0 (Accelrys Inc., San Diego, USA, 2015). The combination of different molecular properties, *ECFP_6* and *FCFP_6* were included in this study to obtain the best combination. Various statistical parameters corresponding to different Bayesian classification models developed in the example study are summarized in Table 2. Multiple combinations of molecular descriptors were systematically evaluated to determine the most appropriate descriptor set for effectively distinguishing HDAC11 inhibitors from non-inhibitors. As evident from Table 2, the model developed using molecular properties (*MPs*) in combination with *ECFP_6* descriptors exhibited superior

Table 2 Various statistical parameters of the developed Bayesian classification models.

Descriptors	Set	Acc	Pre	Sen	Spec	ROC
MPs	Train (5fold CV)	0.731	0.771	0.544	0.875	0.720
	Test	0.747	0.724	0.636	0.826	0.795
MPs + ECFP_6	**Train (5-fold CV)**	**0.845**	**0.824**	**0.817**	**0.866**	**0.812**
	Train (10-fold CV)	**0.850**	**0.829**	**0.826**	**0.869**	**0.822**
	Train (15-fold CV)	**0.850**	**0.826**	**0.830**	**0.866**	**0.819**
	Train (20-fold CV)	**0.854**	**0.833**	**0.830**	**0.872**	**0.823**
	Test	**0.766**	**0.716**	**0.727**	**0.793**	**0.850**
ECFP_6 + FC-FP_6	Train (5fold CV)	0.839	0.817	0.813	0.859	0.808
	Test	0.759	0.684	0.788	0.739	0.845
MPs + FCFP_6	Train (5fold CV)	0.830	0.786	0.838	0.824	0.807
	Test	0.778	0.738	0.727	0.815	0.843
MPs + ECFP_6- + FCFP_6	Train (5fold CV)	0.843	0.818	0.822	0.859	0.809
	Test	0.766	0.693	0.788	0.750	0.847

performance across the evaluated statistical parameters for both the training and test sets. Consequently, this model was identified as the best model and selected for further analysis.

Depending on the best model, the top 20 favourable and 20 unfavourable Bayesian fingerprints were identified. By mapping these positive and negative fingerprints onto the molecular structures of inhibitors and non-inhibitors, the most critical structural features governing HDAC11 inhibitory activity were elucidated (Fig. 1).

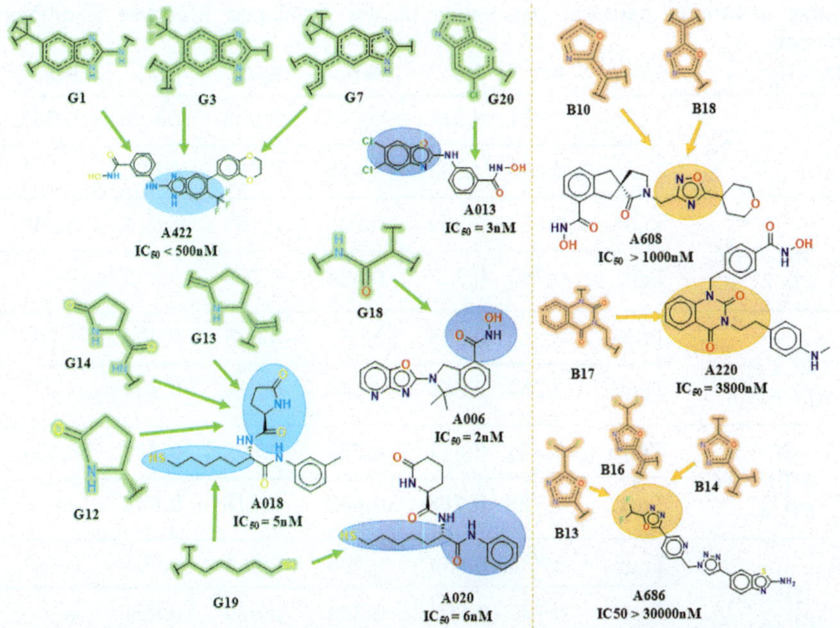

Fig. 1 Representative molecular structures of active and inactive HDAC11 inhibitors illustrating the contribution of Bayesian fingerprints to model predictions.

Multiple recursive partitioning models were also developed using the "Create RP Model" module of DS 3.0 (Accelrys Inc., San Diego, USA, 2015) for the present case study. Various combinations of molecular properties along with *ECFP_6* and *FCFP_6* fingerprints were systematically evaluated to identify the most robust and predictive model. Multiple combinations of molecular descriptor sets (*MPs*, *ECFPs* and *FCFPs*) were systematically examined to identify the most appropriate feature representation for effectively discriminating HDAC11 inhibitors from non-inhibitors. For each descriptor combination, several decision tree models were generated, and the statistical parameters corresponding to the best-performing tree were selected and reported in Table 3. Comparative analysis of these models revealed that the combination of molecular properties (*MPs*) with *FCFP_6* fingerprints consistently outperformed the other descriptor sets across all evaluated statistical metrics in both the training and test sets. To further assess the predictability of the models, fivefold cross-validation was also employed during model evaluation. Decision tree-1 (Fig. 2), developed using the combined molecular properties (*MPs*) and *FCFP_6* descriptors, consists of 31 leaves with a maximum depth of 17. This decision tree comprises six

Table 3 Various statistical parameters of the developed Recursive partitioning models.

Descriptors	Set	Acc	Pre	Sen	Spec	ROC	ROC_{cv}
MPs	Train	0.771	0.747	0.732	0.802	0.832	0.735
	Test	0.690	0.682	0.616	0.753	0.778	—
MPs + EC-FP_6	Train	0.832	0.809	0.806	0.853	0.911	0.769
	Test	0.753	0.727	0.696	0.798	0.835	—
ECFP_6 + FCFP_6	Train	0.816	0.855	0.755	0.875	0.906	0.761
	Test	0.734	0.803	0.646	0.829	0.804	—
MPs + FC-FP_6	**Train**	**0.841**	**0.871**	**0.787**	**0.892**	**0.927**	**0.791**
	Test	**0.728**	**0.788**	**0.642**	**0.818**	**0.844**	—
MPs + EC-FP_6 + FC-FP_6	Train	0.834	0.834	0.794	0.867	0.899	0.762
	Test	0.797	0.803	0.736	0.849	0.804	—

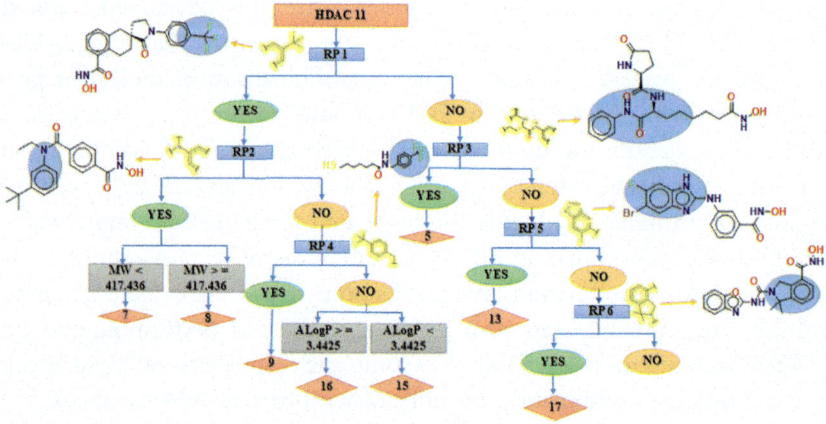

Fig. 2 Decision tree illustrating the classification of compounds into HDAC11 inhibitors and non-inhibitors.

molecular properties and twenty-four structural fingerprints (FP-1 to FP-24). These features collectively contribute to the effective discrimination between HDAC11 inhibitors and non-inhibitors.

For the development of different ML-based classification models, *ECFP_4* fingerprints with a length of 2048 bits and *MACCS* structural keys comprising 167 bits were used as descriptors. The LASSO method was employed for feature selection to identify the most relevant descriptors for model development. The required hyperparameters were systematically optimized using the grid search (GridSearchCV) approach. A fivefold cross-validated accuracy was employed as the objective function to identify the optimal hyperparameter combinations and to ensure improved model performance. The details of the optimized hyperparameters are summarized in Table 4. Using these optimized hyperparameters, two distinct sets of ML-based classification models were developed employing *ECFP_4* fingerprints and *MACCS* keys, respectively.

Based on a comparative analysis of these metrics, the models constructed with *MACCS* fingerprints demonstrated superior capability in discriminating potent HDAC11 inhibitors from non-inhibitors. Among the *ECFP_4*-based models, the Random Forest classifier (RFC) exhibited

Table 4 Various optimized hyperparameters of the developed ML-based classification models.

Descriptors	Models	Hyperparameters
ECFP_4 fingerprints	RFC	criterion: gini, max_depth: 5, min_samples_leaf: 1, min_samples_split: 5, n_estimators: 70
	SVC	C: 0.1, gamma: scale, kernel: rbf
	GBC	criterion: friedman_mse, learning_rate: 0.1, max_depth: 3 min_samples_leaf: 1, min_samples_split: 4,n_estimators: 60
	XGBC	colsample_bytree: 0.8, learning_rate: 0.1, max_depth: 3, n_estimators: 50, subsample: 0.8
MACCS fingerprints	RFC	criterion: entropy, max_depth: 4, min_samples_leaf: 4, min_samples_split: 2, n_estimators: 60
	SVC	C: 1.0, gamma: scale, kernel: rbf
	GBC	criterion: friedman_mse, learning_rate: 1.0, max_depth: 1, min_samples_leaf: 1, min_samples_split: 2, n_estimators: 100
	XGBC	n_estimators: 70, max_depth: 4, learning_rate: 0.01, subsample: 1.0, colsample_bytree: 0.9

Table 5 Various statistical parameters of the developed ML-based classification models.

Descriptors	Models	Set	Accuracy	Precision	Sensitivity	F1 score	MCC	Cohens kappa	AUC-ROC
ECFP_4 fingerprints	**RFC**	**Train**	**0.807**	**0.872**	**0.652**	**0.746**	**0.612**	**0.595**	**0.885**
		Test	**0.804**	**0.830**	**0.667**	**0.740**	**0.594**	**0.585**	**0.840**
	SVC	Train	0.771	0.817	0.610	0.698	0.534	0.520	0.846
		Test	0.804	0.818	0.682	0.744	0.593	0.587	0.830
	GBC	Train	0.839	0.869	0.743	0.801	0.673	0.668	0.921
		Test	0.798	0.815	0.667	0.733	0.580	0.573	0.863
	XGBC	Train	0.807	0.849	0.676	0.753	0.608	0.598	0.888
		Test	0.797	0.804	0.682	0.738	0.580	0.575	0.854
MACCS fingerprints	RFC	Train	0.787	0.840	0.631	0.720	0.569	0.554	0.870
		Test	0.804	0.830	0.667	0.740	0.594	0.585	0.844
	SVC	**Train**	**0.843**	**0.863**	**0.759**	**0.808**	**0.680**	**0.676**	**0.921**
		Test	**0.817**	**0.794**	**0.758**	**0.775**	**0.621**	**0.620**	**0.862**
	GBC	Train	0.796	0.771	0.755	0.763	0.584	0.584	0.877
		Test	0.798	0.750	0.773	0.761	0.586	0.585	0.827
	XGBC	Train	0.780	0.844	0.606	0.705	0.556	0.537	0.876
		Test	0.816	0.849	0.682	0.756	0.621	0.612	0.861

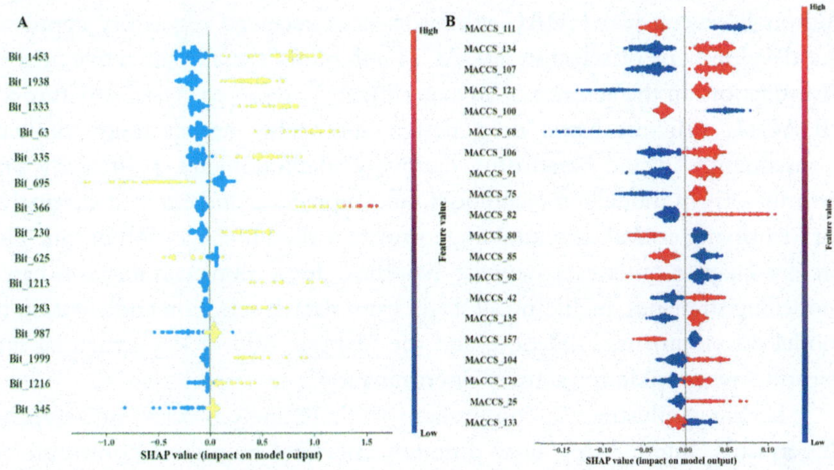

Fig. 3 (A) SHAP summary plot for the RFC model using ECFP_4 fingerprints (B) SHAP summary plot for the SVC model using MACCS fingerprints.

the best overall performance, whereas the Support Vector Classifier (SVC) achieved the highest predictive accuracy among the *MACCS* fingerprint-based classification models. A detailed summary of the statistical performance of all developed models is presented in Table 5.

These two models were selected for further mechanistic interpretation to identify the specific fingerprint bits that contribute most significantly to the discrimination of HDAC11 inhibitors from non-inhibitors. To elucidate the key structural features encoded within the fingerprints, SHAP analysis was performed for both optimal models. The corresponding SHAP summary plots, highlighting the relative importance of individual fingerprint bits, are presented in Fig. 3A and B, respectively. By mapping the fingerprint bit information onto the molecular structures of HDAC11 inhibitors, it becomes possible to identify the key structural features that contribute to either enhancement or reduction of inhibitory potency of HDAC11 inhibitors.

5. Conclusion

HDAC11, the only member of class IV HDACs, has gained significant attention due to its involvement in a wide range of physiological functions and disease-associated pathways. Based on the regulatory significance, HDAC11 has become an important therapeutic target in recent years.

Although several pan HDAC inhibitors have received regulatory approval, the development of selective HDAC11 inhibitors remains an active area of investigation. In the absence of an experimentally resolved crystal structure for HDAC11, structure-based drug design approaches are currently limited. Consequently, ligand-based drug discovery methodologies, particularly fingerprint-driven modelling techniques, offer a practical and efficient alternative for identifying critical structural fingerprints for the HDAC11 inhibition. The present chapter outlines a stepwise workflow for various classification-based modelling strategies, including all stages from dataset selection and curation to model development, validation, and the identification of key structural fingerprints responsible for activity discrimination.

The Fig. 1 illustrates a comprehensive SAR analysis based on Bayesian fingerprints, highlighting how different chemical fragments contribute to the potent inhibition of HDAC11. The green fragments represent the critical structural fingerprints that are responsible for the potent inhibition of HDAC11. On the contrary, orange colour represents bad fingerprints, which may not be promising for HDAC11 inhibition. Various favourable fragments (G1, G3, G7, and G20) are identified, and their presence is systematically highlighted within the promising structures of HDAC11 inhibitors. Some unfavourable fingerprints (B10, B13, B14, B16, B17 and B18) are also represented in the diagram, and these are correlated with the structures of weak inhibitors.

The Fig. 2 illustrates a decision tree–based model for distinguishing HDAC11 inhibitors from non-inhibitors. It integrates various structural fingerprints and molecular properties to effectively discriminate between inhibitors and non-inhibitors. The key fingerprints are highlighted within the structures of the inhibitors, emphasizing their role in inhibition. Overall, the model demonstrates how specific critical structures and physicochemical properties collectively determine the inhibitory potential of compounds against HDAC11.

The two figures (Fig. 3A and 3B) present the SHAP analysis of the best-performing ML models based on *ECFP_4* and *MACCS* fingerprints, illustrating how individual features contribute to the prediction of HDAC11 inhibitory activity. Positively contributing features are critical for identifying potent HDAC11 inhibitors, while those with negative contributions are associated with weak inhibitors. Fig. 3A displays the impact of highly contributing fingerprint bits (e.g., Bit_1453, Bit_1938, Bit_1333, Bit_63 and Bit_335) on the predictions of the model. Each dot represents a compound, and the position along the x-axis indicates the SHAP value, which reflects how much that feature contributes to increasing or decreasing the predicted inhibitory activity. Fig. 3B highlights a similar SHAP analysis but for *MACCS* structural keys, which are predefined chemical substructures. Features such as *MACCS*_111, *MACCS*_134, *MACCS*_107, *MACCS*_121, and *MACCS*_100 show prominent contributions, where the presence of these keys strongly influences the HDAC11 inhibitory activity.

Acknowledgements

SG would like to thank Anusandhan National Research Foundation (ANRF), Govt. of India, for financial assistance under the MATRICS scheme (MTR/2022/000286). The authors thankfully acknowledge Prof. Tarun Jha of Jadavpur University, India, for providing the facilities to use D.S 3.0. We are grateful to the Department of Pharmaceutical Technology, Jadavpur University, Kolkata, for providing research facilities.

Author's contribution

I.D and S.S. wrote the original draft. S.G. supervision and conceptualization. I.D. and S.G. writing – review & editing. S.G. project administration and funding acquisition. The work was supported by funds from the Anusandhan National Research Foundation (ANRF), Govt. of India, under the MATRICS scheme (MTR/2022/000286).

Conflict of interest

The authors declare that they have no conflicts of interest with the contents of this article.

Data availability statement

Data will be made available upon request.

References

Akiba, T., Sano, S., Yanase, T., Ohta, T., & Koyama, M.(2019). Optuna: A next-generation hyperparameter optimization framework. In Proceedings of the 25th ACM SIGKDD international conference on knowledge discovery & data mining (pp. 2623–2631). https://doi.org/10.1145/3292500.3330701.

Amin, S. A., Adhikari, N., & Jha, T. (2021). Development of decision trees to discriminate HDAC8 inhibitors and non-inhibitors using recursive partitioning. *Journal of Biomolecular Structure and Dynamics, 39*(1), 1–8. https://doi.org/10.1080/07391102.2019.1661876.

Amin, S. A., Banerjee, S., Adhikari, N., & Jha, T. (2020). Discriminations of active from inactive HDAC8 inhibitors Part II: Bayesian classification study to find molecular fingerprints. *SAR and QSAR in Environmental Research, 31*(4), 245–260. https://doi.org/10.1080/1062936X.2020.1723136.

Awad, M., & Fraihat, S. (2023). Recursive feature elimination with cross-validation with decision tree: Feature selection method for machine learning-based intrusion detection systems. *Journal of Sensor and Actuator Networks, 12*(5), 67. https://doi.org/10.3390/jsan12050067.

Ballabio, D., Consonni, V., Mauri, A., Claeys-Bruno, M., Sergent, M., & Todeschini, R. (2014). A novel variable reduction method adapted from space-filling designs. *Chemometrics and Intelligent Laboratory Systems, 136*, 147–154. https://doi.org/10.1016/j.chemolab.2014.05.010.

Belete, D. M., & Huchaiah, M. D. (2022). Grid search in hyperparameter optimization of machine learning models for prediction of HIV/AIDS test results. *International Journal of Computers and Applications, 44*(9), 875–886. https://doi.org/10.1080/1206212X.2021.1974663.

Bhagat, R. P., Jyotisha, Dasgupta, I., Amin, S. A., Jakkula, P., Bhattacharya, A., & Gayen, S. (2025). First report on analysis of chemical space, scaffold diversity, critical structural features of HDAC11 inhibitors. *Molecular Diversity,* 1–24. https://doi.org/10.1007/s11030-025-11217-3.

Bisong, E. (2019). Introduction to Scikit-learn. In Building machine learning and deep learning models on google cloud platform: a comprehensive guide for beginners (pp. 215–229). Berkeley, CA: Apress. https://doi.org/10.1007/978–1-4842–4470-8_18.

Bora-Singhal, N., Mohankumar, D., Saha, B., Colin, C. M., Lee, J. Y., Martin, M. W., & Chellappan, S. (2020). Novel HDAC11 inhibitors suppress lung adenocarcinoma stem cell self-renewal and overcome drug resistance by suppressing Sox2. *Scientific Reports, 10*(1), 4722. https://doi.org/10.1038/s41598-020-61295-6.

Chander, B., & Gopalakrishnan, K. (2023). Data clustering using unsupervised machine learning. *Statistical modeling in machine learning*. Academic Press,179–204. https://doi.org/10.1016/B978-0-323-91776-6.00015-4.

Cherrington, M., Thabtah, F., Lu, J., & Xu, Q. (2019). Feature selection: filter methods performance challenges. In 2019 International Conference on Computer and Information Sciences (ICCIS) (pp. 1–4). IEEE. https://doi.org/10.1109/ICCISci.2019.8716478.

de Sousa, J. M. A. (2017). Processing of SMILES, InChI, and hashed fingerprints. *Tutorials in Chemoinformatics,* 75–81. https://doi.org/10.1002/9781119161110.ch4.

Falkenberg, K. J., & Johnstone, R. W. (2014). Histone deacetylases and their inhibitors in cancer, neurological diseases and immune disorders. *Nature Reviews. Drug Discovery, 13*(9), 673–691. https://doi.org/10.1038/nrd4360.

Fernández-de Gortari, E., García-Jacas, C. R., Martinez-Mayorga, K., & Medina-Franco, J. L. (2017). Database fingerprint (DFP): An approach to represent molecular databases. *Journal of Cheminformatics, 9*(1), 9. https://doi.org/10.1186/s13321-017-0195-1.

Gao, L., Cueto, M. A., Asselbergs, F., & Atadja, P. (2002). Cloning and functional characterization of HDAC11, a novel member of the human histone deacetylase family. *Journal of Biological Chemistry, 277*(28), 25748–25755. https://doi.org/10.1074/jbc.M111871200.

Khatun, S., Amin, S. A., Choudhury, D., Chowdhury, B., Jha, T., & Gayen, S. (2024a). Advances in structure-activity relationships of HDAC inhibitors as HIV latency-reversing agents. *Expert Opinion on Drug Discovery, 19*(3), 353–368. https://doi.org/10.1080/17460441.2024.2305730.

Khatun, S., Bhagat, R. P., Dutta, R., Datta, A., Jaiswal, A., Halder, S., & Gayen, S. (2024b). Unraveling HDAC11: Epigenetic orchestra in different diseases and structural insights for inhibitor design. *Biochemical Pharmacology, 225*, 116312. https://doi.org/10.1016/j.bcp.2024.116312.

Khatun, S., Dasgupta, I., Sen, S., Amin, S. A., Qureshi, I. A., Jha, T., & Gayen, S. (2025). Histone deacetylase 8 in focus: Decoding structural prerequisites for innovative epigenetic intervention beyond hydroxamates. *International Journal of Biological Macromolecules, 284*, 138119. https://doi.org/10.1016/j.ijbiomac.2024.138119.

Khatun, S., Islam, R., Amin, S. A., Bhattacharya, A., Dhaked, D. K., Jha, T., & Gayen, S. (2024c). Nitrogen-containing heterocycles as important scaffold for selective and potent HDAC8 inhibition: A step towards effective, non-toxic and selective HDAC8 inhibitor discovery. *Journal of Biomolecular Structure and Dynamics,* 1–19. https://doi.org/10.1080/07391102.2024.2442756.

LeCun, Y., Bengio, Y., & Hinton, G. (2015). Deep learning. *Nature, 521*(7553), 436–444. https://doi.org/10.1038/nature14539.

Liu, L. L., Lu, J., Lu, Y., Zheng, M. Y., Luo, X. M., Zhu, W. L., & Chen, K. X. (2014). Novel Bayesian classification models for predicting compounds blocking hERG potassium channels. *Acta Pharmacologica Sinica, 35*(8), 1093–1102. https://doi.org/10.1038/aps.2014.35.

Martin, T. M., Harten, P., Young, D. M., Muratov, E. N., Golbraikh, A., Zhu, H., & Tropsha, A. (2012). Does rational selection of training and test sets improve the outcome of QSAR modeling? *Journal of Chemical Information and Modeling, 52*(10), 2570–2578. https://doi.org/10.1021/ci300338w.

Mauri, A. (2020). alvaDesc: A tool to calculate and analyze molecular descriptors and fingerprints. *Ecotoxicological QSARs*. New York, NY: Springer US. 801–820. https://doi.org/10.1007/978-1-0716-0150-1_32.

Moriwaki, H., Tian, Y. S., Kawashita, N., & Takagi, T. (2018). Mordred: A molecular descriptor calculator. *Journal of Cheminformatics, 10*(1), 4. https://doi.org/10.1186/s13321-018-0258-y.

Muthukrishnan, R., & Rohini, R. (2016). LASSO: A feature selection technique in predictive modeling for machine learning. In 2016 IEEE international conference on advances in computer applications (ICACA) (pp. 18–20). Ieee. https://doi.org/10.1109/ICACA.2016.7887916.

Nugroho, A., & Suhartanto, H. (2020). Hyper-parameter tuning based on random search for densenet optimization. In 2020 7th International conference on information technology, computer, and electrical engineering (ICITACEE) (pp. 96–99). IEEE. https://doi.org/10.1109/ICITACEE50144.2020.9239164.

Ponce-Bobadilla, A. V., Schmitt, V., Maier, C. S., Mensing, S., & Stodtmann, S. (2024). Practical guide to SHAP analysis: Explaining supervised machine learning model predictions in drug development. *Clinical and Translational Science, 17*(11), e70056. https://doi.org/10.1111/cts.70056.

Rogers, D., & Hahn, M. (2010). Extended-connectivity fingerprints. *Journal of Chemical Information and Modeling, 50*(5), 742–754. https://doi.org/10.1021/ci100050t.

Rose, P. W., Bi, C., Bluhm, W. F., Christie, C. H., Dimitropoulos, D., Dutta, S., ... Bourne, P. E. (2012). The RCSB Protein Data Bank: New resources for research and education. *Nucleic Acids Research, 41*(D1), D475–D482. https://doi.org/10.1093/nar/gks1200.

Roy, K., Kar, S., & Das, R. E. (2015). Validation of QSAR models. *Understanding the Basics of QSAR for Applications in Pharmaceutical Sciences and Risk Assessment,* 231–289. https://doi.org/10.1016/B978-0-12-801505-6.00007-7.

Saravanan, R., & Sujatha, P. (2018). A state of art techniques on machine learning algorithms: a perspective of supervised learning approaches in data classification. In 2018 s international conference on intelligent computing and control systems (ICICCS) (pp. 945–949). IEEE. https://doi.org/10.1109/ICCONS.2018.8663155.

Sarker, I. H. (2021). Machine learning: Algorithms, real-world applications and research directions. *SN Computer Science, 2*(3), 160. https://doi.org/10.1007/s42979-021-00592-x.

Shetty, S. H., Shetty, S., Singh, C., & Rao, A. (2022). Supervised machine learning: Algorithms and applications. *Fundamentals and Methods of Machine and Deep Learning: Algorithms, Tools and Applications,* 1–16. https://doi.org/10.1002/9781119821908.ch1.

Sterling, T., & Irwin, J. J. (2015). ZINC 15–ligand discovery for everyone. *Journal of Chemical Information and Modeling, 55*(11), 2324–2337. https://doi.org/10.1021/acs.jcim.5b00559.

Studio, D. (2008). Discovery studio. *Accelrys [2.1], 420,* 1–9.

Sun, H. (2005). A naive Bayes classifier for prediction of multidrug resistance reversal activity on the basis of atom typing. *Journal of Medicinal Chemistry, 48*(12), 4031–4039. https://doi.org/10.1021/jm050180t.

Sushko, I., Novotarskyi, S., Körner, R., Pandey, A. K., Rupp, M., Teetz, W., & Tetko, I. V. (2011). Online chemical modeling environment (OCHEM): Web platform for data storage, model development and publishing of chemical information. *Journal of Computer-Aided Molecular Design, 25*(6), 533–554. https://doi.org/10.1007/s10822-011-9440-2.

Verma, J., Khedkar, V. M., & Coutinho, E. C. (2010). 3D-QSAR in drug design-a review. *Current Topics in Medicinal Chemistry, 10*(1), 95–115. https://doi.org/10.2174/156802610790232260.

Wassermann, A. M., & Bajorath, J. (2011). BindingDB and ChEMBL: Online compound databases for drug discovery. *Expert Opinion on Drug Discovery, 6*(7), 683–687. https://doi.org/10.1517/17460441.2011.579100.

Wen, Y., Zhang, X., Li, X., Tian, L., Shen, S., Ma, J., & Ai, F. (2022). Histone deacetylase (HDAC) 11 inhibits matrix metalloproteinase (MMP) 3 expression to suppress colorectal cancer metastasis. *Journal of Cancer, 13*(6), 1923. https://doi.org/10.7150/jca.66914.

Xu, W. S., Parmigiani, R. B., & Marks, P. A. (2007). Histone deacetylase inhibitors: Molecular mechanisms of action. *Oncogene, 26*(37), 5541–5552. https://doi.org/10.1038/sj.onc.1210620.

Yang, Y., Yuan, Y., Han, Z., & Liu, G. (2022). Interpretability analysis for thermal sensation machine learning models: An exploration based on the SHAP approach. *Indoor Air, 32*(2), e12984. https://doi.org/10.1111/ina.12984.

Yap, C. W. (2011). PaDEL-descriptor: An open source software to calculate molecular descriptors and fingerprints. *Journal of Computational Chemistry, 32*(7), 1466–1474. https://doi.org/10.1002/jcc.21707.

Yoshida, M., Kudo, N., Kosono, S., & Ito, A. (2017). Chemical and structural biology of protein lysine deacetylases. *Proceedings of the Japan Academy, Series B, 93*(5), 297–321. https://doi.org/10.2183/pjab.93.019.

Zeileis, A., Hothorn, T., & Hornik, K. (2008). Model-based recursive partitioning. *Journal of Computational and Graphical Statistics, 17*(2), 492–514. https://doi.org/10.1198/106186008×319331.

CHAPTER NINE

Mass spectrometry-based mapping of the ubiquitin chaperone code

Ayush Nigam[1], Prasun Kumar Bhunia[1], Trayambak Basak*, and Prasad Kasturi*

School of Biosciences and Bioengineering, Indian Institute of Technology Mandi, Kamand, Himachal Pradesh, India

*Corresponding authors. e-mail address: trayambak@iitmandi.ac.in; prasadkasturi@iitmandi.ac.in

Contents

Abstract

Maintenance of proteome integrity is essential for cellular homeostasis and organismal health. This integrity depends on proteostasis, a coordinated network of protein quality control systems that regulate protein folding, stabilization, and degradation. Molecular chaperones, together with proteolytic pathways such as the ubiquitin–proteasome system (UPS) and the autophagy–lysosomal pathway, prevent the accumulation of misfolded and aggregation-prone proteins. Perturbations, including genetic mutations, environmental stress, and aging challenge protein folding fidelity, leading to proteotoxic stress and contributing to the pathogenesis of neurodegenerative disorders. Among the chaperone machinery, the HSP70 and HSP90 families play central roles in maintaining protein conformational homeostasis and directing damaged or misfolded substrates toward refolding or degradation

[1] Equal first authors

Methods in Enzymology, Volume 734
ISSN 0076-6879, https://doi.org/10.1016/bs.mie.2026.05.047

pathways. Recent studies show that chaperone activity is dynamically regulated by diverse post-translational modifications (PTMs), including phosphorylation, acetylation, and ubiquitination, collectively termed the "chaperone code." These modifications modulate chaperone–client interactions, enzymatic activity, localization, and coordination with protein degradation systems. Mass spectrometry (MS)-based proteomics has emerged as a powerful approach for mapping ubiquitination sites and quantifying ubiquitin signaling dynamics. This chapter outlines experimental and computational strategies for MS-based analysis of the ubiquitin chaperone code, including di-glycine peptide enrichment, site identification, quantitative analysis, and validation.

1. Introduction

Maintenance of proteome integrity is fundamental to cellular homeostasis and organismal health. Proteome integrity refers to the preservation of a functionally balanced and correctly folded proteome, a state termed as proteostasis. To maintain proper proteostasis, cells maintain a coordinated protein quality control systems comprising molecular chaperones, and protein degradation systems that together prevent the accumulation of toxic, misfolded protein species (Balchin et al., 2016; Hartl et al., 2011). Perturbations such as genetic mutations, environmental stress, and aging challenge protein folding fidelity, resulting in protein misfolding and aggregation. Failure to refold or efficiently clear such aberrant proteins leads to loss of function, proteotoxic stress, and is a hallmark of aging and numerous age-associated neurodegenerative disorders (Hipp et al., 2014; López-Otín et al., 2023).

Molecular chaperones assist in the folding of nascent polypeptides, refolding of stress-denatured proteins, and stabilization of metastable conformations. When folding attempts fail, chaperones direct misfolded substrates toward degradation pathways. Two major proteolytic systems operate in eukaryotic cells: the ubiquitin–proteasome system (UPS) and the autophagy–lysosomal pathway (Pohl & Dikic, 2019). Selective degradation is typically mediated by ubiquitination, a reversible post-translational modification (PTM) in which ubiquitin is covalently attached to substrate lysine residues. Importantly, ubiquitination not only labels proteins for proteasomal degradation but also influences their trafficking toward autophagy, sequestration into quality control compartments, or participation in phase-separated assemblies (Baker & Bernardini, 2021; Popovic et al., 2014; Rape, 2018; Sheng et al., 2024).

Among the molecular chaperones, the HSP70 and HSP90 families form the core chaperone machinery. These ATP-dependent chaperones regulate client protein conformational states without becoming part of the final folded complex (Hartl et al., 2011). Their function is fine-tuned by an extensive network of co-chaperones and regulatory factors. Disruption of this chaperone machinery has been strongly linked to proteotoxicity and disease. Indeed, altered chaperone activity contributes to the pathogenesis of major neurodegenerative disorders, including Alzheimer's disease, Parkinson's disease, Huntington's disease, and amyotrophic lateral sclerosis (Duncan et al., 2015; Soto & Pritzkow, 2018). In cellular and animal models, the HSP70/HSP90 systems modulate misfolded protein toxicity, inclusion formation, and cell survival (Lackie et al., 2017; Prodromou et al., 2023).

Emerging evidence indicates that chaperone activity is dynamically regulated by a complex array of PTMs, including phosphorylation, acetylation, and ubiquitination. Collectively, these modifications constitute the "chaperone code," a regulatory layer that controls chaperone–client interactions, subcellular localization, ATPase activity, and cooperation with degradation pathways (Backe et al., 2025; Cloutier & Coulombe, 2013; Nitika & Truman, 2020). Notably, HSP70 and HSP90 are extensively modified at multiple lysine, serine, threonine, and tyrosine residues. Ubiquitination of chaperones themselves can regulate their stability, modulate their interaction networks, or alter their ability to triage client proteins between folding, degradation, sequestration, and phase separation (Baker & Bernardini, 2021; Panda et al., 2023; Shiber et al., 2013).

Functional interplay between chaperones and the UPS is particularly critical for maintaining proteostasis. HSP70, together with its co-chaperones and E3 ubiquitin ligases such as CHIP, coordinates the ubiquitination and proteasomal targeting of misfolded substrates (Kevei et al., 2017; Shiber and Ravid, 2014). Chaperones also participate in selective autophagy pathways, including chaperone-mediated autophagy and aggrephagy (Bauer et al., 2023). Under conditions of limited chaperone capacity, ubiquitin conjugation can promote the sequestration of misfolded proteins into quality control foci, thereby limiting proteotoxic damage (Shiber et al., 2013). Thus, ubiquitination acts not only as a degradation signal but as a central determinant of protein fate within the chaperone-guided triage system.

Given the complexity and site-specific nature of ubiquitin modifications, comprehensive mapping of the ubiquitin chaperone code requires sensitive and high-resolution analytical approaches. Mass spectrometry (MS)-based proteomics has emerged as the method of choice for the identification and quantification of ubiquitination sites at proteome-wide scale (Sahu et al., 2023). In bottom-up proteomics workflows, ubiquitinated lysines are typically detected via the di-glycine (diGly) remnant left on lysine residues after tryptic digestion. Both global analyses without enrichment and antibody-based enrichment of diGly-modified peptides are employed, depending on experimental goals and sample complexity (Fulzele & Bennett, 2018). Advances in instrumentation, fragmentation methods, and quantitative strategies such as label-free quantification, SILAC, and isobaric labeling enable precise measurement of peptide abundance and calculation of ubiquitination occupancy (Prus et al., 2024).

However, accurate mapping of ubiquitination sites requires rigorous data processing and validation. Database searching with appropriate variable modifications, control of false discovery rates at peptide and site levels, localization probability scoring, and careful statistical analysis are essential to ensure confident site assignment. Quantitative assessment further demands normalization strategies, correction for protein abundance changes, and appropriate statistical modeling to identify biologically meaningful alterations in ubiquitination dynamics. In this chapter, we describe methodological approaches for mass spectrometry–based mapping of the ubiquitin chaperone code, with emphasis on experimental design, sample preparation, acquisition of MS data, enrichment strategies for diGly-modified peptides, computational data analysis, site validation, quantification of peptide abundance, calculation of ubiquitination occupancy, and statistical evaluation. By integrating biochemical and proteomic strategies, these methods provide a powerful framework to decipher how ubiquitination shapes chaperone function in health, aging, and disease.

2. Chaperone code

The concept of the chaperone code refers to the collective set of post-translational modifications (PTMs) that occur on molecular chaperones and regulate their activity in a coordinated and context-dependent manner. Analogous to the "histone code," the chaperone code proposes that specific PTM patterns on chaperones act as a molecular language that dictates their conformational states, interaction networks, and client specificity (Truman et al., 2012). Rather than functioning as simple on/off switches, these modifications act combinatorially to fine-tune chaperone behavior in response to stress conditions, metabolic changes, and disease states.

Over the past decade, rapid advances in mass spectrometry–based proteomics have uncovered an extensive array of PTMs on chaperones, including phosphorylation, acetylation, methylation, and ubiquitination. Although hundreds of modification sites have been identified on Hsp70, Hsp90, and their co-chaperones, the functional consequences of many of these PTMs remain poorly characterized (Omkar et al., 2023). Increasing evidence indicates that PTMs do not merely adjust canonical folding cycles but can redirect chaperones toward distinct functional outputs, including degradation, sequestration into inclusions, modulation of signaling complexes, and participation in biomolecular condensates (Fig. 1).

2.1 Molecular chaperones

Molecular chaperones are responsible for maintaining proteostasis by assisting in folding, refolding, assembly, disassembly, and degradation of proteins. Chaperones are broadly categorized based on molecular weight and mechanism of action into several conserved families, including Hsp60, Hsp70, Hsp90, Hsp100, and small heat shock proteins. Among these, Hsp70 and Hsp90 represent the primary ATP-dependent chaperone systems in eukaryotic cells and function as central hubs coordinating protein folding and signaling pathways (Genest et al., 2019; Morán Luengo et al., 2019).

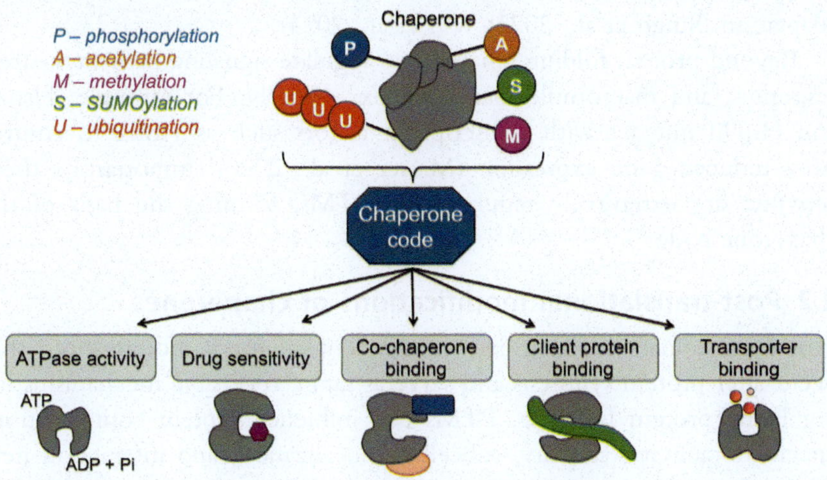

Fig. 1 The chaperone code controls chaperone function via combinatorial PTMs that regulate ATPase activity, co-chaperone/client binding, transport, and drug sensitivity.

The Hsp70 family plays a pivotal role in recognizing short hydrophobic segments exposed in unfolded or partially folded polypeptides. Through ATP-regulated cycles, Hsp70 binds and releases substrates in a process tightly controlled by co-chaperones. J-domain proteins (Hsp40s) stimulate ATP hydrolysis and facilitate substrate delivery, whereas nucleotide exchange factors (NEFs) promote ADP release and substrate dissociation. In humans, multiple Hsp70 isoforms exist, including constitutive and stress-inducible forms, enabling adaptation to diverse cellular contexts. There are 13 types of Hsp70 proteins (HSPA1A-HSPA14), which are categorized by their cellular localization and whether they are constitutively expressed or stress-inducible. HSPA1A is a stress-inducible member of the Hsp70 family, whereas HSPA8 (Hsc70) is constitutively expressed under normal cellular conditions. Beyond assisting folding, Hsp70 participates in disaggregation, cooperates with degradation pathways, and acts as a triage factor that determines whether a client protein is refolded or targeted for degradation (Wentink et al., 2026).

Hsp90 primarily functions downstream of Hsp70 and specializes in the maturation and activation of signaling proteins, including kinases, steroid hormone receptors, and transcription factors (Schopf et al., 2017). Its ATP-driven conformational cycle is regulated by a large set of co-chaperones that modulate client selection, complex stability, and ATPase activity. Through these interactions, Hsp90 influences numerous signaling cascades and plays critical roles in cancer biology, neurodegeneration, and stress adaptation (Singh et al., 2025; Wei et al., 2024).

Beyond protein folding, chaperones regulate signaling pathways, stress responses, and macromolecular complex assembly. For example, Hsp70 and Hsp90 interact with transcription factors such as HSF1 to control stress-induced gene expression (Masser et al., 2020). Importantly, these activities are extensively regulated by PTMs, forming the basis of the chaperone code.

2.2 Post-translational modifications of chaperones

Post-translational modifications are covalent chemical modifications that occur after protein synthesis and serve as rapid, reversible mechanisms for regulating protein function. PTMs can influence protein conformation, enzymatic activity, stability, subcellular localization, and interaction networks. In molecular chaperones, PTMs provide an additional regulatory layer that complements co-chaperone-mediated control and allows rapid adaptation to environmental and cellular changes (Chu et al., 2025).

Common PTMs on chaperones include phosphorylation, acetylation, methylation, and ubiquitination.

Phosphorylation is one of the most extensively studied PTMs in chaperones. Site-specific phosphorylation events can alter ATPase activity, conformational transitions, and co-chaperone binding. In Hsp70, phosphorylation within nucleotide-binding or substrate-binding domains can shift the balance between folding and degradation pathways, particularly under proteotoxic stress Stress-induced phosphorylation enhances interactions with NEFs and J-domain proteins, enabling rapid adaptation to fluctuating proteostasis demands (Velasco et al., 2019). In Hsp90, phosphorylation modulates its conformational cycle and affects recruitment of specific co-chaperones, thereby influencing client maturation (Mollapour et al., 2011).

Acetylation of chaperones on lysine residues modulates chaperone activity, stability, and interaction with client proteins (Liu et al., 2021). Acetylation of Hsp70 has been implicated in modulating its interaction with E3 ubiquitin ligases and promoting degradation of specific clients (Seo et al., 2016). In Hsp90, acetylation within the middle and C-terminal domains reduces ATPase activity and client binding, thereby altering functional output (Scroggins et al., 2007).

Ubiquitination represents a more complex and versatile PTM, involving the covalent attachment of ubiquitin to lysine residues. Chaperones can themselves be ubiquitinated, influencing their turnover, stability, and interaction networks (Kundrat & Regan, 2010a). Moreover, ubiquitination of client proteins provides a direct link between folding attempts and degradation pathways. Importantly, crosstalk between phosphorylation, acetylation, and ubiquitination can determine whether a substrate is refolded, sequestered, or degraded (Muller et al., 2013), highlighting the combinatorial nature of the chaperone code.

Importantly, PTMs often function combinatorially; crosstalk between PTMs can determine whether a substrate is refolded, sequestered, or degraded (Griffith & Holmes, 2019). For example, phosphorylation and acetylation can cooperate or antagonize each other to direct Hsp70-bound clients toward folding or ubiquitin-mediated degradation (Seo et al., 2016).

2.3 Ubiquitination chaperone code

Ubiquitination is a highly versatile post-translational modification characterized by the covalent attachment of the 8-kDa ubiquitin protein to substrate lysine residues. Unlike small chemical modifications, ubiquitination can generate structurally diverse signals through mono-ubiquitination or polyubiquitin chains of varying linkage types. The topology, linkage (e.g., K48, K63, K11, K6), and chain length encode distinct functional outcomes (Liao et al., 2024; Rape, 2018). (Fig. 2).

In the context of proteostasis, K48-linked polyubiquitin chains typically signal for proteasomal degradation, whereas K63-linked and other atypical linkages often mediate non-proteolytic roles such as signaling, trafficking, or regulation of macromolecular assemblies (Komander & Rape, 2012). This diversity allows ubiquitination to function not only as a degradation signal but also as a regulatory modification shaping protein interactions and cellular architecture. The E3 ubiquitin ligase CHIP (C-terminus of Hsc70-interacting protein) provides a direct molecular bridge between chaperones and the UPS. When Hsp70-bound clients fail to achieve proper folding, CHIP ubiquitinates these substrates, committing them to proteasomal degradation (Kundrat & Regan, 2010b). Notably, Hsp70 and Hsp90 themselves are ubiquitinated at multiple lysine residues. CHIP-mediated polyubiquitination includes K6-, K11-, K48-, and K63-linked chains, suggesting both degradative and regulatory roles (Tracz & Bialek, 2021). Such modifications influence chaperone stability, complex assembly, and functional transitions. Ubiquitination also plays important roles in stress adaptation and phase separation (Baker & Bernardini, 2021; Maxwell et al., 2021). During proteotoxic stress, global ubiquitin levels increase, facilitating the sequestration of misfolded proteins into quality-control compartments or inclusions (Lee & Goldberg, 2022; Shiber et al., 2013). These structures are often associated with liquid–liquid phase separation (LLPS). Chaperones and co-chaperones are recruited to these biomolecular condensates, where ubiquitin signaling regulates assembly and disassembly dynamics (Dao et al., 2022; Hou & Tang, 2023; Krause et al., 2022).

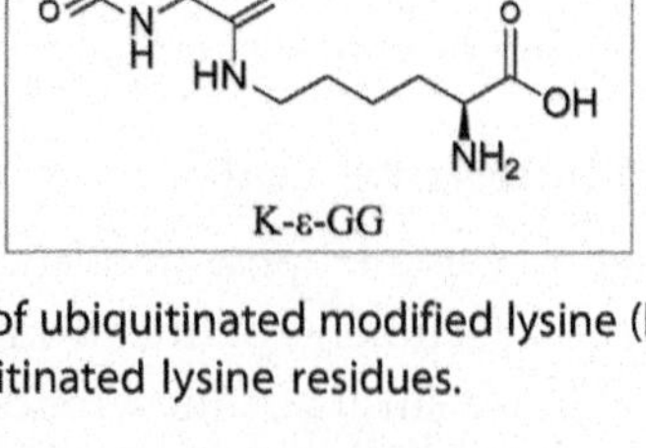

Fig. 2 Chemical structure of ubiquitinated modified lysine (K), trypsin digestion leaves a diGly remnant on ubiquitinated lysine residues.

2.4 Functional roles of the ubiquitin chaperone code

A principal function of the ubiquitin chaperone code is the regulation of the balance between protein folding and degradation. Hsp70 and Hsp90 operate sequentially during client maturation, with Hsp90 acting downstream of Hsp70 in many cases. Co-chaperones coordinate substrate transfer between these systems. Ubiquitination influences these transitions by modulating complex stability, ATPase activity, and client retention time. Thus, ubiquitin modifications dynamically reprogram chaperone complexes to favor either folding or degradation pathways (Kevei et al., 2017). Another critical role of the ubiquitin chaperone code lies in regulating conformational states and interaction networks (Chu et al., 2025). In summary, the ubiquitin chaperone code represents a sophisticated regulatory framework that integrates folding machinery with degradation systems and stress responses. Decoding this molecular language is essential for understanding how cells maintain proteome integrity and how its dysregulation contributes to disease.

The ability to quantify site-specific ubiquitination is critical for understanding how ubiquitin signaling regulates various cellular functions including chaperone and proteostasis. However, this remains technically challenging due to the low abundance, transient nature, and heterogeneity of ubiquitin modifications. This chapter is therefore motivated by the need for a clear, standardized, and rigorous methodological framework to accurately detect and quantify ubiquitination at the peptide level. In particular, advances in MS-based proteomics such as diGly remnant enrichment, high-resolution data acquisition, and refined database search strategies have enabled precise mapping of ubiquitination sites. This chapter systematically addresses these challenges by outlining best practices for enriching ubiquitinated peptides, ensuring confident site localization, and performing reliable quantification, including approaches to estimate ubiquitination occupancy and assess data quality.

3. Quantifying site-specific ubiquitination by mass spectrometry and statistical analysis

The strategy used during the sample preparation step directly determines the confidence and quantitative robustness of site-specific ubiquitination analysis. (Fig. 3).

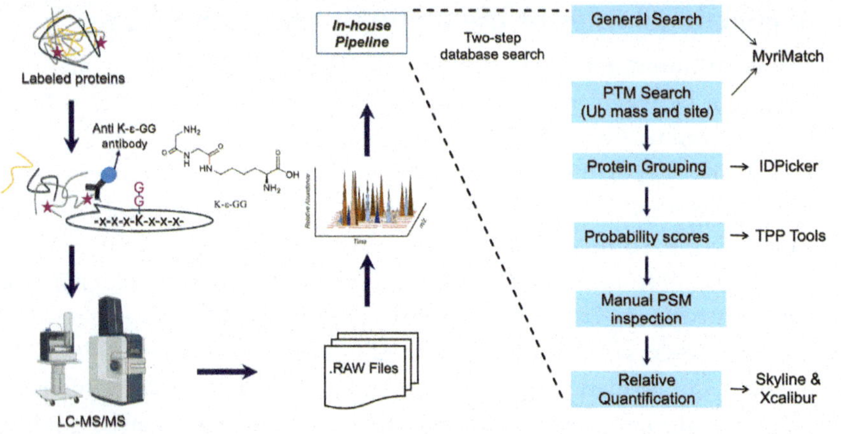

Fig. 3 Pipeline for identification of ubiquitination sites by diGly remnant mass spectrometry.

3.1 Sample preparation

1. Under denaturing conditions, protein samples should be prepared to ensure enough solubilization and to prevent post-lysis deubiquitylation.
2. Followed by the reduction and alkylation of cysteine (C) residues and then subjected to trypsin digestion.
3. Trypsin cleaves the C-terminal to lysine (K) and Arginine (R) residues; however, the ubiquitinated modified lysine (K) was unaffected and retained a characteristic di-glycine (K-ε-GG) remnant (Udeshi et al., 2013).
4. Due to this modification, there is a defined mass shift (+114.0429 Da) which enables site-specific identification at MS/MS level.
5. The peptides produced are purified by solid phase extraction using C18 chemistry and re-suspended in LC-MS compatible solvent.

3.1.1 Without enrichment

6. In this case, cleaned peptide mixtures will be directly injected into LC-MS/MS.

3.1.2 With antibody-based enrichment

7. While here the diGly-modified peptides were enriched using antibodies specific to the K-ε-GG motif.
8. Under acidic conditions, enriched diGly containing peptides were eluted and then desalted to remove contaminants derived from antibodies.

Notes:
- The ubiquitinylated lysine modifications are characterized by the presence of a K-ε-GG tag after tryptic digestion.
- Without enrichment approach, the ubiquitination sites detection is biased towards higher abundance events because of the low stoichiometry of this modification.
- This enrichment approach substantially enhances the sensitivity of ubiquitination site identification (Bezstarosti et al., 2020).

3.2 MS data acquisition

1. After this, peptide samples were analyzed by nano LC-MS/MS using high-resolution Orbitrap instrumentation in Data Dependent Acquisition (DDA) mode.
2. Full MS scans were acquired at high mass accuracy, and sequentially, the top 20 most intense precursor ions were selected for Higher-energy Collisional Dissociation (HCD) (Kalli et al., 2013).
3. All acquired spectra were stored in instrument specific .raw files.

Notes:
- HCD fragmentation generated fragment ions that confidently enable the localization of diGly-modified lysine residues.

3.3 MS data processing and analysis

1. After LC-MS/MS acquisition, MS data are generated in vendor-specific raw formats, which includes, .raw files (Thermo), .d directories (Bruker, which often contains multiple sub-files), .wiff and .wiff.scan (Sciex), .d directories (Agilent, with profile data) and .RAW directories (Waters). These files serve as the primary input for downstream database searches and quantitative analyses.
2. If the MS data is in .RAW format, this format is not directly interchangeable, so the files are needed to be converted them into a standard, open format which can be easily read by the downstream analysis tools such as mzML (XML-based, vendor neutral, retains complete spectral information), mzXML (older but still in use in some of the pipelines). For this purpose, MSConvert (GUI only works in Windows system), a ProteoWizard's tool can be used. This gives us a spectrum file for each run (Chambers et al., 2012).
3. The proteomic database search was performed on two levels using MyriMatch search engine (Tabb et al., 2007), (user can explore other search engines such as Mascot, SEQUEST, Fragpipe, etc.): The first

level search (General search) followed by the second-level search (PTM-specific search).

4. After the search, a list of peptide-spectrum matches (PSMs), along with the information about which lysine is modified by the diGly (hence ubiquitinated) for each peptide will be retrieved.

Notes:

- Database search engine basically mimics the experiment: the software digests every protein in the database in silico, predicting the masses of the final peptides and their fragments, and gives scores to each spectrum against all these predictions.
- In actual use, each observed fragment mass is compared with the predicted fragments of candidate peptides. High scores are given for good matches (enough matching fragment ions within the given tolerance).

3.3.1 First (general) database search

For this level search, some specific files are in the same folder for convenience.

I. .RAW/.wiff/.d files (based on different commercial mass-spectrometry formats)
II. .config file
III. .fasta file

1. Firstly, download the ProteoWizard Suite (Kessner et al., 2008) to get all the tools required for the search and the commercial format of the raw data should be converted to standard mzML format using the MSConvert standard pipelines.
2. Then, a protein FASTA database can be downloaded from UniProt to run this database search.
3. At last, a configuration file (.cfg file format) mentioning all the technical parameters such as tolerance for precursor and fragment ions was set at $\pm$ 10 ppm, $\pm$ 20 ppm respectively, type of cleavage (e.g. trypsin cleavage), fixed modification such as carbamidomethyl on cysteine (+57.0236), dynamic modifications such as methionine oxidation (+15.9949), allow a diGly (+114.0429) modification on lysine site (which corresponds to the ubiquitin remnant), maximum dynamic modifications per peptide were set to 4 needed to be created (Panda et al., 2023).
4. The output of this search will be in.pepXML format, from which the subset fasta file will be extracted using IDPicker with $< 1\%$ FDR (for PSM, Peptide and proteins IDs).

Notes:

- A Ubiquitin-modified Lysine often resists trypsin cleavage, and thus at least up to 2 missed cleavages are allowed.

3.3.2 Second (PTM-specific) database search

Like the previous search, some specific files, again in the same folder (but not same as the previous search) are needed for convenience,

I. .mzML files
II. .config file
III. .fasta (subset) file

1. In this case, the.fasta (subset) file is the subset of proteins identified in the previous search and this search will be performed against this subset protein database for further in-depth ubiquitination site identification.
2. In this search, the technical parameters were, same precursor and fragment tolerances were set, a maximum of 6 modifications per peptide and up to 4 missed cleavages were allowed.

3.3.3 Ensuring accuracy of Identified sites

1. After obtaining the final.pepXML output files, will be imported them to IDPicker (Ma et al., 2009) for parsimonious protein grouping and to control FDR.
2. The output from the database search consists of many candidate peptides, including false positives. A false discovery rate (FDR) filter (which often uses a target-decoy strategy) needs to be applied to keep only high confidence PSMs (generally $< 1\%$ at peptide, protein and PSMs levels) using IDPicker.
3. Now, manually inspect and cross-verify the spectra for key peptides to see whether the major MS/MS peaks line up with the predicted fragment series around the modified lysine.

Notes:

- The scoring algorithms are good but not perfect, hence there is a utmost need to verify that the spectrum has a clear pattern (b and y ions) supporting the sequence and the location of the diglycine tag.

3.4 Measurement of peptide abundance (quantification)

1. Once the validation of the ubiquitinated peptide is ensured by high-quality PSM, and its corresponding unmodified peptide, measurement of its abundance is performed from the MS1 signal.

2. If the proteomic workflow is label-free, it is recommended to integrate the extracted ion chromatogram (XIC) for that specific peptide's precursor m/z value (across its elution time, take the area under the peak).
3. TPP tools can be used to calculate probability scores which is required during spectral library generation and then Skyline can be used to perform the quantification.
4. If the proteomic workflow is label-based (e.g., SILAC), then it is required to calculate the heavy/light ratio from the MS1 scan intensities.
5. Both approaches will yield numerical intensities for each peptide in each sample.
6. Sometimes, overlapping scans for a peptide may be obtained; these peptides are typically co-eluting peptides that cannot be considered for quantitation.
7. In any sample, these intensities tell us about how much that peptide (modified or unmodified) is present.

3.4.1 Calculation of ubiquitination occupancy

1. To calculate the site-specific occupancy of a ubiquitination-modified lysine, a comparison of the peptide intensities of the modified peptide vs. the unmodified peptide forms across replicates. Let's say, L_mod is the intensity value of the diGly-modified peptide and L_unmod is the intensity value of the corresponding unmodified peptide (or sum of the intensities of multiple unmodified peptides) covering that site. One of the simplest ways to calculate occupancy is [100 % x (L_mod)/(L_mod + L_unmod)]. Thus, the level of lysine ubiquitinylation can be estimated.

Notes:

- One point to note is to consider only those peptides that are modified and their respective unmodified peptide versions, as it's not possible to calculate occupancy without both versions of the peptide being detected. As there are only two possibilities for lysine, either ubiquitinated or not, the percentage sum should be 100 %.
- Please note that both the relative abundance changes of the modified peptide and the occupancy change should be reported. Let's say, if a site's occupancy goes from 5 % to 30 % from healthy condition samples to diseased condition samples, this sixfold change is more meaningful information than the intensity values.

3.4.2 Statistical analysis

If multiple biological replicates are used, standard statistical (nonparametric and/or parametric) should be conducted within groups using site-specific occupancy values for modified peptides.

4. Conclusions

This chapter provides a comprehensive workflow for the identification and quantification of site-specific ubiquitination using mass spectrometry (MS) coupled with rigorous statistical analysis. It begins with introduction to proteostasis, focusing on molecular chaperones and their PTMs. Next it highlights importance of chaperone code with focus on ubiquitination chaperone code. Then it presents an integrated workflow for quantifying site-specific ubiquitination using mass spectrometry and statistical analysis, covering key steps from sample preparation to data interpretation. It outlines strategies with and without enrichment of ubiquitinated peptides, including antibody-based isolation of diGly-modified peptides to improve detection sensitivity, followed by considerations for MS data acquisition and processing. A two-stage database search approach combining general and PTM-specific searches is described to enhance identification accuracy, along with methods to validate site localization. The chapter further details approach for peptide quantification and enabling calculation of ubiquitination occupancy at specific residues. Together, this workflow provides a practical and reliable guide for studying ubiquitination dynamics at site-specific resolution.

References

Backe, S. J., Heritz, J. A., & Mollapour, M. (2025). Hsp70 and Hsp90 post-translational modifications and translating the chaperone code. *Cell Stress & Chaperones, 30*(6), 100118.

Baker, H. A., & Bernardini, J. P. (2021). It's not just a phase; ubiquitination in cytosolic protein quality control. *Biochemical Society Transactions, 49*(1), 365–377.

Balchin, D., Hayer-Hartl, M., & Hartl, F. U. (2016). In vivo aspects of protein folding and quality control. *Science (New York, N. Y.), 353*(6294), aac4354.

Bauer, B., Martens, S., & Ferrari, L. (2023). Aggrephagy at a glance. *Journal of Cell Science, 136*(10), jcs260888.

Bezstarosti, K., van der Wal, L., & Demmers, J. A. A. (2020). Detection of protein ubiquitination sites by peptide enrichment and mass spectrometry. *Journal of Visualized Experiments: JoVE*(157), https://doi.org/10.3791/59079.

Chambers, M. C., Maclean, B., Burke, R., Amodei, D., Ruderman, D. L., Neumann, S., ... Mallick, P. (2012). A cross-platform toolkit for mass spectrometry and proteomics. *Nature Biotechnology, 30*(10), 918–920.

Chu, F., Sharma, S., Ginsberg, S. D., & Chiosis, G. (2025). PTMs as molecular encoders: Reprogramming chaperones into epichaperomes for network control in disease. *Trends in Biochemical Sciences, 50*(10), 892–905.

Cloutier, P., & Coulombe, B. (2013). Regulation of molecular chaperones through post-translational modifications: decrypting the chaperone code. *Biochimica et Biophysica Acta, 1829*(5), 443–454.

Dao, T. P., Yang, Y., Presti, M. F., Cosgrove, M. S., Hopkins, J. B., Ma, W., ... Castañeda, C. A. (2022). Mechanistic insights into enhancement or inhibition of phase separation by different polyubiquitin chains. *EMBO Reports, 23*(8), e55056.

Duncan, E. J., Cheetham, M. E., Chapple, J. P., & van der Spuy, J. (2015). The role of HSP70 and its co-chaperones in protein misfolding, aggregation and disease. *Sub-cellular Biochemistry, 78*, 243–273.

Fulzele, A., & Bennett, E. J. (2018). Ubiquitin diGLY Proteomics as an Approach to Identify and Quantify the Ubiquitin-Modified Proteome. *Methods in Molecular Biology (Clifton, N. J.), 1844*, 363–384.

Genest, O., Wickner, S., & Doyle, S. M. (2019). Hsp90 and Hsp70 chaperones: Collaborators in protein remodeling. *Journal of Biological Chemistry, 294*(6), 2109–2120.

Griffith, A. A., & Holmes, W. (2019). Fine tuning: Effects of post-translational modification on Hsp70 chaperones. *International Journal of Molecular Sciences, 20*(17), 4207.

Hartl, F. U., Bracher, A., & Hayer-Hartl, M. (2011). Molecular chaperones in protein folding and proteostasis. *Nature, 475*(7356), 324–332.

Hipp, M. S., Park, S. H., & Hartl, F. U. (2014). Proteostasis impairment in protein-misfolding and -aggregation diseases. *Trends in Cell Biology, 24*(9), 506–514.

Hou, X. N., & Tang, C. (2023). The pros and cons of ubiquitination on the formation of protein condensates. *Acta Biochimica et Biophysica Sinica, 55*(7), 1084–1098.

Kalli, A., Smith, G. T., Sweredoski, M. J., & Hess, S. (2013). Evaluation and optimization of mass spectrometric settings during data-dependent acquisition mode: Focus on LTQ-Orbitrap mass analyzers. *Journal of Proteome Research, 12*(7), 3071–3086.

Kessner, D., Chambers, M., Burke, R., Agus, D., & Mallick, P. (2008). ProteoWizard: Open source software for rapid proteomics tools development. *Bioinformatics (Oxford, England), 24*(21), 2534–2536.

Kevei, É., Pokrzywa, W., & Hoppe, T. (2017). Repair or destruction-an intimate liaison between ubiquitin ligases and molecular chaperones in proteostasis. *FEBS Letters, 591*(17), 2616–2635.

Komander, D., & Rape, M. (2012). The ubiquitin code. *Annual Review of Biochemistry, 81*, 203–229.

Krause, L. J., Herrera, M. G., & Winklhofer, K. F. (2022). The role of ubiquitin in regulating stress granule dynamics. *Frontiers in Physiology, 13*, 910759.

Kundrat, L., & Regan, L. (2010a). Identification of residues on Hsp70 and Hsp90 ubiquitinated by the cochaperone CHIP. *Journal of Molecular Biology, 395*(3), 587–594.

Kundrat, L., & Regan, L. (2010b). Balance between folding and degradation for Hsp90-dependent client proteins: A key role for CHIP. *Biochemistry, 49*(35), 7428–7438.

Lackie, R. E., Maciejewski, A., Ostapchenko, V. G., Marques-Lopes, J., Choy, W. Y., Duennwald, M. L., ... Prado, M. A. M. (2017). The Hsp70/Hsp90 chaperone machinery in neurodegenerative diseases. *Frontiers in Neuroscience, 11*, 254.

Lee, D., & Goldberg, A. L. (2022). 26S proteasomes become stably activated upon heat shock when ubiquitination and protein degradation increase. *Proceedings of the National Academy of Sciences of the United States of America, 119*(25), e2122482119.

Liao, Y., Zhang, W., Liu, Y., Zhu, C., & Zou, Z. (2024). The role of ubiquitination in health and disease. *MedComm, 5*(10), e736.

Liu, P., Xiao, J., Wang, Y., Song, X., Huang, L., Ren, Z., ... Wang, Y. (2021). Posttranslational modification and beyond: interplay between histone deacetylase 6 and heat-shock protein 90. *Molecular Medicine, 27*(1), 110.

López-Otín, C., Blasco, M. A., Partridge, L., Serrano, M., & Kroemer, G. (2023). Hallmarks of aging: An expanding universe. *Cell, 186*(2), 243–278.

Ma, Z. Q., Dasari, S., Chambers, M. C., Litton, M. D., Sobecki, S. M., Zimmerman, L. J., ... Tabb, D. L. (2009). IDPicker 2.0: Improved protein assembly with high discrimination peptide identification filtering. *Journal of Proteome Research, 8*(8), 3872–3881.

Masser, A. E., Ciccarelli, M., & Andréasson, C. (2020). Hsf1 on a leash - controlling the heat shock response by chaperone titration. *Experimental Cell Research, 396*(1), 112246.

Maxwell, B. A., Gwon, Y., Mishra, A., Peng, J., Nakamura, H., Zhang, K., ... Taylor, J. P. (2021). Ubiquitination is essential for recovery of cellular activities after heat shock. *Science (New York, N. Y.), 372*(6549), eabc3593.

Mollapour, M., Tsutsumi, S., Donnelly, A. C., Beebe, K., Tokita, M. J., Lee, M. J., ... Neckers, L. (2011). Swe1-dependent phosphorylation of Hsp90 regulates distinct facets of chaperone function. *Molecular Cell, 42*(5), 555–566.

Morán Luengo, T., Mayer, M. P., & Rüdiger, S. G. D. (2019). The Hsp70-Hsp90 chaperone cascade in protein folding. *Trends in Cell Biology, 29*(2), 164–177.

Muller, P., Ruckova, E., Halada, P., Coates, P. J., Hrstka, R., Lane, D. P., & Vojtesek, B. (2013). C-terminal phosphorylation of Hsp70 and Hsp90 regulates alternate binding to co-chaperones CHIP and HOP to determine cellular protein folding/degradation balances. *Oncogene, 32*(25), 3101–3110.

Nitika, & Truman, A. W. (2020). The expanding universe of the chaperone code: Interplay of post-translational modifications of Hsp70 and Hsp90. *Trends in Biochemical Sciences, 45*(11), 943–960.

Omkar, S., Rysbayeva, A., & Truman, A. W. (2023). Understanding chaperone specificity: Evidence for a 'client code'. *Trends in Biochemical Sciences, 48*(8), 662–664.

Panda, P., Sarohi, V., Basak, T., & Kasturi, P. (2023). Elucidation of site-specific ubiquitination on chaperones in response to mutant huntingtin. *Cellular and Molecular Neurobiology, 44*(1), 3.

Pohl, C., & Dikic, I. (2019). Cellular quality control by the ubiquitin-proteasome system and autophagy. *Science (New York, N. Y.), 366*(6467), 818–822.

Popovic, D., Vucic, D., & Dikic, I. (2014). Ubiquitination in disease pathogenesis and treatment. *Nature Medicine, 20*(11), 1242–1253.

Prodromou, C., Aran-Guiu, X., Oberoi, J., Perna, L., Chapple, J. P., & van der Spuy, J. (2023). HSP70-HSP90 chaperone networking in protein-misfolding disease. *Sub-Cellular Biochemistry, 101*, 389–425.

Prus, G., Satpathy, S., Weinert, B. T., Narita, T., & Choudhary, C. (2024). Global, site-resolved analysis of ubiquitylation occupancy and turnover rate reveals systems properties. *Cell, 187*(11), 2875–2892.e21.

Rape, M. (2018). Ubiquitylation at the crossroads of development and disease. *Nature Reviews. Molecular Cell Biology, 19*(1), 59–70.

Sahu, I., Zhu, H., Buhrlage, S. J., & Marto, J. A. (2023). Proteomic approaches to study ubiquitinomics. *Biochimica et Biophysica Acta. Gene Regulatory Mechanisms, 1866*(2), 194940.

Schopf, F. H., Biebl, M. M., & Buchner, J. (2017). The HSP90 chaperone machinery. *Nature Reviews. Molecular Cell Biology, 18*(6), 345–360.

Scroggins, B. T., Robzyk, K., Wang, D., Marcu, M. G., Tsutsumi, S., Beebe, K., ... Neckers, L. (2007). An acetylation site in the middle domain of Hsp90 regulates chaperone function. *Molecular Cell, 25*(1), 151–159.

Seo, J. H., Park, J. H., Lee, E. J., Vo, T. T., Choi, H., Kim, J. Y., ... Kim, K. W. (2016). ARD1-mediated Hsp70 acetylation balances stress-induced protein refolding and degradation. *Nature Communications, 7*, 12882.

Sheng, X., Xia, Z., Yang, H., & Hu, R. (2024). The ubiquitin codes in cellular stress responses. *Protein & Cell, 15*(3), 157–190.

Shiber, A., Breuer, W., Brandeis, M., & Ravid, T. (2013). Ubiquitin conjugation triggers misfolded protein sequestration into quality control foci when Hsp70 chaperone levels are limiting. *Molecular Biology of the Cell, 24*(13), 2076–2087.

Shiber, A., & Ravid, T. (2014). Chaperoning proteins for destruction: Diverse roles of Hsp70 chaperones and their co-chaperones in targeting misfolded proteins to the proteasome. *Biomolecules, 4*(3), 704–724.

Singh, M. K., Ranbhise, J. S., Fu, M., Ju, S., Han, S., Yun, H. R., ... Kang, I. (2025). Beyond folding: Expanding the functional landscape of Hsp90 chaperone machinery in health and disease. *International Journal of Molecular Sciences, 26*(21), 10279.

Soto, C., & Pritzkow, S. (2018). Protein misfolding, aggregation, and conformational strains in neurodegenerative diseases. *Nature Neuroscience, 21*(10), 1332–1340.

Tabb, D. L., Fernando, C. G., & Chambers, M. C. (2007). MyriMatch: Highly accurate tandem mass spectral peptide identification by multivariate hypergeometric analysis. *Journal of Proteome Research, 6*(2), 654–661.

Tracz, M., & Bialek, W. (2021). Beyond K48 and K63: Non-canonical protein ubiquitination. *Cellular & Molecular Biology Letters, 26*(1), 1.

Truman, A. W., Kristensen, O., Wolfgeher, D., Hasin, N., & Frydman, J. (2012). Post-translational modification of the molecular chaperone Hsp70 regulates its interaction with co-chaperones and clients. *Molecular Cell, 48*(1), 118–129.

Udeshi, N. D., Mertins, P., Svinkina, T., & Carr, S. A. (2013). Large-scale identification of ubiquitination sites by mass spectrometry. *Nature Protocols, 8*(10), 1950–1960.

Velasco, L., Dublang, L., Moro, F., & Muga, A. (2019). The complex phosphorylation patterns that regulate the activity of Hsp70 and its cochaperones. *International Journal of Molecular Sciences, 20*(17), 4122.

Wei, H., Zhang, Y., Jia, Y., Chen, X., Niu, T., Chatterjee, A., He, P., & Hou, G. (2024). Heat shock protein 90: Biological functions, diseases, and therapeutic targets. *MedComm, 5*(2), e470.

Wentink, A., Rosenzweig, R., Kampinga, H., & Bukau, B. (2026). Mechanisms and regulation of the Hsp70 chaperone network. *Nature Reviews. Molecular Cell Biology, 27*(2), 110–128.

CHAPTER TEN

Evaluation of lysine modified peptides and their cell penetrating peptide conjugates with sirtuin activity inhibition properties and intracellular delivery

Kamal D. Patel[a], Shalini Aricthota[c], Gayan Heruka De Zoysa[b], Vijayalekshmi Sarojini[a], and Devyani Haldar[c,]*

[a]School of Chemical Sciences and the Centre for Green Chemical Science, The University of Auckland, Auckland, New Zealand

[b]The MacDiarmid Institute for Advanced Materials and Nanotechnology, Victoria University of Wellington, Wellington, New Zealand

[c]Laboratory of Chromatin Biology and Epigenetics, BRIC-Centre for DNA Fingerprinting and Diagnostics, Uppal, Hyderabad, India

*Corresponding author. e-mail address: devyani@cdfd.org.in

Contents

Abstract

Sirtuins are NAD$^+$ dependent class of histone deacetylase enzymes responsible for post-translational modifications regulating transcription, cell cycle, metabolism, DNA repair and apoptosis. Substrates of sirtuins are histones, α-tubulin, p65, FOXO1 etc. all

Methods in Enzymology, Volume 734
ISSN 0076-6879, https://doi.org/10.1016/bs.mie.2026.06.016

of which have multiple N^{ε}-acetyl lysine residues taking part in catalytic reactions of sirtuins with the help of NAD^+. In this chapter, we present methods for Solid Phase Peptide Synthesis and solution phase cyclization of a series of substrate-based peptide inhibitors of sirtuins and their cyclic and CPP conjugated derivatives **KP 1**, **Cyc KP 1**, **Tat KP 1, KP 2**, **Cyc KP 2**, **Tat KP 2, KP 3**, **Cyc KP 3** and **Tat KP 3**. Further, describe methods that could be used to study sirtuin inhibition activity and cytotoxicity properties of peptides and small molecules. We present data of three peptides from this family **Tat KP 1**, **Tat KP 2** and **Tat KP 3** that show promising yeast sirtuin and mammalian SIRT1 inhibition potential (IC_{50} 6–12 μM) comparable to known sirtuin inhibitors suramin and splitomicin and also showed cytotoxicity against HeLa and BE(2)-C cells. Finally, the morphological studies using SEM and TEM showed membrane disruption potential and evidence of apoptosis of Tat conjugated derivatives. This study documents the one of the initial reports of Tat conjugation modification to enhance the sirtuin inhibition potential. Collectively, this chapter provides comprehensive protocols and practical guidance for the characterization and biological evaluation of sirtuin inhibitors which could be used to identify and characterize novel peptides inhibitors of sirtuins with cell penetrating properties.

1. Introduction

Post-translational modifications (PTMs) of lysine residues in proteins play crucial conserved roles in normal cellular functions and deregulation causes disease (Nakayasu et al., 2017). Lysine acetylation and deacetylation is one of the most prevalent epigenetic modification responsible for regulating important cellular processes including, gene expression, DNA repair, metabolism, protein-protein interactions and cellular homeostasis which is regulated by two sets of enzymes known as histone acetyl transferase (HATs) and histone deacetylases (HDACs) (Zencheck et al., 2012). Sirtuins (Silent Information Regulator) are a family of NAD^+ dependent class of HDACs having seven members in humans (SIRT1-SIRT7). Although identified as HDACs, Sirtuins deacetylate histones as well as non-histone proteins. HATs and HDACs act in a systematic fashion to regulate protein acetylation dynamics effectively, depending on the signals and needs of the cell (Seto & Yoshida, 2014). However, stress and certain diseases cause upregulated levels of these enzymes, which can alter biological functions (Sanaei & Kavoosi, 2019). Elevated levels of these SIRT enzymes cause cascade of events leading to increased prevalence of several diseases. For e.g., SIRT1 has been linked to diseases including diabetes, cancer and heart diseases, (Karbasforooshan & Karimi, 2017) while SIRT2 to tumour development and Parkinson's disease (Kiviranta et al., 2009).

Interestingly, sirtuins also act on non-histone proteins such as p53, acetyl co-A synthetase and human α-tubulin, Tau, α-synuclein (α-Syn) etc. (Moniot et al., 2012). The human sirtuin ortholog SIRT2 is reported to deacetylate lysine at position 40 of the human α-tubulin protein in-vivo and in vitro (North et al., 2003; Skoge & Ziegler, 2016). Knockdown of SIRT2 has resulted in hyperacetylation of perinuclear microtubules, justifying the catalytic role of sirtuins in human α-tubulin (Skoge & Ziegler, 2016). Human α-tubulin is 451 amino acids long, and the acetylation of Lys-40 is correlated with increased microtubule stability, particularly the sequence position 36-GGIT**K(Ac)**DSPM-44 is responsible for sirtuin binding (Szyk et al., 2014). Various substrate-based inhibitors designed by replacing lysine with trifluoroacetyl lysine (K^{tfa}) and thioacetyl lysine (K^{tac}) modifications are recognised by the catalytic grove of sirtuins and block the NAD^+ dependant catalytic reaction (Huhtiniemi et al., 2010; Fatkins & Zheng, 2008). The human SIRT2 levels increase with aging, and several studies provide evidence that SIRT2 activity mediates various processes involved in PD pathogenesis, including aggregation of α-synuclein (α-syn), microtubule function, oxidative stress, inflammation, and autophagy. SIRT2 deacetylates α-synuclein at K6 and K10 to promote its aggregation propensity and cytotoxicity promoting Parkinson's disease. Inhibiting SIRT2 decreases the α-Syn mediated toxicity and increase acetylation of α-tubulin, making it a promising therapeutic target for Parkinson's disease (Liu et al., 2020).

Potent SIRT2 selective peptide based inhibitors with the core sequence **RIK^{tfa}RY** were previously discovered by a genetic code reprogramming system known as flexible tRNA acylation ribozyme technology (FIT) and random nonstandard peptide integrated discovery (RaPID) (Morimoto et al., 2012). Both the selectivity of these peptides against SIRT2 and their relatively potent enzyme inhibitory activities at nanomolar concentrations (280 nM) suggest that these peptides might interact with the SIRT2 active site. Unfortunately, small peptide inhibitors like **RIK^{tfa}RY** lack cell-permeability properties and protease stability, making it difficult to harness their beneficial properties. The lack of cell permeability makes the testing on cell models difficult despite having potent activity for the target within the cell. In the literature, FIT assisted synthesis has been used to generate to macrocyclic peptides with a thioester bond. Macrocyclic peptides like S2iL5, S2iL8 and S2iD7 generated using the FIT system have shown selectivity towards SIRT2 with higher potency of 32–47 nM compared to their linear counterparts which are 10 fold less potent (Morimoto et al., 2012; Yamagata et al., 2014).

Cell penetrating peptides (CPP) are widely explored for their drug delivery properties (Snyder & Dowdy, 2004). Many anticancer drug-CPP combinations have helped improve the cellular localization of drugs such as protein kinase inhibitors (van Wandelen et al., 2013), CPP-paclitaxel (Duan et al., 2017) and CPP-gemcitabine conjugates (Zakeri-Milani et al., 2017). CPPs, either alone or in conjugation with nano Particles has been used to alleviate the pathogenic effects of neurodegenerative diseases Particularly AD and PD (38). The CPP conjugated sirtuin inhibitors have been very less explored previously. In this investigation, a K^{tfa} modified linear peptide - **KP 1**, its cyclic counterpart **Cyc KP 1**, and its CPP conjugated derivative **Tat KP 1** of the **RIKRY** sequence were synthesized and assessed for their in vitro sirtuin inhibitory potential and cytotoxicity. For comparative purposes, K^{tac} modified analogues of **RIKRY** (**KP 2**, **Cyc KP 2** and **Tat KP 2**) were also synthesized. In addition, K^{tfa} modified analogues of the human α-tubulin core inhibitor sequence GGIT**K(Ac)** DSPM responsible for sirtuin binding in the linear, cyclic and Tat-conjugated forms (**KP 3**, **Cyc KP 3** and **Tat KP 3**) were also included in our study (Fig. 1). A yeast cell-based reporter transcriptional silencing assay was used for primary screening of the peptides (Grozinger et al., 2001). Peptides were tested for their ability to inhibit yeast sirtuin family NAD-dependent Sir2 protein at 50 μM concentration, and the hit peptide **Tat KP 1** was

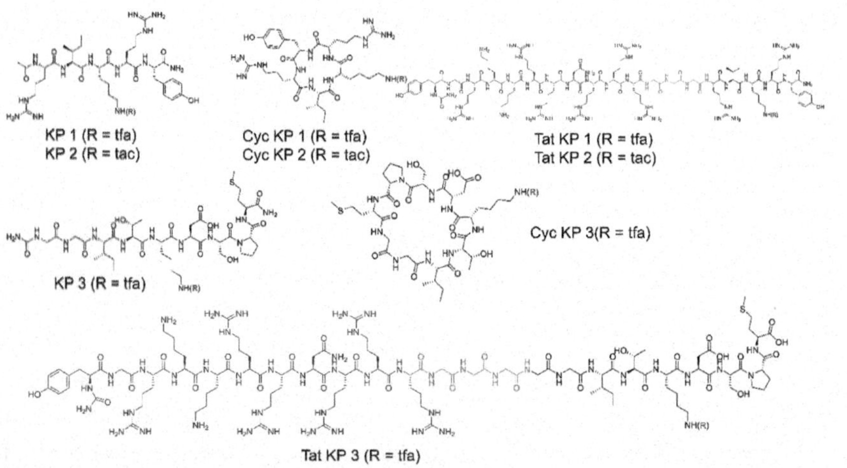

Fig. 1 Chemical structures of designed peptide based sirtuin inhibitors. CPP Tat (highlighted in brick red) is attached to the peptide sequence via triglycine linker (highlighted in green).

further studied for its percentage inhibition properties on yeast sirtuins. Cytotoxicity against HeLa and neuroblastoma (BE(2)-C) cells was also explored. Further investigation of their apoptosis properties and morphological study forms additional aspects of this study.

Here, we provide method for synthesis and purification of linear and cyclic cell penetrating peptides. Describe methods for their chemical and biological characterization. Furthermore, we have given the methodologies that could be very useful for testing molecules of this class to develop better therapeutics for diseases where sirtuins are over-expressed such as Parkinson's disease where increase human sirtuin, SIRT2 contribute to the disease development via regulation of α-synuclein and tubulin. This study also shows that cellular internalisation of the inhibitors can be achieved through conjugation to short cell penetrating peptides.

Rationale:

Cell penetrating peptides (CPP) are widely explored for their drug delivery properties (Snyder & Dowdy, 2004). Protein kinase inhibitors (van Wandelen et al., 2013), CPP-paclitaxel (Duan et al., 2017) and CPP-gemcitabine conjugates and several anticancer drug-CPP combinations have helped improve the cellular localization of these drugs (Zakeri-Milani et al., 2017). CPPs, either alone or in conjugation with nano Particles has been used to alleviate the pathogenic effects of neurodegenerative diseases Particularly AD and PD (38). The CPP conjugated sirtuin inhibitors have been very less explored previously. Therefore, whether CPPs increase the activity of sirtuin inhibitors was checked. In this investigation, a K^{tfa} modified linear peptide - **KP 1**, its cyclic counterpart **Cyc KP 1**, and its CPP conjugated derivative **Tat KP 1** of the **RIKRY** sequence were synthesized and assessed for their in vitro sirtuin inhibitory potential and cytotoxicity. For comparative purposes, K^{tac} modified analogues of **RIKRY** (**KP 2**, **Cyc KP 2** and **Tat KP 2**) were also synthesized. In addition, K^{tfa} modified analogues of the human α-tubulin core inhibitor sequence GGIT**K(Ac)**DSPM responsible for sirtuin binding in the linear, cyclic and Tat-conjugated forms (**KP 3**, **Cyc KP 3** and **Tat KP 3**) were also included in our study (Fig. 1). A yeast cell-based reporter transcriptional silencing assay was used for primary screening of the peptides (Bedalov et al., 2001; Grozinger et al., 2001). This method could be very useful for testing intra-cellular effects of molecules of this class to develop better therapeutics for diseases where sirtuins are over-expressed.

2. Synthesis of sirtuin inhibitor peptides: Linear, cyclic and CPP conjugated

All linear peptides were acetylated at the *N*-terminus and amidated at the *C*-terminus to mimic the native protein, which is an important parameter to consider when designing substrate-based inhibitors. Conjugation of the Tat sequence to peptides was carried out using triglycine as a linker, as this does not affect the peptide secondary structure and impart flexibility to the peptide chain. Side chain protected linear peptides for the synthesis of head to tail cyclic peptides were generated using 1 % v/v TFA during cleavage from the resin. Selective, head to tail cyclisation was attempted in DMF under high dilution using either a mixture of HCTU/HOAt or PyAOP coupling reagents. Cyclisation using PyAOP produced higher yield (20 %) than HCTU/HOAt mixture (6 %).

2.1 Chemicals and reagents

Fmoc-protected amino acids, 2-chlorotrityl chloride resin, and Fmoc-rink amide linker were purchased from GL Biochem. Fmoc-Lys-OH-HCl was purchased from ChemPep Inc. Ethyl dithioacetate was purchased from Sigma Aldrich. Tentagel S NH_2 was purchased from Peptide International. Human cervical carcinoma cells (HeLa), human neuroblastoma cells (BE(2)-C), cisplatin and Splitomicin was purchased from Sigma-Aldrich. Annexin V alexa fluor-488, SYTO9 and propidium iodide (PI) was purchased from Thermo Fisher Scientific Solvents used for synthesis and purification were of analytical grade and were used without further purification.

2.2 Synthesis of Fmoc-N-thioacetyl lysine

1. Fmoc-Lys-OH HCl (5.43 mmol) was dissolved in EtOH (30 mL) and was cooled to 0 °C in an ice bath, followed by dropwise addition of 10 mL of 10 % (w/v) Na_2CO_3.
2. The reaction mixture was allowed to warm to room temperature and stirred extensively.
3. Ethyl dithioacetate (5.98 mmol) was added and the reaction mixture was stirred extensively for 3 h at room temperature.
4. Solvent was evaporated; the crude product was acidified with 3 M HCl (aq.) and extracted 3 times with DCM.
5. The organic phase was washed twice with brine, dried with Na_2SO_4, filtered and evaporated to obtain desired Fmoc-*N*-thioacetyl lysine as viscous product. This was used directly without further purification (Fatkins et al., 2006).

2.3 Synthesis and purification of linear peptides

2.3.1 Linear peptides

The linear peptides were synthesized according to conventional Fmoc/tBu SPPS methodology.

1. The amino acids were attached to Tentagel S NH_2 resin as C terminal amides using Rink amide linker (0.25 mmol/g substitution, 0.1 mmol scale).
2. Amino acids were coupled to Rink amide linked resin in DMF using HCTU as coupling reagent, HoAt as additive and DIPEA as the base.
3. The *N*-terminal was acetylated using acetic anhydride/pyridine/DMF (1:2:3).
4. The final peptides were cleaved from resin using 10 mL TFA cocktail mixture (TFA/TIS/H_2O 9.5/0.25/0.25).
5. The cleavage solution was filtered and concentrated under a flow of nitrogen and the crude peptides were precipitated using an excess of cold diethyl ether.
6. The solid was washed with cold ether thrice, discarding the supernatant each time.
7. Finally centrifuged and the pellet obtained was dissolved in water and lyophilized to obtain crude peptide.
8. The crude peptides were purified by RP-HPLC and C_{18} 100 A° column, using gradient system of 0.1% TFA in water as solvent A and acetonitrile/water/TFA mixture (99/1/0.1) as solvent B.
9. Linear gradient of 10–60 % of solvent B over a period of 35 mins was followed.
10. The peptides were confirmed using Bruker MALDI-TOF and ESI-MS.

2.3.2 Cyclic peptides

1. The cyclic peptides were assembled on acid sensitive 2-chlorotrityl chloride resin (1.7 mmol/g substitution, 0.1 mmol scale) using the same coupling conditions as described above. 2. The linear protected peptides were cleaved from the resin using 1 % TFA in DCM 2 for minutes repeated once.
2. The cleavage mixture was combined and was evaporated under nitrogen and was lyophilized to yield the protected peptide as a white fluffy solid.
3. Head-to-tail cyclisation was carried out using 3 eq of PyAOP in the presence of DIPEA (1 % v/v in DMF) for 4 h.
4. The reaction mixture was concentrated under reduced pressure.

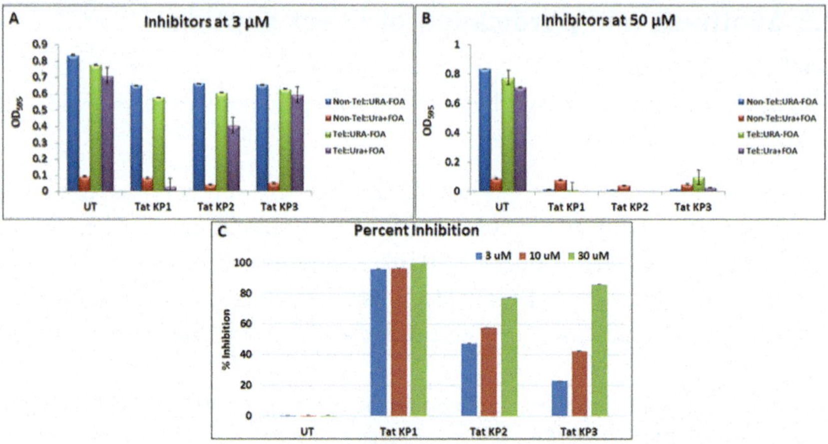

Fig. 2 Inhibition of Sir2 mediated transcriptional silencing at the telomeric locus in yeast by **Tat KP 1, Tat KP 2** and **Tat KP 3**. (A) and (B) represent growth inhibition of yeast in the presence of peptides at concentration of and 50 µM respectively. (C) Representative % inhibitory activity of peptides at 3 µM, 10 µM and 30 µM against yeast.

5. It was followed by global deprotection using 10 mL of TFA cocktail mixture and lyophilisation to obtain the crude cyclic peptides.
6. The cyclic crude peptides were purified to homogeneity using RP-HPLC.
7. Peptides were confirmed using MALDI-TOF and ESI-MS as discussed in section (Fig. 2 and supplementary data) 2.3.1

3. Yeast sirtuin inhibition assays

i) *In vivo* screening for inhibitors of Sirtuins using a cell-based reporter silencing assay in yeast model system. To identify potent inhibitors of Sirtuins, budding yeast *Saccharomyces cerevisiae* can be used as a model system for primary screening. Yeast can be very useful and effective model system for drug screening as it requires simple and inexpensive media for growth and assay can be high throughput in which large number of compounds can be screened at less cost (Grozinger et al., 2001; Nakhi et al., 2013; Rambabu, et al., 2013). Various types of drug like molecules including, small molecules as well as peptides can be screened whether these possess NAD+ dependent HDAC inhibitory activity using this yeast cell based assay. Compounds designed as described above were screened for HDAC inhibitor activity of sirtuins using the cell based reporter silencing assay in yeast model system.

Cell-based phenotypic reporter silencing assay in yeast:

This assay is based on the fact that histone deacetylases Sir2 is required for silencing of any gene inserted into transcriptionally silent telomeric region of yeast chromosome. In this case, a yeast strain in which URA3 reporter gene has been inserted at the telomeric locus (**TEL::URA3 Strain** MATa ura3-52 lys2-801 ade2-101 trpD63 his3D200 leu2-D1 TEL adh4::URA3) and control **Non-TEL::URA3 strain (**MATa ura3-52 lys2-801 ade2-101 trpD63 his3D200 leu2-D1 adh4::URA3) (Bedalov et al., 2001; Grozinger et al., 2001) will be grown in YPDA medium in presence and absence of 5-FOA (5-Fluroorotic acid) in the presence of various concentrations of proposed small molecule inhibitors or other potential inhibitors.

Growth will be monitored by absorbance at 590 nM of the cultures using a spectrophotometer or by spotting serial dilutions of strains on YPDA (rich growth medium) plates in absence and presence of 5-FOA.

A control experiment will be carried out where a wild-type yeast strain will be grown only in presence of DMSO and in presence of the compounds to identify small molecules which are cytotoxic to normal cells.

Sir2 will repress URA3 gene expression at this locus and as a result yeast cells will be able to grow in the presence of thymidine analog 5-FOA.

However, in the presence of a Sir2 inhibitor, cells die because of the inhibition of silencing of URA3 gene or activation of URA3 gene and subsequent conversion of 5-FOA to toxic fluorouracil by URA3. This assay can be performed on plates (YPDA with and without FOA) using serial dilution of yeast strains as **shown in** Figs. 3 and 4 or can be used for high-throughput screening in 96 well plates using liquid growth media as described above.

3.1 Yeast sirtuin inhibition assay (cell based in vivo assay)

Yeast strain (TEL::URA3 strain (MATα ura3-52 lys2-801 ade2-101 trpΔ63 his3Δ200 leu3Δ200 leu2-Δ1 TEL adh4::URA) was used in this assay with a reporter gene URA3 inserted in the silenced telomeric region silenced by yeast Sir2 protein.

1. The yeast strain was inoculated in 5.0 mL of YPDA media.
2. The cells were grown at the exponential phase and pipetted into round bottom 96-well plates.
3. A Stock concentration of 10 % 5-FOA was used to make a final concentration of 0.3 % 5-FOA in the wells of the 96-well plates.

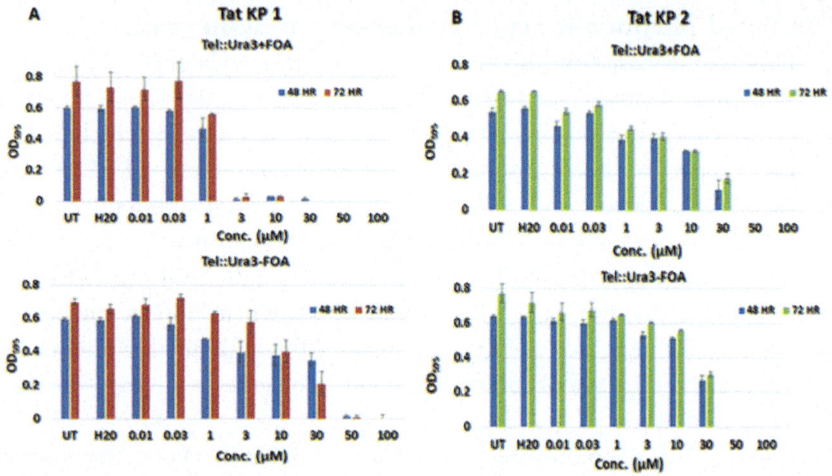

Fig. 3 Dose dependent % inhibition of Ura3 + FOA and Ura3-FOA strains upon treatment with (A) **Tat KP 1**, (B) **Tat KP 2** and (C) **Tat KP 3** with an IC_{50} of 1.06 μM, 5.52 μM and 9.65 μM respectively.

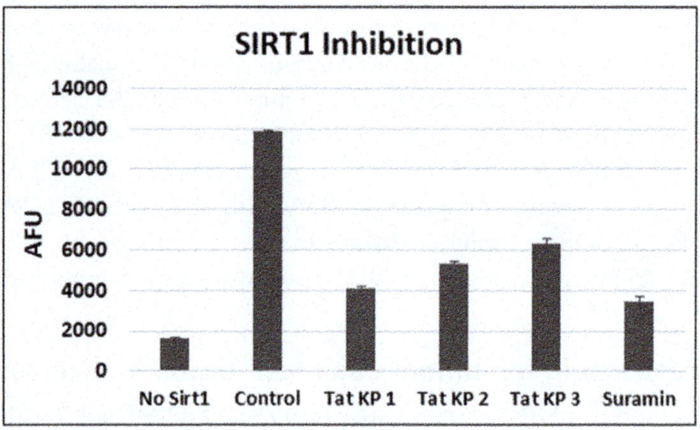

Fig. 4 Sirtuin inhibition assay. Graph showing the sirtuin inhibition activity of the tat conjugates and suramin used at 100 μM represented in fluorescence units AFU.

4. The peptides at a concentration of 3 μM, 30 and 50 μM were added to each well and the plates were incubated at 30 ^{0}C.
5. Absorbance at 595 was measured using 96 well plate reader after 24 and 48 h.
6. The inhibitory effect of compounds was analyzed after plotting the OD vs concentration of the compound.

3.2 *In vitro* deacetylase activity assay for SIRT1

The in vitro deacetylase assay was performed according to manufacturer's protocol (Biomol SIRT1 Deacetylase kit) and the human SIRT1 activity inhibition by CPPs was checked.

The deacetylase activity was calculated by measuring fluorescence intensity.

1. The reaction mixture consisting of recombinant protein SIRT1, FLUOR DE LYS SIRT1 substrate (amino acids 379–382 of human p53 peptide), potential inhibitors and NAD+ were incubated at 37 °C for 60 min as per manufacturer's protocol.
2. The fluorescence intensity was measured using an opaque microtiter plate fluorimeter with an excitation at 360 nM and emission 460 nM.
3. The results were expressed as a mean and SEM of three independent experiments (as **shown in** Fig. 5).

Cytotoxicity assay against Hela and BE(2)-C cell lines.

The detailed procedure for cell viability as determined by IC_{50} is described in our previous publication (Patel et al., 2020).

1. Cells were seeded at a seeding density of 5×10^4 cells/mL in 96-well tissue culture plates for 24 h at 37 °C in 5 % CO_2.
2. On the subsequent day, the media was carefully discarded and the cells were treated with serial concentration of the peptide inhibitors and anticancer cisplatin (0.4–200 μM) for 48 h at 37 °C in 5 % CO_2.

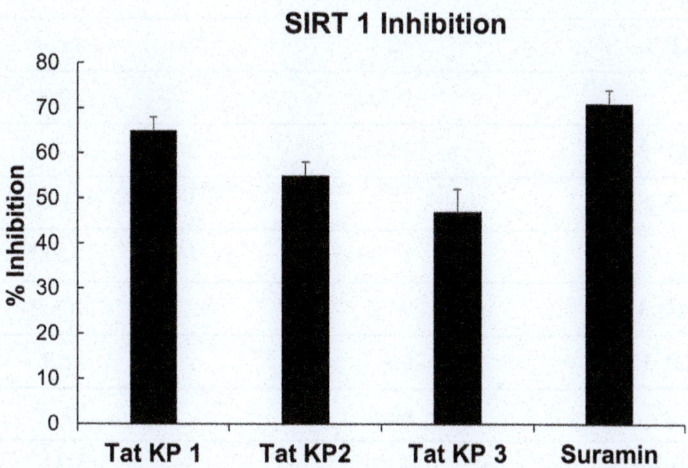

Fig. 5 Graph showing percentage SIRT 1 inhibitory activity of the tat conjugates and suramin used at 100 μM.

3. After the treatment, cells were stained with MTT (20 μL, 0.5 mg/mL) for 2–4 h at 37 °C in 5 % CO_2.
4. MTT stain was washed and 100 μL of DMSO was added to each well to solubilize formazan crystals.
5. The absorbance was measured at 570 nm using an Enspire Multimode plate reader.
6. The IC_{50} values were calculated using Sigma Plot 11.0.

Sirtuin inhibitors are widely studied for their apoptosis inducing properties against HeLa cells and neuroblastoma cells (Hu et al., 2014). The synthetic peptides were tested for cytotoxicity against HeLa and BE(2)-C cells (Table 1). Linear peptides **KP 1**, **KP 2** and **KP 3** were non-cytotoxic toward these cell lines, up to 200 μM. The cyclic versions (**Cyc KP 1**, **Cyc KP 2** and **Cyc KP 3**) were slightly more potent and had approximately two-fold lower IC_{50} values (95- 120 μM against HeLa cells and 105–140 μM, against BE(2)-C cells) (Fig. 6).

The CPP conjugated derivatives **Tat KP 1**, **Tat KP 2** and **Tat KP 3** were more cytotoxic with IC_{50} of 12.4, 13.5 and 6.4 μM respectively against

Table 1 Cytotoxicity of peptides against HeLa and BE(2)-C cells.

Peptide	IC_{50} (μM)*	
	HeLa cells	**BE (2)-C cells**
KP 1	>200	>200
Cyc KP 1	120.2	140.5
Tat KP 1	12.4	35.9
KP 2	>200	>200
Cyc KP 2	115.6	136.8
Tat KP 2	13.5	33.6
KP 3	>200	>200
Cyc KP 3	95.2	105.3
Tat KP 3	6.4	12.8
Tat	62.5	94
Cisplation	6.3	5.3

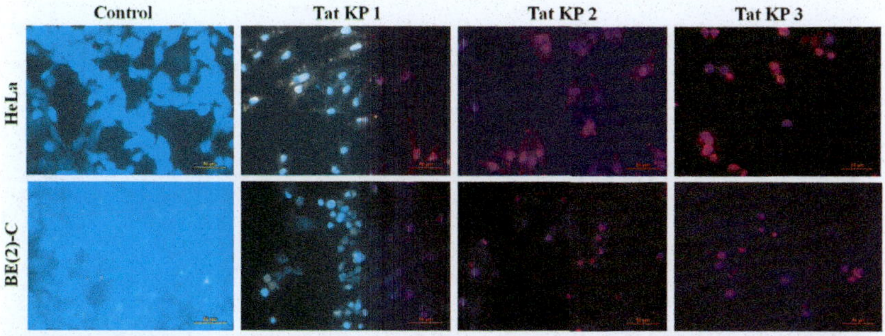

Fig. 6 Annexin V alexa fluor 488/PI staining of HeLa and BE(2)-C cells. Scale bar represent 50 μm.

HeLa cells and 35.9, 33.6 and 12.8 μM against BE(2)-C cells respectively. **Tat KP 1** and **Tat KP 2** were equally potent against both the cell lines, indicating that both tfa and tac modifications have the same effect on cytotoxicity. Particularly important to note that **Tat KP 3** is >30 times potent than **KP 3**. The increased cytotoxicity the Tat conjugates can possibly be attributed to the increased cell permeability of the conjugates because of the known cell permeable nature of Tat arising particularly from its cationic nature. The Tat peptide itself had lower potency than the conjugates and showed moderate activity with IC_{50} of 62.5 μM against HeLa (5–10 fold lower potency) and 94 μM (3–8 old lower potency) against BE(2)-C cells. The known standard drug cisplatin which was used as a reference showed IC_{50} of 6.34 μM against HeLa and 5.30 μM against BE(2)-C cells which are in agreement with the IC_{50} of cisplatin reported in the literature (Mohanty et al., 2005; Gill et al., 2016; Minagawa et al., 1999). These results confirm that **Tat KP 3** has the same potency as cisplatin against HeLa cell lines (Fig. 7).

3.3 Annexin V alexa fluor 488/PI staining assay'

The detailed procedure used for this assay is reported in our previous publication (Patel et al., 2020; Crowley et al., 2016).

1. The cells were seeded at density of 5×10^4 cells/mL on 6-well plates with poly lysine coated coverslips for 24 h.
2. The inhibitory compounds at $2 \times IC_{50}$ were added and incubated for further 24 h.
3. The coverslips were carefully removed, washed with phosphate buffer and stained with 5 μL of Alexa Fluor 488 annexin V and 1 μL of 100 μg/mL PI solution.

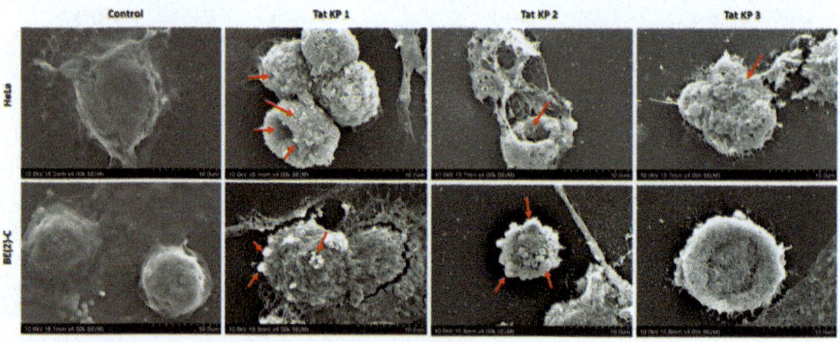

Fig. 7 SEM images of HeLa and BE(2)-C in the presence and absence of 2 x IC_{50} concentration of **Tat KP 1, Tat KP 2** and **Tat KP 3.** Membrane blebs are marked in red which are characteristic of apoptosis. Scale bar represent 10 μM.

4. The stained coverslips were incubated for 15 min in the dark and viewed under fluorescence microscopy (Nikon eclipse TE2000) at 400 x magnification using a filter with excitation and emission of 488 nm and 530 nm respectively.
5. The images were captured using a built-in DS-UI camera.

3.4 Morphology of cells using SEM and TEM

3.4.1 SEM (Scanning Electron Microscopy)

Equipment:

Termonax coverslip

FEI Quanta 200 F ESEM microscope

Procedure:

1. The cells were seeded (5×10^4 cells/mL) and incubated for 24 h on 12 well plates with cell culture termonax coverslip.
2. After 24 h, the coverslips were treated with the peptide inhibitors at $2 \times IC_{50}$ for 12 h.
3. The treated coverslips were washed with phosphate buffer, fixed with 4 % glutaldehyde for 1 h and dehydrated with graded ethanol (10−100 %) prior to drying at 60 °C for 10 min.
4. The coverslips were platinum coated for 2 min at 20 mA before viewing under high vacuum with FEI Quanta 200F ESEM microscope at 10 kV (Patel et al., 2020) (Fig. 8).

3.4.2 TEM (transmission electron microscopy)

Equipment:

Microtome.

Transmission Electron Microscope (FEI Tecnai 12).

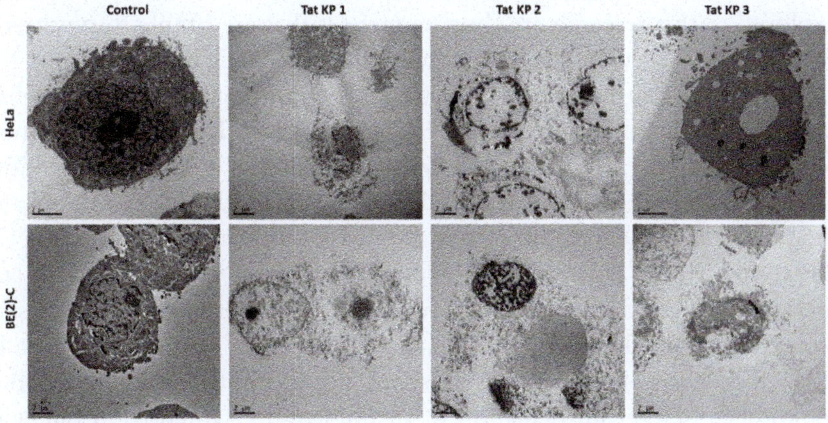

Fig. 8 TEM images of HeLa and BE(2)-C in the presence and absence of 2 x IC_{50} concentration of **Tat KP 1, Tat KP 2** and **Tat KP 3.** Scale bar represent 2 µm.

Procedure:

1. The cells were seeded (5×10^4 cells/mL) and incubated for 24 h and treated with 0.25 % trypsin-EDTA for 2 min to detach the cells.
2. The detached cells were centrifuged at 4600 rpm for 10 min to obtain a cell pellet which was then treated with peptide inhibitors at $2 \times IC_{50}$ for 12 h.
3. The treated cells were washed with phosphate buffer, fixed with fixed with 4 % glutaldehyde for 1 h and postfixed with 1 % OSO_4 in PBS for 1 h.
4. The postfixed cells were washed with PBS followed by dehydration with graded ethanol and acetone.
5. The dehydrated pellets were submerged in 1:1 mixture of acetone and 812 epoxy resin for 1 h and then pure epoxy resin at 60 °C for 48 h.
6. The dried samples were sectioned using a microtome and stained with uranyl acetate and lead citrate to view under transmission electron microscopy (Tecnai 12) (Patel et al., 2020).

4. Conclusion

This study forms one of the pioneering reports where cell penetrating peptide conjugation has been used to internalise sirtuin inhibitory peptides. CPP conjugation led to 10-fold increase in cytotoxic profile against HeLa and neuroblastoma cell lines. The sirtuin inhibitory peptides explored in this research are short sequences that are simple to synthesize chemically at efficient yield and purity. **TAT-KP3** is as potent as the clinically used drug cisplatin against HeLa cell lines while it was only half less potent than cisplatin against BE(2)-C cell lines. **Tat KP 1** is the most potent peptide at targeting both yeast and human (SIRT1) inhibition. SEM and TEM morphological studies of HeLa and BE(2)-C cell lines in the presence of the Tat conjugates showed significant damage to the cells due to membrane disruptions. The treated cells have nuclear pyknosis and membrane blebs due to apoptosis which was further corroborated using annexin V alexafluor 488/PI staining fluorescent microscopy. Results from this study show the promise of using simple substrate peptide inhibitors to target upregulation of sirtuin activity which is prevalent in certain diseases including Parkinson's disease. This study also shows that cellular internalisation of the inhibitors can be achieved through conjugation to short cell penetrating peptides.

Acknowledgements

This work was supported by the BRIC-Centre for DNA Fingerprinting (CDFD), Hyderabad, India (core grant) to DH, and University of Auckland, New Zealand to VS. We thank Dr. Yuan Tao from Auckland University of Technology for technical assistance with acquiring SEM images and Dr. Adrian Turner from the School of Biological Sciences, University of Auckland for all the technical assistance with acquiring TEM images.

References

Bedalov, A., Gatbonton, T., Irvine, W. P., Gottschling, D. E., & Simon, J. A. (2001). Identification of a small molecule inhibitor of Sir2p. *Proceedings of the National Academy of Sciences of the United States of America, 98*(26), 15113–15118.

Crowley, L. C., Marfell, B. J., Scott, A. P., & Waterhouse, N. J. (2016). Quantitation of apoptosis and necrosis by annexin V binding, propidium iodide uptake, and flow cytometry. *Cold Spring Harb Protocol Cold Spring Harbor Protocols.*

Duan, Z. Q., Chen, C. T., Qin, J., Liu, Q., Wang, Q., Xu, X. C., & Wang, J. X. (2017). Cell-penetrating peptide conjugates to enhance the antitumor effect of paclitaxel on drug-resistant lung cancer. *Drug Delivery, 24*(1), 752–764.

Fatkins, D. G., Monnot, A. D., & Zheng, W. (2006). Nepsilon-thioacetyl-lysine: A multifacet functional probe for enzymatic protein lysine Nepsilon-deacetylation. *Bioorganic & Medicinal Chemistry Letters, 16*(14), 3651–3656.

Fatkins, D. G., & Zheng, W. (2008). Substituting N-epsilon-thioacetyl-lysine for N-epsilon-acetyl-lysine in peptide substrates as a general approach to inhibiting human NAD(+)-dependent protein deacetylases. *International Journal of Molecular Sciences, 9*(1), 1–11.

Gill, M. R., Harun, S. N., Halder, S., Boghozian, R. A., Ramadan, K., Ahmad, H., & Vallis, K. A. (2016). A ruthenium polypyridyl intercalator stalls DNA replication forks, radiosensitizes human cancer cells and is enhanced by Chk1 inhibition. *Scientific Reports, 6*(1), 31973.

Grozinger, C. M., Chao, E. D., Blackwell, H. E., Moazed, D., & Schreiber, S. L. (2001). Identification of a class of small molecule inhibitors of the sirtuin family of NAD-dependent deacetylases by phenotypic screening. *Journal of Biological Chemistry, 276*(42), 38837–38843.

Hu, J., Jing, H., & Lin, H. (2014). Sirtuin inhibitors as anticancer agents. *Future Medicinal Chemistry, 6*(8), 945–966.

Huhtiniemi, T., Suuronen, T., Lahtela-Kakkonen, M., Bruijn, T., Jaaskelainen, S., Poso, A., ... Jarho, E. (2010). N-epsilon-modified lysine containing inhibitors for SIRT1 and SIRT2. *Bioorganic & Medicinal Chemistry, 18*(15), 5616–5625.

Karbasforooshan, H., & Karimi, G. (2017). The role of SIRT1 in diabetic cardiomyopathy. *Biomedicine & Pharmacotherapy, 90*, 386–392.

Kiviranta, P. H., Suuronen, T., Wallen, E. A. A., Leppanen, J., Tervonen, J., Kyrylenko, S., ... Jarho, E. M. (2009). N-epsilon-thioacetyl-lysine-containing Tri-, Tetra-, and pentapeptides as SIRT1 and SIRT2 inhibitors. *Journal of Medicinal Chemistry, 52*(7), 2153–2156.

Liu, Y., Zhang, Y., Zhu, K., Chi, S., Wang, C., & Xie, A. (2020). Emerging role of sirtuin 2 in parkinson's disease. *Frontiers in Aging Neuroscience, 11*, 372.

Minagawa, Y., Kigawa, J., Itamochi, H., Kanamori, Y., Shimada, M., Takahashi, M., & Terakawa, N. (1999). Cisplatin-resistant HeLa cells are resistant to apoptosis via p53-dependent and -independent pathways. *Japanese Journal of Cancer Research, 90*(12), 1373–1379.

Mohanty, S., Huang, J., & Basu, A. (2005). Enhancement of cisplatin sensitivity of cisplatin-resistant human cervical carcinoma cells by bryostatin 1. *Clinical Cancer Research, 11*(18), 6730–6737.

Moniot, S., Weyand, M., & Steegborn, C. (2012). Structures, substrates, and regulators of mammalian Sirtuins - opportunities and challenges for drug development. *Frontiers in Pharmacology, 3*.

Morimoto, J., Hayashi, Y., & Suga, H. (2012). Discovery of macrocyclic peptides armed with a mechanism-based warhead: Isoform-selective inhibition of human deacetylase SIRT2. *Angewandte Chemie (International Ed. in English), 51*(14), 3423–3427.

Nakayasu, E. S., Burnet, M. C., Walukiewicz, H. E., Wilkins, C. S., Shukla, A. K., Brooks, S., ... Payne, S. H. (2017). Ancient regulatory role of lysine acetylation in central metabolism. *Mbio, 8*(6).

Nakhi, A., Archana, S., Seerapu, G. P. K., Chennubhotla, K. S., Kumar, K. L., Kulkarni, P., ... Pal, M. (2013). AlCl3-mediated hydroarylation-heteroarylation in a single pot: A direct access to densely functionalized olefins of pharmacological interest. *Chemical Communications (Cambridge, England), 49*, 6268–6270.

North, B. J., Marshall, B. L., Borra, M. T., Denu, J. M., & Verdin, E. (2003). The human Sir2 ortholog, SIRT2, is an NAD(+)-dependent tubulin deacetylase. *Molecular Cell, 11*(2), 437–444.

Patel, K. D., De Zoysa, G. H., Kanamala, M., Patel, K., Pilkington, L. I., Barker, D., ... Sarojini, V. (2020). Novel cell-penetrating peptide conjugated proteasome inhibitors: anticancer and antifungal investigations. *Journal of Medicinal Chemistry, 63*(1), 334–348.

Rambabu, D., Raja, G., Sreenivas, B. Y., Seerapu, G. P. K., Kumar, K. L., Deora, G. S., ... Pal, M. (2013). Spiro heterocycles as potential inhibitors of SIRT1: Pd/C-mediated synthesis of novel N-indolylmethyl spiroindoline-3,2 '-quinazolines. *Bioorganic & Medicinal Chemistry Letters, 23*(5), 1351–1357.

Sanaei, M., & Kavoosi, F. (2019). Histone deacetylases and histone deacetylase inhibitors: Molecular mechanisms of action in various cancers. *Advanced Biomedical Research, 8*, 63.

Seto, E., & Yoshida, M. (2014). Erasers of histone acetylation: The histone deacetylase enzymes. *Cold Spring Harbor Perspectives in Biology, 6*(4).

Skoge, R. H., & Ziegler, M. (2016). SIRT2 inactivation reveals a subset of hyperacetylated perinuclear microtubules inaccessible to HDAC6. *Journal of Cell Science, 129*(15), 2972–2982.

Snyder, E. L., & Dowdy, S. F. (2004). Cell penetrating peptides in drug delivery. *Pharmaceutical Research, 21*(3), 389–393.

Szyk, A., Deaconescu, A. M., Spector, J., Goodman, B., Valenstein, M. L., Ziolkowska, N. E., ... Roll-Mecak, A. (2014). Molecular basis for age-dependent microtubule acetylation by tubulin acetyltransferase. *Cell, 157*(6), 1405–1415.

van Wandelen, L. T. M., van Ameijde, J., Ismail-Ali, A. F., van Ufford, H. C. Q., Vijftigschild, L. A. W., Beekman, J. M., ... Liskamp, R. M. J. (2013). Cell-penetrating bisubstrate-based protein kinase C inhibitors. *ACS Chemical Biology, 8*(7), 1479–1487.

Yamagata, K., Goto, Y., Nishimasu, H., Morimoto, J., Ishitani, R., Dohmae, N., ... Nureki, O. (2014). Structural basis for potent inhibition of SIRT2 deacetylase by a macrocyclic peptide inducing dynamic structural change. *Structure (London, England: 1993), 22*(2), 345–352.

Zakeri-Milani, P., Farkhani, S. M., Shirani, A., Mohammadi, S., Mojarrad, J. S., Akbari, J., & Valizadeh, H. (2017). Cellular uptake and anti-tumor activity of gemcitabine conjugated with new amphiphilic cell penetrating peptides. *Excli Journal, 16*, 650–662.

Zencheck, W. D., Xiao, H., & Weiss, L. M. (2012). Lysine post-translational modifications and the cytoskeleton. *Lysine-Based Post-Translational Modification of Proteins, 52*, 135–145.

Further reading

Burattini, S., & Falcieri, E. (2013). Analysis of cell death by electron microscopy. *Methods in Molecular Biology, 1004*, 77–89.

Elmore, S. (2007). Apoptosis: A review of programmed cell death. *Toxicologic Pathology, 35*(4), 495–516.

Gartenberg, M. R. (2000). The Sir proteins of Saccharomyces cerevisiae: Mediators of transcriptional silencing and much more. *Current Opinion in Microbiology, 3*(2), 132–137.

Hoffmann, G., Breitenbucher, F., Schuler, M., & Ehrenhofer-Murray, A. E. (2014). A novel sirtuin 2 (SIRT2) inhibitor with p53-dependent pro-apoptotic activity in non-small cell lung cancer. *Journal of Biological Chemistry, 289*(8), 5208–5216.

Kozako, T., Mellini, P., Ohsugi, T., Aikawa, A., Uchida, Y., Honda, S., & Suzuki, T. (2018). Novel small molecule SIRT2 inhibitors induce cell death in leukemic cell lines. *BMC Cancer, 18*.

Pruitt, K., Zinn, R. L., Ohm, J. E., McGarvey, K. M., Kang, S. H. L., Watkins, D. N., ... Baylin, S. B. (2006). Inhibition of SIRT1 reactivates silenced cancer genes without loss of promoter DNA hypermethylation. *PLoS Genetics, 2*(3), 344–352.

CHAPTER ELEVEN

Epigenetic clues: Predicting maternal depression through DNA methylation

Varghese Merin Jerusha, Chinnasamy Thirumoorthy, and Kuppan Gokulakrishnan*

Department of Neurochemistry, National Institute of Mental Health & Neuro Sciences (NIMHANS), Bangalore, India

*Corresponding author. e-mail address: gokul@nimhans.ac.in; gokulnimhans@gmail.com

Contents

Abstract

Perinatal depression (PND) is a prevalent and multifactorial mood disorder affecting approximately 10–20 % of women globally, with higher burdens reported in low- and middle-income countries. Despite the availability of screening tools such as the Edinburgh Postnatal Depression Scale, these approaches primarily identify risk without elucidating underlying biological mechanisms. Emerging evidence highlights the role of epigenetic regulation particularly DNA methylation as a key mediator linking genetic susceptibility and environmental exposures during the perinatal period. This review synthesizes current knowledge on DNA methylation dynamics in

Methods in Enzymology, Volume 734
ISSN 0076-6879, https://doi.org/10.1016/bs.mie.2026.06.020

maternal depression, emphasizing both candidate gene and epigenome-wide association study (EWAS) approaches. Candidate gene studies have identified differential methylation in stress-related pathways, including HPA axis genes (NR3C1, FKBP5), serotonergic signalling (SLC6A4), and oxytocin pathways (OXTR), though findings remain limited by poor reproducibility and small sample sizes. In contrast, EWAS provides a hypothesis-free framework, identifying novel differentially methylated positions and regions associated with PND, including predictive CpG panels with potential diagnostic utility. The review also highlights the importance of tissue specificity, temporal epigenetic remodeling across pregnancy, and the interplay between maternal and fetal epigenomes. Furthermore, methodological challenges such as heterogeneity in study design, lack of replication, and analytical inconsistencies remain barriers to clinical translation. Integrating genetic, epigenetic, and environmental data through multi-omics approaches may enhance predictive accuracy and improve early intervention strategies. Overall, DNA methylation represents a promising avenue for understanding the biological underpinnings of PND and developing robust biomarkers for risk prediction and personalized care.

1. Introduction

The transition into motherhood, also referred to as the perinatal period, has been recognised for centuries as a vulnerable phase during which mothers are more likely to experience depressive episodes. Symptoms such as persistent low mood, guilt, changes in sleep and appetite, lack of worth, and suicidal ideation may be exhibited. However, this phenomenon was formally recognised in Diagnostic and Statistical Manual of Mental Disorders (DSM) IV in 1994, which diagnosed women with postpartum onset of depression. Further, the concept was expanded in DSM-5 as a phenomenon occurring during the onset of pregnancy and childbirth (Kroh & Lim, 2021). Therefore, Perinatal Depression (PND), as characterized in DSM-5, is a mood disorder associated with a major depressive episode during the onset of pregnancy or within 4 weeks of delivery, thereby encompassing both antenatal and postpartum depression (AP, 2000; Carlson et al., 2025).

Globally, 10–20% of women are affected by PND (Ganjiwale & Nimbalkar, 2025). A recent meta-analysis reported a disproportionately higher mean adjusted pooled prevalence of PND in Low- and Middle-Income countries (LMICs) than in High-Income countries, at 13.1 % vs 11.4 % (Bluett-Duncan et al., 2021). In India, the prevalence of PND was around 15–45 % (Raghavan et al., 2021; Upadhyay et al., 2017). Despite the availability of validated screening tools, such as the Edinburgh Postnatal Depression Scale (EPDS), Hamilton Depression Rating scale (HAM-D), which facilitate early identification of women who are at high risk of developing PND. These screening tools restricted to preliminary risk identification and are limited in understanding the underlying biological mechanisms of the disorder.

Perinatal depression (PND) arises from a complex, multifactorial aetiology shaped by the interaction of psychological, socioeconomic, obstetric, lifestyle, and biological factors (Carlson et al., 2025; Ghaedrahmati et al., 2017; Koleva et al., 2011; Konjevod et al., 2023; Lancaster et al., 2010; Rodríguez et al., 2010). More recently, studies have increasingly focused on the polygenic basis of PND to better understand its heritability and the extent to which its genetic profile differs from that of non-perinatal depression. Estimates suggest that heritability ranges from approximately 37 % in antenatal depression to 40 % in postnatal depression(Viktorin et al., 2016). Another study reported that sibling concordance for postpartum depression is highest among female siblings during the early postpartum period, particularly between 6 and 8 weeks after childbirth (Forty et al., 2006).

Early genetic studies adopted a candidate-gene association study that relied on a hypothesis-driven approach to explain the association of the genes with the traits of interest. However, poor reproducibility generated high false positive associations, small sample sizes, and heterogeneity in result generation, leading to inconsistent reporting on biological significance and low statistical power (Bosker et al., 2011; Duncan et al., 2019; Sullivan et al., 2001). The above limitations led to the adoption of Genome-Wide Association Studies (GWAS), which is an agnostic approach in large sample sizes that identifies all trait-associated variants (SNPs) that contribute to the polygenic architecture of a psychiatric condition and constructs an individual aggregated risk score using Polygenic Risk Score (PRS) (McIntosh et al., 2019). However, the largest genome-wide association studies of PND identified only one significant locus, mapped to GULP1, which was not previously reported to be associated with MDD, and none in a 2023 meta-analysis of 20 multi-ancestry cohorts.

While genetics provides the susceptibility to PND, it is insufficient to explain the heterogeneous expression of PND, i.e., the gene-centric paradigm overlooks the phenotypic variability or the plastic nature of the genetic sequence in response to environmental cues (Bernstein et al., 2007). Therefore, epigenomics accounts for the dynamic regulation of gene expression in response to environmental, hormonal, and psychosocial factors. Epigenetic modifications refer to heritable, reversible chemical tags that regulate the accessibility of DNA without altering the underlying DNA sequence. The perinatal period represents time-sensitive windows that are modulated by alterations in environmental, immune, and endocrine factors that, in turn, modulate the epigenome. Therefore, integrating genetic susceptibility with epigenomic regulation in PND can provide comprehensive insight into the disease pathophysiology. This review aims to outline the key epigenetic regulation, DNA methylation, in association with perinatal depression, and the overview of the methodology to study DNA methylation changes is provided, followed by a comprehensive view of the heterogeneity of expression in different tissues for potential DNA Methylation signatures in PND.

2. Materials and methods

Methods such as microarray, whole genome bisulfite sequencing, Pyrosequencing, and methyl-sensitive PCR could be used to study DNA methylation changes. In this review, we have presented the EPIC array protocol, DNA isolation from whole blood using the phenol-chloroform method and a DNA bisulfite conversion method for epigenome-wide association studies (EWAS).

2.1 Genomic DNA isolation from whole blood

Genomic DNA isolation from whole blood workflow for sequencing studies is provided in Fig. 1.

Equipment: Refrigerated microcentrifuge (capable of 12,000 g), Tabletop centrifuge, Vortex mixer, Dry bath/incubator, Pipettes (P10, P20, P200, P1000), NanoDrop ND-1000 Spectrophotometer (NanoDrop Technologies, Wilmington, USA), Qubit Fluorometer, 1.5 mL and 2.0 mL microcentrifuge tubes, Tube rotator/mixer, Biosafety cabinet, Ice bucket.

Reagents:

Erythrocyte Lysis Buffer: M Tris-HCl (pH 8.0), 320 mM sucrose, 5 mM $MgCl_2$, 1 % Triton X-100; **Extraction Buffer:** 400 mM Tris-HCl, 60 mM EDTA, 150 mM NaCl, 1 % SDS; **Additional Reagents:** 5 M sodium perchlorate, Tris-saturated phenol, Chloroform: Isoamyl alcohol (24:1), Absolute ethanol (100 %), 70 % ethanol, TE buffer (10 mM Tris-HCl, pH 8.0; 1 mM EDTA), Nuclease-free water.

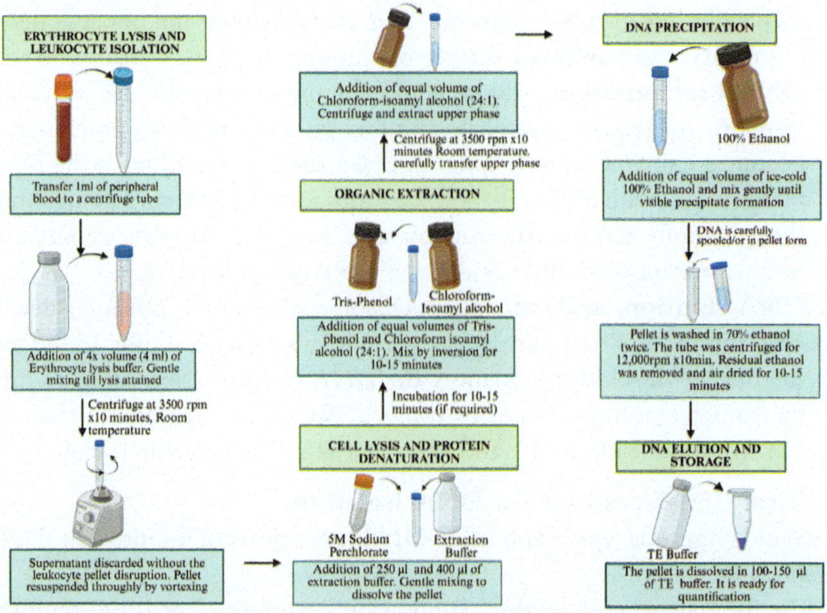

Fig. 1 Genomic DNA isolation from whole blood workflow for sequencing studies.

Procedure.

1. **Erythrocyte Lysis and Leukocyte Isolation:** Transfer 1.0 mL of peripheral whole blood into a sterile centrifuge tube and add four volumes (4 mL) of erythrocyte lysis buffer. Mix gently by inversion until complete erythrocyte lysis is achieved. Centrifuge the sample at 3500 rpm for 10 min at room temperature, then carefully discard the supernatant without disturbing the leukocyte pellet. Finally, thoroughly disturb the leukocyte pellet by vortexing. The overview of the protocol is given in Fig. 1.
2. **Cell Lysis and Protein Denaturation:** Add 250 µL of 5 M sodium perchlorate to the leukocyte pellet, followed by 500 µL of extraction buffer. Mix gently until the pellet is completely dissolved. If required, incubate the mixture at room temperature for 10–15 min to ensure complete cell lysis.
3. **Organic Extraction:** Add an equal volume of Tris-phenol to the lysed sample, followed by an equal volume of chloroform:isoamyl alcohol (24:1). Mix gently by inversion for 5–10 min and centrifuge at 3500 rpm for 10 min. Carefully transfer the upper aqueous phase into a fresh tube, then add an equal volume of chloroform:isoamyl alcohol (24:1).

Centrifuge again at 3500 rpm for 10 min and transfer the upper aqueous layer into a new tube for further processing.

4. **DNA Precipitation:** Add an equal volume of ice-cold 100 % ethanol to the recovered aqueous phase and mix gently until DNA precipitation becomes visible. Carefully spool or pellet the DNA, and wash the DNA pellet twice with 70% ethanol. Centrifuge at 12,000 rpm for 10 min, then carefully remove the supernatant, ensuring complete removal of residual ethanol. Finally, air-dry the DNA pellet for 1−2 h.
5. **DNA Elution and Storage:** Dissolve the dried DNA pellet in 100–150 μL of TE buffer. If necessary, incubate at 55 °C for 15–30 min to enhance dissolution. Quantify the DNA concentration using a Qubit fluorometer and assess purity using a NanoDrop spectrophotometer. Finally, aliquot the DNA and store at −80 °C until further analysis.

Critical Considerations for DNA Isolation:

- Avoid vigorous vortexing after DNA precipitation to prevent DNA shearing.
- Use freshly prepared erythrocyte lysis buffer for efficient RBC removal.
- Ensure complete removal of phenol residues during chloroform extraction.
- Residual ethanol can inhibit downstream enzymatic reactions.
- Use nuclease-free consumables to prevent DNA degradation.
- Perform phenol handling under a fume hood with appropriate PPE.

2.2 Bisulfite conversion of genomic DNA

Equipment: Thermocycler, Microcentrifuge (capable of ≥10,000 g), Vortex mixer, Pipettes (P10, P20, P200, P1000), NanoDrop spectrophotometer, Ice bucket, Zymo-Spin™ IC Columns with collection tubes, 1.5 mL microcentrifuge tubes.

Reagents: EZ DNA Methylation™ Kit (D5001/D5002; Zymo Research, USA), CT Conversion Reagent, M-Dilution Buffer, M-Binding Buffer, M-Wash Buffer, M-Desulphonation Buffer, M-Elution Buffer, Nuclease-free water, 100 % Ethanol.

Procedure.

1. **DNA Denaturation:** Add 500 ng of genomic DNA into a PCR tube and add 5 μL of M-dilution buffer. Adjust the final volume to 50 μL using nuclease-free water and mix gently. Incubate the reaction at 37 °C for 15 min. DNA bisulfite conversion of methylated cytosine to uracil by sulphonation, deamination and desulphonation is given in Fig. 2.

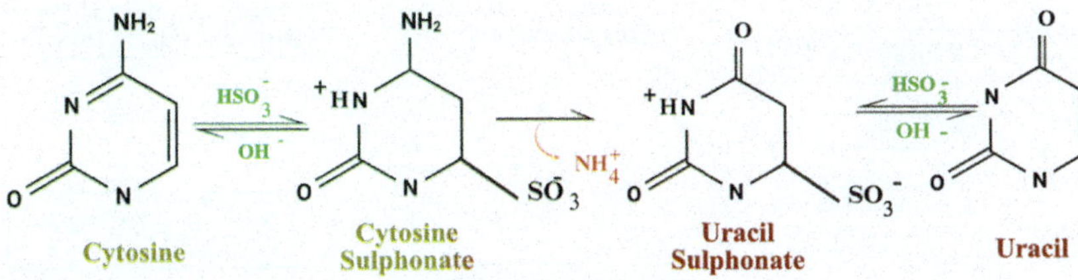

Fig. 2 Bisulfite conversion of methylated cytosine to uracil by sulphonation, deamination and desulphonation.

2. **Bisulfite Conversion:** Prepare fresh CT conversion reagent according to the manufacturer's instructions. Add 100 μL of CT conversion reagent to each sample and mix thoroughly by gentle vortexing followed by brief pulse centrifugation. Perform bisulfite conversion in a thermal cycler using the following program: 16 cycles consisting of 30 s at 95 °C, followed by 60 min at 50 °C, and a hold at 4 °C indefinitely after completion. After cycling, place the samples on ice for 10 min. The protocol overview is given in Fig. 3.
3. **DNA Binding and Purification:** Add 400 μL of M-Binding Buffer to a Zymo-Spin IC column, then transfer the entire reaction mixture onto the column. Mix by gentle inversion several times to ensure proper binding. Centrifuge at > 10,000 g for 30 s and discard the flow-through.
4. **Column Washing:** Add 100 μL of M-Wash Buffer to the column and centrifuge at full speed for 30 s. Discard the flow-through.
5. **Desulphonation:** Add 200 μL of M-Desulphonation Buffer to the column and incubate at room temperature for 15–20 min. After incubation, centrifuge at full speed for 30 s. Discard the flow-through.
6. **Final Washing:** Wash the column twice with 200 μL of M-Wash Buffer, centrifuging at full speed for 30 s after each wash.
7. **DNA Elution:** Transfer the column into a clean 1.5 mL tube and add 10 μL of M-Elution Buffer directly onto the membrane. Incubate at room temperature for 2–5 min, then centrifuge at full speed for 30 s. The eluted bisulfite-converted DNA can be stored at −20 °C until downstream analysis.

Storage Conditions: Bisulfite-converted DNA can be stored at −20 °C for up to 1 week.

Critical Considerations for Bisulfite Conversion.

- Fresh preparation of the CT conversion reagent is essential.
- Protect samples from prolonged light exposure.
- Ensure ethanol is added to the M-Wash buffer before use.

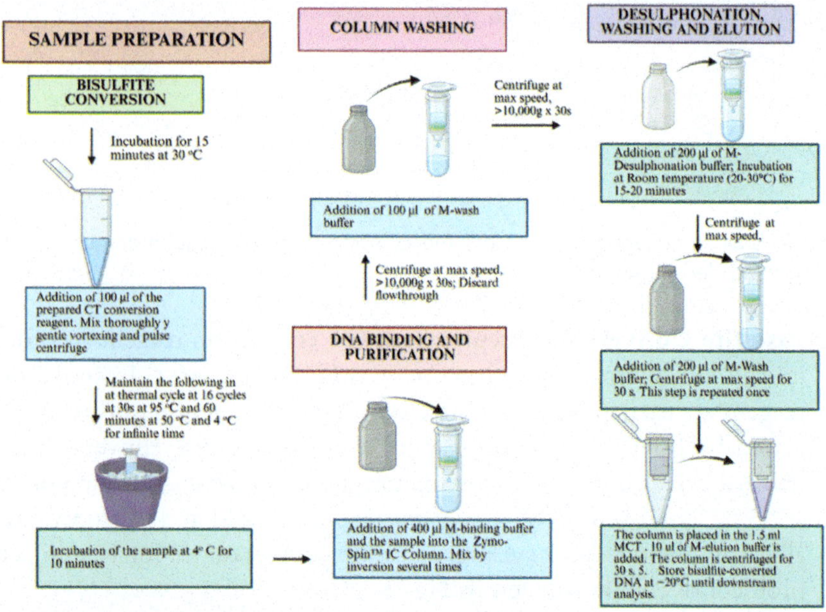

Fig. 3 Bisulfite conversion workflow.

- Use high-quality, intact genomic DNA for optimal conversion efficiency.
- Avoid repeated freeze-thaw cycles of bisulfite-converted DNA.
- Bisulfite treatment causes DNA fragmentation; therefore, downstream PCR amplicons should ideally be <300 bp.
- Complete desulphonation is critical for efficient PCR amplification.

2.3 DNA methylation profiling using the infinium methylation epic beadchip

Equipment: Illumina Hybridization Oven, Illumina iScan System, Refrigerated centrifuge, Microplate centrifuge, Heat block/incubator, Thermocycler, Vortex mixer, Vacuum desiccator, Water circulator with temperature control, Pipettes (P10, P20, P200, P1000), 96-well plates and cap mats, Hybridization chamber and gasket slides, Flow-through chamber assembly, BeadChip alignment fixture, Tube racks, Ice bucket.

Reagents: Infinium MethylationEPIC BeadChip Kit Components: MA1 buffer, RPM buffer, MSM buffer, FMS fragmentation reagent, PM1 precipitation reagent, RA1 resuspension buffer, PB1 wash buffer, PB2 buffer, XC1 buffer, XC2 buffer, XC3 buffer, XC4 buffer, TEM reagent, STM reagent, ATM reagent.

Additional Reagents: N NaOH, 100 % 2-propanol, 95 % formamide/1 mM EDTA, 100 % ethanol, Deionized water.

Experimental Procedure.

1. **Whole-Genome Amplification:** Add 20 μL of MA1 buffer to each well of the MSA4 plate, then transfer 4 μL of bisulfite-converted DNA from the BCD plate into each corresponding well. Add 4 μL of 0.1 N NaOH to each reaction. Seal the plate using a 96-well cap mat, shake at 1600 rpm for 1 min, and centrifuge at 280g for 1 min. Finally, incubate the plate at room temperature for 10 min. An overview of genome-wide methylation profiling by the Illumina Infinium MethylationEPIC BeadChip Kit is given in Fig. 4.

   Amplification Reaction: Remove the cap mat and add 68 μL of RPM buffer, followed by 75 μL of MSM buffer to each well. Reseal the plate, then vortex at 1600 rpm for 1 min and pulse centrifuge at 280 g for 1 min. **Incubation:** Incubate plates at 37 °C for 20–24 h for DNA amplification.
2. **DNA Fragmentation:** Briefly centrifuge the plate at 280 g, then add 50 μL of FMS buffer to each well. Seal the plate and vortex at 1600 rpm for 1 min, then centrifuge at 280 g for 1 min. Finally, incubate at 37 °C for 1 h to fragment the amplified DNA.

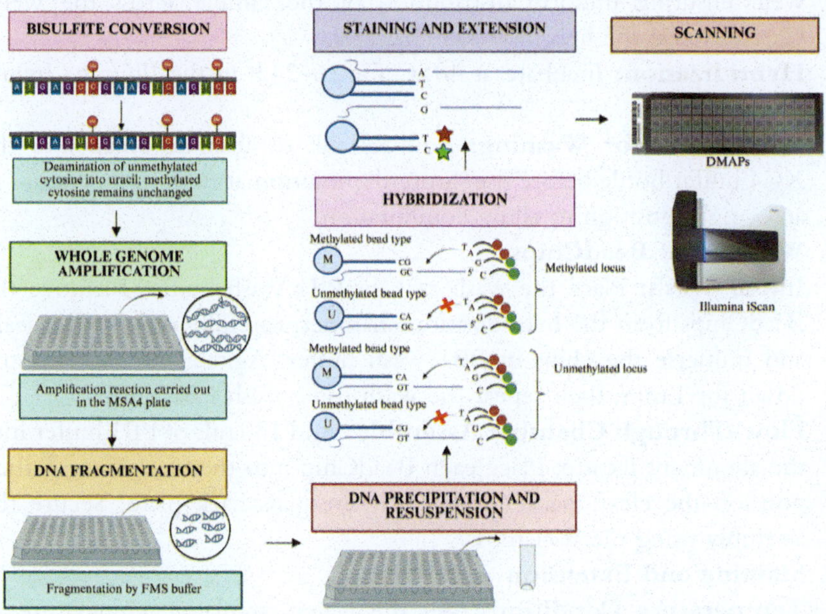

Fig. 4 Overview of genome-wide methylation profiling by methylation EPIC BeadChip Kit.

3. **DNA Precipitation:** Add 100 µL of PM1 buffer to each well, seal the plate, and vortex for 1 min. Incubate at 37 °C for 5 min, then centrifuge at 280 g for 1 min. Carefully remove the cap mat and add 300 µL of 100 % 2-propanol to each well. Reseal the plate and invert it 10 times to mix thoroughly, then incubate at 4 °C for 30 min. Finally, centrifuge at 3000 g at 4 °C for 20 min. **Pellet Recovery:** Confirm the formation of a visible blue pellet, then carefully discard the supernatant by inversion without disturbing the pellet. Air-dry the pellet at room temperature for approximately 1 h, ensuring that pellet integrity is maintained.
4. **DNA Resuspension:** Add 46 µL of RA1 buffer to each well and seal the plate using a foil heat seal. Incubate at 48 °C for 1 h, then vortex at 1800 rpm for 1 min followed by a brief centrifugation at 280 *g* for 1 min.
5. **Hybridization to BeadChip:**
   DNA Denaturation: Denature the resuspended DNA at 95 °C for 20 min, then allow it to equilibrate at room temperature for 30 min. Finally, perform a brief pulse centrifugation. **Hybridization Setup:** Assemble the hybridization chamber using gasket slides. Add 400 µL of PB2 buffer to both the upper and lower reservoirs. Remove the BeadChips from their packaging and place them into the designated inserts.
   Sample Loading: Load 26 µL of denatured DNA into each BeadChip well, ensuring uniform distribution of the sample across the wells. Carefully seal the hybridization chamber.
   Hybridization: Incubate at 48 °C for 16–24 h in the Illumina hybridization oven.
6. **Preparation for Washing:** Add 330 mL of absolute ethanol to the XC4 buffer bottle before use. Store the solution at room temperature in an upright position to allow equilibration.
7. **Washing of BeadChips**
   Initial Wash: Place the wash rack in PB1 wash buffer. Remove the BeadChips from the hybridization chamber, carefully remove the seals, and immerse the chips in PB1 wash buffer. Agitate the rack up and down for 1 min, then repeat the wash using fresh PB1 buffer.
   Flow-Through Chamber Assembly: Add 150 mL of PB1 buffer into the alignment fixture. Place each BeadChip into the black frames, then position the clear spacers and attach the glass back plate. Secure the assembly using metal clamps.
8. **Staining and Extension**
   Temperature Conditions: Set the water circulator temperature to 44 °C.

Reagent Flow Sequence

Initial Steps: Add 150 μL of RA1 and wait for 30 s; repeat this step four times. Then add 450 μL of XC1 and incubate for 10 min, followed by 450 μL of XC2 with an additional 10-minute incubation. Set the water circulator temperature to 32 °C. Next, add 200 μL of TEM and incubate for 15 min, and finally add 450 μL of 95 % formamide/1 mM EDTA and incubate for 5 min.

Washing: Add 450 μL of XC3 and incubate for 1 min; repeat this step once. Then wait until the water circulator temperature reaches 32 °C.

Staining and Extension Cycles

Perform the following sequence: Add 250 μL of STM and incubate for 10 min. Then add 450 μL of XC3 and incubate for 1 min; repeat this step once. Wait for 5 min, then add 250 μL of ATM and incubate for 10 min. Next, add 450 μL of XC3 and incubate for 1 min, repeating the step once again. Finally, wait for 5 min.

Repeat until: 3 STM cycles, 2 ATM cycles

9. **Final Washing and Drying**

   Washing: Remove the flow-through chambers. Prepare XC4 and PB1 wash dishes. Wash the BeadChips by gently agitating them 10 times, then soak them for 5 min in each wash solution.

   Drying: Arrange the BeadChips with the barcode side facing upward, and dry them in a vacuum desiccator for 50–55 min at 675 mm Hg (0.9 bar).

10. **BeadChip Scanning**

    Scan the BeadChips using the Illumina iScan System, and extract probe-level fluorescence intensities using GenomeStudio software (Illumina).

2.4 DNA methylation data processing and quality control

Overview: The EPIC array interrogates approximately 850,000–930,000 CpG loci across the genome. Each CpG site is quantified using methylated (M) and unmethylated (U) probe intensities.

Calculation of β-values.

DNA methylation levels were represented as β-values calculated using:

$$\beta = \frac{M}{M + U + 100}$$

Where:

- **M** = methylated signal intensity
- **U** = unmethylated signal intensity

The constant value 100 was included to stabilize β-values when fluorescence intensities were low and to minimize background noise.
β-values range from:
- 0 = completely unmethylated
- 1 = fully methylated

Intermediate values indicate partial methylation.

2.5 Bioinformatics processing and statistical analysis

Software Environment: All preprocessing and statistical analyses were performed in:

- R version 4.3.1 or higher version
- Bioconductor-based workflows

Data Import: Raw.idat files were imported as RGChannelSet objects using methylation array analysis workflows.

Quality Control Filtering: The following probes were excluded:

1. Probes with poor detection p-values
2. Probes located on sex chromosomes
3. Cross-reactive probes
4. Probes overlapping common SNPs

These filtering steps minimized technical variability and reduced false-positive findings.

Normalization and Differential Methylation Analysis: Differential methylation analysis was performed using the CHAMP package in R.

Statistical Model:
- Linear regression models were applied to compare methylation levels between groups.
- Fold change and p-values were calculated for each CpG site.

Multiple Testing Correction: P-values were adjusted using the Benjamini–Hochberg false discovery rate (FDR) correction method.

Critical Considerations.
Wet-Lab Considerations.
- Ensure complete bisulfite conversion prior to amplification.
- Avoid repeated freeze–thaw cycles of bisulfite-converted DNA.

- Blue pellet formation after precipitation is critical for successful downstream processing.
- Hybridization chambers must remain tightly sealed to prevent evaporation.
- Ensure uniform sample loading onto BeadChips.

Instrumentation Considerations.
- Maintain accurate hybridization oven temperature.
- Prevent bubble formation during reagent loading.
- Ensure BeadChips are completely dry before scanning.

Bioinformatics Considerations.
- Detection p-value filtering is essential for reliable methylation measurements.
- Removal of SNP-overlapping probes reduces genotype-associated bias.
- Exclusion of sex chromosome probes minimizes confounding by sex-specific methylation.

3. Epigenetic alterations in the whole genome – mechanisms and profiling approaches

Epigenetic modification provides a mechanistic framework that explores the malleability of the epigenetic marks in response to the external and internal milieu on the genome (Mill & Heijmans, 2013). However, the variations in the epigenome can also be random or stochastic in nature, which are predominantly present in the vicinity of genes involved in the development and morphogenesis of the foetus (Feinberg & Irizarry, 2010; Landan et al., 2012). Furthermore, epigenetic alterations are stably inherited during early development, these changes, thereby supporting the concept of the Developmental Origin of Health and Disease (DoHaD) and explaining the discordance of phenotypic variance across identical twins and differences in sex.

A major challenge in epigenomic studies is that it is dynamic and tissue-specific. Therefore, cross-tissue comparison of epigenetic marks within individuals must be analysed to categorize the variation at specific genomic regions (Mill & Heijmans, 2013). Epigenetic inheritance can take the following forms: DNA methylation, non-coding RNAs, and histone tail modifications. DNA methylation involves the addition of a methyl group to a sequence. Histone modifications can introduce acetylation, phosphorylation, and ubiquitination at the amino acid sequence of the histone region, which may modify chromatin accessibility or activate or suppress transcription. Post-transcription modifications are regulated by non-coding RNAs (Álvarez-Mejía et al., 2025; Ma et al., 2025).

High-throughput techniques have been employed to characterize the whole genome methylation profile. Among which, whole-genome bisulphite sequencing is considered to be the gold standard technique that interrogates approx. 28 million CpG sites across the human genome. The other widely used technique is methylation array that targets < 2 % of the CpG sites present in the whole genome, thereby predominantly focusing on CpG islands and promoters (Mill & Heijmans, 2013; Sandoval et al., 2011).

3.1 Mechanism of DNA methylation

DNA Methylation is the covalent addition of a methyl group to the 5th carbon of the cytosine residues that reside within the DNA sequence. The methylation process is catalysed by a family of methyl transferases (Dnmts) that transfer a methyl group from the S-adenyl methionine (SAM) to the fifth position of the cytosine region, which leads to the formation of 5mC. Each enzyme is associated with a distinct role. The maintenance of the methylation pattern from the parental DNA strand to the daughter strand is handled by DNMT1, thereby preserving the cell identity during replication. DNMT3A and DNMT3B mediate the de novo methylation by the introduction of the methylation to an unmodified DNA sequence, whereas DNMT3L serves as an accessory to DNMT3A and DNMT3B (Moore et al., 2013).

Despite its association with cell development and growth, the post-mitotic neuronal cells of the mature mammalian brain show a persistent expression of Dnmts, suggesting the dynamic involvement not just in developmental programming but also in brain function (Feng et al., 2005; Goto et al., 1994). The entire process is dynamic and reversible. The demethylation process can occur by the passive inhibition of DNMT1 or by the active intervention of the TET-eleven translocation (TETim) enzyme that removes the methylated sites by oxidation (Gujar et al., 2019). Methylation predominantly occurs at the CpG sites in the mammalian genome, which are interspersed in various genomic regions. CpG islands are characterized as vast stretches of roughly 1000 bp long DNA that have a higher CpG density than the rest of the genome (Bird et al., 1985). 70 % of the gene promoters reside within the islands that are associated with housekeeping genes (Illingworth et al., 2010; Saxonov et al., 2006). These islands are evolutionarily conserved and maintain an open chromatin state with low nucleosome density that promotes the accessibility of the sequence to transcription factors despite the lack of promoter regions (Mikkelsen et al., 2007; Moore et al., 2013; Tazi & Bird, 1990). While CpG islands are predominantly unmethylated, their methylation can result in a stable expression of gene silencing (Mohn et al., 2008) which were observed to establish imprinting (Caspary et al., 1998; Choi et al., 2005; Wutz et al., 1997; Zwart et al., 2001). In contrast, the intergenic regions are bulk methylated regions (Schulz et al., 2006) that silence transposable and viral elements that cause deleterious effects to the genome (Gwynn et al., 1998; Kuster et al., 1997; Michaud et al., 1994; Ukai et al., 2003; Wu et al., 1997).

3.2 DNA methylation in PND

DNA methylation patterns associated with perinatal depression (PND) reflect the complex interaction between environmental exposures and neuroendocrine system dysregulation, highlighting how these factors influence gene expression plasticity. Current research has largely focused on DNA methylation changes in fetal and placental tissues in response to maternal stress. At the same time, there is growing interest in examining maternal peripheral tissues as potential predictive biomarkers for PND. However, findings across studies remain inconsistent, partly due to tissue-specific differences in methylation patterns and limited replication in multi-ancestry cohorts, both at the candidate gene and genome-wide levels. Decoding Perinatal Depression Through DNA Methylation Signatures is depicted in the graphical abstract.

3.2.1 Candidate gene DNA methylation in maternal depression

A hypothesis-driven approach is adopted to target genes associated with biological pathways, particularly in the Hypothalamus-Pituitary Axis, Serotonergic Transport, and Oxytocin signalling. Within the maternal-focused methylation framework, NR3C1, CRH, and FKBP5, which have been reported to have differential methylation expression, are associated with altered expression of cortisol levels. Furthermore, the hypermethylation of serotonin transporter SLC6A4 was associated with lower serotonin levels exhibited in PND symptomology (Carnevali & Buoli, 2021). The regulation of the HPA axis by oxytocin was reported to have a significant increase in the methylation of OXTR exon 3 during pre-and postpartum, thereby suppressing the OXTR expression (King et al., 2017). Despite the biological significance, candidate gene approaches are limited by the small sample sizes and poor reproducibility, which may generate high false positive associations. A summary of candidate gene methylation studies in perinatal depression are presented in Table 1.

3.2.2 Tissue specificity of epigenetic markers and the link between maternal peripheral tissues and fetal epigenetic programming

Evidence indicates substantial variability in DNA methylation patterns across different biological tissues, including maternal blood, saliva, buccal cells, placenta, and umbilical cord blood. This heterogeneity has been observed in several candidate gene studies. For example, a longitudinal study involving 49 pregnant women that analysed matched placental and umbilical cord samples found that increased placental expression of stress-related genes such as SLC6A4, HSD11B2, and CR2H2 was associated with hypermethylation at the IGF2/H19 imprinting region under conditions of maternal stress. The same study also reported reduced global methylation in placental tissue and low concordance between placental and umbilical cord blood methylation at these loci (Chen et al., 2014). Similarly, studies integrating maternal salivary cortisol levels (measured during the second and third trimesters) with oxytocin receptor (OXTR) methylation in cord blood showed that higher maternal cortisol linked to elevated depressive symptoms on the EPDS was associated with lower OXTR methylation. This finding suggests that fetal epigenetic patterns are sensitive to maternal stress-related biological signals (Unternaehrer et al., 2016). Collectively, these findings point to a relative independence between fetal epigenetic profiles and maternal peripheral methylation markers, underscoring the importance of tissue-specific effects in epigenetic studies.

3.2.3 Animal models as complementary models to elucidate methylation in perinatal depression

Animal models provide causal evidence that can recapitulate key features within a controlled framework of environmental exposure, genetic background, and developmental timings. These models explore the dysregulation in the glucocorticoid signalling pathway and the disruption of the intrauterine environment in response to the exposure. thereby providing an excellent framework for investigating foetal reprogramming. The causality is tested by the quantification of the DNA Methylation and gene expression in regions pertaining to maternal stress, such as the frontal cortex, amygdala, and hippocampus (Nemoda & Szyf, 2017). For instance, in a prenatal restraint mouse model, increased levels of DNMT1 and

Table 1 Summary of candidate gene methylation studies in perinatal depression.

Study	Study design	Sample	DNA methylation analysis	Biomarkers	Results
Jahnke et al. (2021) Galapagos cohort	Longitudinal, 24 Mother-Child Dyads Sample collection: 3rd Trimester: 34–36 gestation weeks Postpartum: 2 months	Maternal salivary cortisol [3 time points- Awakening, +30 min post awakening (CAR), Evening before sleep] Infant salivary cortisol [2 time points- pre and post stressors] Placenta (sampling at 4 quadrants of the placenta) Perceived Stress Scale and PHQ-8	Bisulphite Conversion (EZ DNA Methylation™ kit), Pyrosequencing	Placental methylation HSD11B2 (Measures at 4 CpG sites) HSD11B2 expression in maternal and infant cortisol	Maternal HPA dysregulation is associated with lower HSD11B2 expression, which leads to exaggerated cortisol reactivity in infants
Nazzari et al. (2022) MOM-COPE cohort	Longitudinal, multi-centric cohort N = 283 mother-infant dyads	Maternal and infant buccal cells collected within 24 h post-delivery BDI-II STAI-Y	Bisulphite-treated DNA + NEXTSeq-500 (Illumina, San Diego, California, USA)	NR3C1 (44 CpGs), and SLC6A4 (13 CpGs) Principal Component Analysis to reduce the methylation data to one dimension between Maternal and	Mothers and infants exposed to COVID-19 were measured across three trimesters. Mothers during 1st trimester showed significant lower methylation in LSC6A4 and NR3C1. Replicated in infants too.

(continued)

Table 1 Summary of candidate gene methylation studies in perinatal depression. (*cont'd*)

Study	Study design	Sample	DNA methylation analysis	Biomarkers	Results
				Infant PC gene methylation	Therefore, early pregnancy appears to be a critical window wherein stress induces stronger epigenetic modifications in both mother and infant
Braithwaite et al. (2015)	Longitudinal 57 mother-infant pairs Control = 37 Case = 20	Maternal Salivary cortisol (5 time points pre and post stress diurnal 6 samples) Infant buccal swabs EPDS	Bisulphite Conversion Pyrosequencing	Infant NR3C1 exon 1 F (10 CpG sites), and BDNF promoter IV	Sex specific response of male infants with a decrease in BDNF IV and increased NR3C1 exon 1 F methylation. Therefore, a positive association between EPDS and NR3C1 and a negative association with BDNF. However, no association of EPDS and maternal cortisol was observed
Chen et al. (2014) Stress in	Longitudinal Study 49 mother-infant pairs	Placental biopsy samples (4 quadrants) and Umbilical Cord Blood samples	Bisulphite pyrosequencing PyroMark Q24 (Qiagen)	HPA axis: 6 genes-HSD11B2, SLCA4, NR3C1, NR3C2, CRH1, and CRH2	Maternal stress correlated with higher levels of HPA genes, such as elevated SLC6A4 and HSD11B2,

Pregnancy Cohort		(Uncoagulated Blood in PAX tubes) PSS-14 STAI SCID-I GAF		Imprinting genes: IGF2, H9 Neurodevelopmental gene: EGR1 Global Methylation levels and analysis of methylation status of IGF1/H19	CRHR2 expression, and elevated EGR1, with a lower GAF score in the placenta Higher gene expression in Placenta 7/9. Low concordance was exhibited between the two tissues, except for H19 Moreover, methylation of imprinting genes was lower in the placenta than in UCB. Global methylation is higher in the placenta
Bell et al. (2015) ALSPAC cohort	Singleton Caucasian women N = 545 PPD cases = 269 Controls = 276 Adjusted analysis N = 500	Blood samples in pregnant individuals (between 7 and 41 weeks of gestation) EPDS	Bisulphite treatment (Kit MECOV50, Invitrogen, Carlsbad, CA, USA) + Pyromark PCR kit (Qiagen, Valencia, CA, USA)	Correlation of OXTR DNA methylation (CpG −934) with OXTR SNPs (rs53576, rs2254298)	No significant association between OXTR methylation and SNPs (rs53576, rs2254298) and PPD risk pre and post adjustment. Significant 3-way interaction between OXTR genotype, OXTR methylation, and depression

(continued)

Table 1 Summary of candidate gene methylation studies in perinatal depression. (*cont'd*)

Study	Study design	Sample	DNA methylation analysis	Biomarkers	Results
	(5 psychosocial confounding factors)				in both adjusted and unadjusted models. Women who didn't exhibit depression during pregnancy had the rs53576 GG genotype and higher methylation, exhibited a higher risk for PPD despite its non-association with methylation

Abbreviations used in Table 1:
BDNF: Brain-derived Neurotrophic factor
BDI-II: Beck's Depression Inventory Scale−II
CAR: Cortisol Awakening Response
CRHR1: Corticotropin-Releasing Hormone Receptor 1
CRHR2: Corticotropin-Releasing Hormone Receptor 2
EGR1: Early growth response protein 1
EPDS: Edinburgh Postnatal Depression Scale
GAF: Global Assessment of Functioning Scale
HSD11B2: 11β-hydroxysteroid dehydrogenase type 2
OXTR: Oxytocin Receptor
NR3C1: Nuclear Receptor Subfamily 3 Group C Member 1
NR3C2: Nuclear Receptor Subfamily 3 Group C Member 2
PSS-14: Perceived Stress Scale (PSS-14)
SCID-I: Structured Clinical Interview for DSM-IV Axis I Disorders
SNP: Single Nucleotide Polymorphism
SLC6A4: Solute Carrier Family 6 Member 4
STAI-Y State-Trait Anxiety Inventory (STAI)- Form Y
UCB: Umbilical Cord Blood

HDACs resulted in an increased methylation of BDNF I, IV, VI, and IX promoters in the hippocampus that reduced the BDNF mRNA expression (Zheng et al., 2016). Table 2 summarizes studies that have employed rodent models to examine locus-specific alterations in methylations induced by perinatal stress and hormonal exposure.

3.3 Epigenome-wide association studies (EWAS)

Epigenome-wide association studies examine epigenetic variations genome-wide and identify novel associations between the phenotype (i.e., maternal depression) and alterations in the epigenome. EWAS adopts a GWAS-like pipeline to independently assess the methylation patterns expressed at individual CpGs to identify differentially methylated regions (DMRs) (Bock, 2012). The premise of the approach is that the methylation of cytosine is not localised to a position but rather a proportion of methylated cytosines that thereby controls the chromatin conformation and the transcriptomic potential of the gene (Mill & Heijmans, 2013). Therefore, EWAS captures a global outlook on the epigenome across a large sample size set that neglects hypothesis-driven biases present in candidate genes.

3.3.1 Predictive potential of peripheral DNA methylation biomarkers in perinatal depression

A study on an Indian population from a longitudinal pregnancy cohort of 201 participants investigated the variation in methylation associated with antenatal depression. A panel of 7 CpG sites was identified with their respective annotated genes and 591 DMPs. The panel identified the CpG sites with the best discriminatory power were annotated to the following genes: RPL31, MAP3K7IP2, TMEM132D, C9orf129, ADAM8, BCL2, and PLA2G16 genes (Thirumoorthy et al., 2025). The panel was subsequently tested to assess its potential to identify postpartum risk, and the same 7 CpGs were associated with postpartum depression. The study also explored the association of 3 major pathways: inositol phosphate metabolism, calcium signalling, and notch signalling, respectively, that were critical for neurodevelopment and adult brain homeostasis (notch signalling) (Ables et al., 2011); progression of depression with lower levels of inositol (Barkai et al., 1978); and alteration of various cellular processes in bipolar disorder (Bowden et al., 1988; Harrison et al., 2021). Similarly, Guintivano et al. proposed a Postpartum Depression (PPD) prediction model from antenatal blood DNA methylation by the rationale that the

Table 2 DNA methylation signature of perinatal stress and hormonal modulation in animal and cross-species studies.

Study	Study design	DNA methylation analysis	Biomarkers	Results
Boersma et al. (2014)	Sprague Dawley rats (Gestation day 2), 12 prenatally stressed dams, and 12 controls (male only) Stress exposure at variable time points within the 3rd gestation week. Tissue collection PND21 (weaning), PND80 (adulthood), and analysis of hippocampus, PFC, and amygdala for gene expression analysis (qPCR)	Bisulphite Pyrosequencing	BDNF exon IV and BDNF VI	Increased methylation of BDNF Exon IV in amygdala and hippocampus during PND20 timepoint, and persistent increase in amygdala in PND80 timepoint. Therefore, resulting in a decreased BDNF expression. Increased expression of Dnmt1 and 3a supports the epigenetic regression model of Bdnf.
Plank et al. (2021)	C57Bl/6N mice, subjected to acute trauma by fear conditioning at day 12 using a foot shock paradigm (ss, 1.5 mA). Adult and Male offspring (PND150 and GD18) Behavioural Assessment: Adults-Elevated Plus Maize ELISA- corticosterone Tissue collection- dHPC	Bisulphite Conversion + MassARRAY (EpiTYPER)	Crhr1, Fkbp5, Nr3c1, Nr3c2	Elevated basal and stress-induced corticosterone levels in both dams and offspring, followed by the hemispheric-specific changes at the left dHPC-hypomethylation of the promoter regions of Crhr1, and an increased gene expression of Crhr1 in the offspring. Hypermethylation of Fkbp5 was observed. No changes in the Nr3c1 and Nr3c2 were observed

Guintivano et al. (2014)	Cross-species study wherein 15 out of the total 30 Ovariectomised C57BL6/6J mice received 17β-E2 by a subcutaneous implantation of a silastic capsule (randomised process, control received an empty capsule) across three time points (1,2,and 4 weeks) Tissue: Hippocampus Integration with a human longitudinal cohort of 93 pregnant women at risk of PPD	Sodium Bisulphite pyrosequencing EZ DNA Methylation Gold Kit (Zymo Research, Irvine, CA, USA)	EWAS on human blood samples, and methylation in hippocampal promoter regions	On estradiol exposure, 891 DMRs were identified that were enriched in transcription binding motifs, indicating epigenetic remodelling. An overlap of CpG sites was rendered with the integration of the human methylation profile involved in PPD. Two loci, HP1BP3 and TTC9B, were indicated as robust methylation markers with high accuracy in prediction of PPD risk.

Abbreviations used in Table 2:
BDNF: Brain-Derived Neurotrophic Factor
dHPC: Dorsal Hippocampus
Crhr1: Corticotropin-Releasing Hormone Receptor 1 gene
Crhr2: Corticotropin-Releasing Hormone Receptor 2 gene
Fkpb5: FK506 Binding Protein 5
PND 20: Postnatal Day 80
PND 80: Postnatal Day 80
PND150: Postnatal Day 150
E2: 17β-estradiol

women who eventually develop PPD have an increased sensitivity to oestrogen E2-mediated epigenetic changes through DNA methylation that can be detected in blood. The study employed a cross-species approach in which 1578 overlapping CpG sites from the antenatal depression period of 93 women were compared with mouse hippocampal 891 DMRs associated with the estrogenic response in response to the induction of 17β-oestradiol. A 103 loci subsequently depicted a nominal association with PPD, but only two loci within HP1BP3 and TTC9B with high accuracy were identified as predictive markers in euthymic women. (Guintivano et al., 2014). Both markers are implicated in oestrogen signalling pathways; however, TTC9B has also been associated with hippocampal synaptic plasticity and has been shown to inhibit AMPA receptor activity, a process essential for mechanisms such as long-term potentiation and depression (Gerges et al., 2004). Subsequent studies have replicated these findings, supporting their reliability (Osborne et al., 2016; Payne et al., 2020). Notably, these proposed biomarkers are present in approximately 80% of postpartum depression (PPD) cases. This has led to the suggestion that PPD may be classified into distinct subtypes, including those sensitive to oestrogen fluctuations and others influenced primarily by non-hormonal factors (Wenzel et al., 2025).

3.3.2 Dynamic regulation of methylation across trimesters

Fradin et al. demonstrated the dynamic state of methylation across gestation by employing a longitudinal sampling across trimesters that exhibited remodelling of the maternal epigenome, progressively from the 1st to 3rd trimester. 39 CpGs across 157 regions were identified that identified the temporal shift in immune regulation, neuroendocrine function, and tissue and placental remodelling. The 1st trimester was characterised by broad epigenetic changes that were followed by dynamic remodelling associated with placental neurodevelopment in mid-pregnancy and reinforcement of pathways linked to maternal-foetal communication and parturition in the third trimester. The study further demonstrates the distinction in stable and highly dynamic loci that maintains the core regulatory functions and pregnancy-responsive epigenetic plasticity. Therefore, the study collectively provided evidence of a time-dependent epigenomic regulation in a developmental progression (Fradin et al., 2023).

3.3.3 Synergistic influence of intrauterine environment and maternal DNA methylation patterns on infant epigenome

The placenta serves as a barrier to protect the developing foetus; the barrier can be compromised on exposure to maternal stress (Lund et al., 2021). Elevation of neuroendocrine levels, particularly glucocorticoid levels, was reported to cross the placenta and can, in turn, modify the foetal epigenome (Zucchi et al., 2013). For example, the most extensively studied gene, the NR3C1 gene (glucocorticoid receptor gene), was reported to be higher in the third trimester of pregnancy and was associated with increased methylation in the promoter region of the cord blood sample of the foetus (Oberlander et al., 2008). Similar outcomes were observed in other candidate genes, such as FKBP5, SLC6A4, and their altered response in behaviour (Buss et al., 2012; Oberlander et al., 2008; Zucchi et al., 2013). Therefore, the foetus is subjected to a concept of time-sensitive windows during growth and differentiation, wherein the foetus is vulnerable to maternal perturbations (Nagy & Turecki, 2012). During this period, the foetus undergoes extensive epigenetic reprogramming wherein the embryo's epigenome is being reset and reestablished at the tissue level. Several cohorts have explored the exposure to prenatal stress during specific time points during the pregnancy and its association with the foetus's methylome.

For instance, the FinnBrain Birth Cohort study of 252 pregnant women demonstrated that the first-trimester maternal symptoms are associated with altered placental DNA methylation marks on genomic and intergenomic regions that are important for the development and long-term growth. Out of the 11 predominantly significant methylation marks, the SHROOM1 gene (intron) and PWWP3A were the most significant (Lund et al., 2021). The protein encoded by the SHROOM1 gene promotes neurulation (Hildebrand & Soriano, 1999) and the promotion of cell survival in response to genotoxic stress (Huen et al., 2010). Furthermore, a multi-ancestry cohort of 301 pregnant women identified a panel of 16 CpG sites associated with the 2nd and 3rd trimesters, and 2 CpGs that were associated with perceived stress (Tesfaye et al., 2021). A functional analysis was also performed to investigate whether placental DNA methylation was associated with the expression of nearby genes, including ADAM23 and CTDP1, which are linked to neurodevelopment and have been implicated in neuropsychiatric conditions. The temporal fluctuations in the ADAM 23 gene expression may not be reflected in the transcriptomic level beyond the neonatal period, and hence, elucidate the crucial time-sensitive window that is involved in utero neurodevelopmental programming (Sun et al., 2004).

3.3.4 Exploration of gene-environmental interactions in PND (G x E) and intervention potential

Perinatal depression is shaped by the interaction of gene and environment, wherein the interaction often manifests at the epigenomic level, such that the genetic susceptibility can influence the outcome of epigenomic changes to environmental exposure (Álvarez-Mejía et al., 2025). Similarly, the environmental factor can modulate the functional role of risk -associated genes via epigenetic modifications (Vaiserman, 2015). The cross-over model can potentially program certain genes in the foetal genetic makeup to become more or less responsive to maternal influences, thereby creating an individualistic trajectory of risk/resilience (Dieckmann & Czamara, 2024). Mitchell et al. provided the framework of the cross-over model in PPD. The study explored two polymorphisms in the serotonin transporter gene (5-HTT)- %-HTTLPR, and Stin2 VNTR, respectively, that demonstrated a dynamic interaction with the environmental factor, i.e., socio-economic status. Women who exhibited a higher number of reactive allele counts displayed a greater sensitivity to the socio-economic status (SES), wherein a low SES was associated with increased PPD risk, and a higher SES exhibited a protective effect (Mitchell et al., 2011). Furthermore, the heightened sensitivity of epigenetic mechanisms due to the influence of extensive hormonal, immune, and developmental remodeling can establish a window wherein environmental inputs can influence the maternal neurobiological pathways and foetal development. This window can be crucial for early intervention strategies, such as diet and lifestyle, to mitigate epigenetic risk in perinatal depression. The extensive role of epigenetic plasticity has been explored during the perinatal period in response to nutritional supplements such as choline, folates, and zinc. These supplements, involved in the one-carbon metabolism, serve as methyl-donors that contribute to the formation of a universal methyl donor- S-adenosylmethionine (SAM) that maintains DNA and histone methylation (Friso et al., 2017; Mentch & Locasale, 2016). However, the existing studies are restricted to covering the impact of these supplements on the offspring's neurodevelopment and provide a limited understanding of their influence on the maternal epigenome.

3.3.5 Methodological and analytical challenges in DNA methylation studies

The development of methylation markers for the prediction of perinatal depression is limited to the methodological challenges during EWAS. The predominant issue lies in the identification of differentially methylated positions vs differentially methylated regions, wherein DMPs assess each CpG site in a GWAS-like fashion, whereas the DMR approach reflects the regional methylation patterns that capture the functional regulation of gene expression. However, both approaches have limitations such that DMPs are prone to more false positives due to multiple testing, whereas DMRs may miss single-site specific CpGs associated with the disease. Secondly, tissue heterogeneity, in most tissues, contains multiple epigenetically distinct cell types that can be directly associated with the disease outcome, such as inflammation. For instance, whole blood tissues do contain distinct cell populations. To counter the variation in routine cell counts, algorithms can be applied. However, these approaches may not be sufficient, and isolation of particular cell types can improve power detection. Furthermore, replication across EWAS studies remains the most significant barrier, wherein the differences due to study design, sample timing, analytical pipelines, and ethnicity may reduce reproducibility. Therefore, standardized analytical workflows and profiling platforms can ensure reproducibility (Mill & Heijmans, 2013).

3.3.6 Risk assessment models in perinatal depression

Prediction models developed to explore perinatal depression risk can aid in timely diagnosis and intervention. The current risk-assessment tool is restricted to the genetic and epigenetic contributions in isolation. The Polygenic Risk Score analyses the cumulative genetic liability by selecting SNPs associated with PND. Similarly, Methylation Risk Score, an analogue to PRS, quantifies the cumulative effects of environmental exposure on differentially methylated CpG sites associated with the particular disease, thereby providing a comprehensive outlook of the epigenetic risk (Ohi et al., 2025). MRS tracks the dynamic shift in expression and demonstrates a model that outperforms PRS's predictive performance (Thompson et al., 2022). Unfortunately, the current literature lacks a unified framework model that utilises the combinatorial performance of the PRS-MRS model in risk assessment, leveraging multi-omics data that thereby links genetic architecture with real-time functional regulation. Furthermore, the integration of peripheral tissue, such as blood-based methylation signatures, mentioned in the previous section, can predict time-point-associated risks as potential diagnostic tools (Payne et al., 2020).

4. Future directions

Future studies should prioritize large-scale, longitudinal, multi-ancestry cohort studies to improve the reproducibility and generalizability of epigenetic findings in PND. Standardization of EWAS methodologies, including sampling time points, tissue selection, and analytical pipelines, is essential to reduce variability across studies. Integrative multi-omics frameworks combining polygenic risk scores (PRS), methylation risk scores (MRS), transcriptomics, and environmental data should be developed to better capture the complex gene–environment interplay underlying PND.

Additionally, more studies are needed to clarify tissue-specific methylation signatures and their concordance across maternal and fetal compartments. Exploring dynamic epigenetic changes across pregnancy trimesters may help identify critical windows for intervention. Translational studies assessing the impact of modifiable factors such as nutrition, stress reduction, and lifestyle interventions on the maternal epigenome are also warranted. Finally, the development of clinically applicable, blood-based methylation biomarkers for early prediction and stratification of PND risk remains a key priority.

5. Conclusion

Perinatal depression is a complex disorder shaped by the interaction of genetic predisposition, environmental exposures, and dynamic epigenetic regulation. DNA methylation plays a central role in mediating these interactions, offering insights into both disease mechanisms and potential biomarkers. While current evidence underscores the promise of epigenetic approaches particularly EWAS in identifying predictive signatures, challenges related to tissue specificity, methodological variability, and limited replication hinder clinical application. Advancing the field will require integrative, standardized, and longitudinal research efforts that bridge genetic and environmental frameworks. Ultimately, leveraging epigenetic insights may enable earlier diagnosis, targeted interventions, and improved maternal and offspring outcomes.

Conflict of interest

Nil.

References

Ables, J. L., Breunig, J. J., Eisch, A. J., & Rakic, P. (2011). Not(ch) just development: Notch signalling in the adult brain. *Nature Reviews. Neuroscience, 12*(5), 269–283. https://doi.org/10.1038/nrn3024.

Álvarez-Mejía, D., Rodas, J. A., & Leon-Rojas, J. E. (2025). From womb to mind: Prenatal epigenetic influences on mental health disorders. *International Journal of Molecular Sciences, 26*(13), 6096. https://doi.org/10.3390/ijms26136096.

AP, A. (2000). Diagnostic and statistical manual of mental disorders (DSM-IV-TR), Fourth ed. Text revision. https://cir.nii.ac.jp/crid/1370285712545076872.

Barkai, A. I., Dunner, D. L., Gross, H. A., Mayo, P., & Fieve, R. R. (1978). Reduced myo-inositol levels in cerebrospinal fluid from patients with affective disorder. *Biological Psychiatry, 13*(1), 65–72.

Bell, A. F., Carter, C. S., Steer, C. D., Golding, J., Davis, J. M., Steffen, A. D., ... Connelly, J. J. (2015). Interaction between oxytocin receptor DNA methylation and genotype is associated with risk of postpartum depression in women without depression in pregnancy. *Frontiers in Genetics, 6*. https://doi.org/10.3389/fgene.2015.00243.

Bernstein, B. E., Meissner, A., & Lander, E. S. (2007). The mammalian epigenome. *Cell, 128*(4), 669–681. https://doi.org/10.1016/j.cell.2007.01.033.

Bird, A., Taggart, M., Frommer, M., Miller, O. J., & Macleod, D. (1985). A fraction of the mouse genome that is derived from islands of nonmethylated, CpG-rich DNA. *Cell, 40*(1), 91–99. https://doi.org/10.1016/0092-8674(85)90312-5.

Bluett-Duncan, M., Kishore, M. T., Patil, D. M., Satyanarayana, V. A., & Sharp, H. (2021). A systematic review of the association between perinatal depression and cognitive development in infancy in low and middle-income countries. *PLoS One, 16*(6), e0253790. https://doi.org/10.1371/journal.pone.0253790.

Bock, C. (2012). Analysing and interpreting DNA methylation data. *Nature Reviews. Genetics, 13*(10), 705–719. https://doi.org/10.1038/nrg3273.

Boersma, G. J., Lee, R. S., Cordner, Z. A., Ewald, E. R., Purcell, R. H., Moghadam, A. A., & Tamashiro, K. L. (2014). Prenatal stress decreases Bdnf expression and increases methylation of Bdnf exon IV in rats. *Epigenetics: Official Journal of the DNA Methylation Society, 9*(3), 437–447. https://doi.org/10.4161/epi.27558.

Bosker, F. J., Hartman, C. A., Nolte, I. M., Prins, B. P., Terpstra, P., Posthuma, D., ... Nolen, W. A. (2011). Poor replication of candidate genes for major depressive disorder using genome-wide association data. *Molecular Psychiatry, 16*(5), 516–532. https://doi.org/10.1038/mp.2010.38.

Bowden, C. L., Huang, L. G., Javors, M. A., Johnson, J. M., Seleshi, E., Mclntyre, K., ... James, W. M. (1988). Calcium function in affective disorders and healthy controls. *Biological Psychiatry, 23*(4), 367–376. https://doi.org/10.1016/0006-3223(88)90287-9.

Braithwaite, E., Kundakovic, M., Ramchandani, P., Murphy, S., & Champagne, F. (2015). Maternal prenatal depressive symptoms predict infant NR3C1 1F and BDNF IV DNA methylation. *Epigenetics: Official Journal of the DNA Methylation Society, 10*(5), 408–417. https://doi.org/10.1080/15592294.2015.1039221.

Buss, C., Davis, E. P., Shahbaba, B., Pruessner, J. C., Head, K., & Sandman, C. A. (2012). Maternal cortisol over the course of pregnancy and subsequent child amygdala and hippocampus volumes and affective problems. *Proceedings of the National Academy of Sciences, 109*(20), E1312–E1319. https://doi.org/10.1073/pnas.1201295109.

Carlson, K., Mughal, S., Azhar, Y., & Siddiqui, W. (2025). *Perinatal depression. StatPearls*. StatPearls Publishing. http://www.ncbi.nlm.nih.gov/books/NBK519070/.

Carnevali, G. S., & Buoli, M. (2021). The role of epigenetics in perinatal depression: Are there any candidate biomarkers? *Journal of Affective Disorders, 280*, 57–67. https://doi.org/10.1016/j.jad.2020.11.056.

Caspary, T., Cleary, M. A., Baker, C. C., Guan, X.-J., & Tilghman, S. M. (1998). Multiple mechanisms regulate imprinting of the mouse distal chromosome 7 gene cLuster. *Molecular and Cellular Biology, 18*(6), 3466–3474. https://doi.org/10.1128/MCB.18.6.3466.

Chen, J., Li, Q., Rialdi, A., Mystal, E., Ly, J., Finik, J., ... Nomura, Y. (2014). Influences of maternal stress during pregnancy on the epi/genome: Comparison of placenta and umbilical cord blood. *Journal of Depression & Anxiety, 3*(2), 152. https://doi.org/10.4172/2167-1044.1000152.

Choi, J. D., Underkoffler, L. A., Wood, A. J., Collins, J. N., Williams, P. T., Golden, J. A., ... Oakey, R. J. (2005). A novel variant of Inpp5f is imprinted in brain, and its expression is correlated with differential methylation of an internal CpG Island. *Molecular and Cellular Biology, 25*(13), 5514–5522. https://doi.org/10.1128/MCB.25.13.5514-5522.2005.

Dieckmann, L., & Czamara, D. (2024). Epigenetics of prenatal stress in humans: The current research landscape. *Clinical Epigenetics, 16*, 20. https://doi.org/10.1186/s13148-024-01635-9.

Duncan, L. E., Ostacher, M., & Ballon, J. (2019). How genome-wide association studies (GWAS) made traditional candidate gene studies obsolete. *Neuropsychopharmacology: Official Publication of the American College of Neuropsychopharmacology, 44*(9), 1518–1523. https://doi.org/10.1038/s41386-019-0389-5.

Feinberg, A. P., & Irizarry, R. A. (2010). Evolution in health and medicine Sackler colloquium: Stochastic epigenetic variation as a driving force of development, evolutionary adaptation, and disease. *Proceedings of the National Academy of Sciences of the United States of America, 107*(1), 1757–1764. https://doi.org/10.1073/pnas.0906183107.

Feng, J., Chang, H., Li, E., & Fan, G. (2005). Dynamic expression of de novo DNA methyltransferases Dnmt3a and Dnmt3b in the central nervous system. *Journal of Neuroscience Research, 79*(6), 734–746. https://doi.org/10.1002/jnr.20404.

Forty, L., Jones, L., Macgregor, S., Caesar, S., Cooper, C., Hough, A., ... Jones, I. (2006). Familiality of postpartum depression in unipolar disorder: Results of a family study. *American Journal of Psychiatry, 163*, 1549–1553. https://doi.org/10.1176/ajp.2006.163.9.1549.

Fradin, D., Tost, J., Busato, F., Mille, C., Lachaux, F., Deleuze, J.-F., ... Benachi, A. (2023). DNA methylation dynamics during pregnancy. *Frontiers in Cell and Developmental Biology, 11*. https://doi.org/10.3389/fcell.2023.1185311.

Friso, S., Udali, S., De Santis, D., & Choi, S.-W. (2017). One-carbon metabolism and epigenetics. *Molecular Aspects of Medicine, Epigenetics and Metabolism, 54*, 28–36. https://doi.org/10.1016/j.mam.2016.11.007.

Ganjiwale, J., & Nimbalkar, S. (2025). Perinatal depression in India: A narrative review of prevalence, risk factors, and healthcare provider awareness. *Journal of Family Medicine and Primary Care, 14*(9), 3630–3636. https://doi.org/10.4103/jfmpc.jfmpc_370_25.

Gerges, N. Z., Tran, I. C., Backos, D. S., Harrell, J. M., Chinkers, M., Pratt, W. B., & Esteban, J. A. (2004). Independent functions of hsp90 in neurotransmitter release and in the continuous synaptic cycling of AMPA receptors. *Journal of Neuroscience, 24*(20), 4758–4766. https://doi.org/10.1523/JNEUROSCI.0594-04.2004.

Ghaedrahmati, M., Kazemi, A., Kheirabadi, G., Ebrahimi, A., & Bahrami, M. (2017). Postpartum depression risk factors: A narrative review. *Journal of Education and Health Promotion, 6*, 60. https://doi.org/10.4103/jehp.jehp_9_16.

Goto, K., Numata, M., Komura, J.-I., Ono, T., Bestor, T. H., & Kondo, H. (1994). Expression of DNA methyltransferase gene in mature and immature neurons as well as proliferating cells in mice. *Differentiation; Research in Biological Diversity, 56*(1), 39–44. https://doi.org/10.1046/j.1432-0436.1994.56120039.x.

Guintivano, J., Arad, M., Gould, T. D., Payne, J. L., & Kaminsky, Z. A. (2014). Antenatal prediction of postpartum depression with blood DNA methylation biomarkers. *Molecular Psychiatry, 19*(5), 560–567. https://doi.org/10.1038/mp.2013.62.

Gujar, H., Weisenberger, D. J., & Liang, G. (2019). The roles of human DNA methyltransferases and their isoforms in shaping the epigenome. *Genes, 10*(2), 172. https://doi.org/10.3390/genes10020172.

Gwynn, B., Lueders, K., Sands, M. S., & Birkenmeier, E. H. (1998). Intracisternal A-particle element transposition into the murine β-glucuronidase gene correlates with loss of enzyme activity: A new model for β-glucuronidase deficiency in the C3H mouse. *Molecular and Cellular Biology, 18*(11), 6474–6481. https://doi.org/10.1128/MCB.18.11.6474.

Harrison, P. J., Hall, N., Mould, A., Al-Juffali, N., & Tunbridge, E. M. (2021). Cellular calcium in bipolar disorder: Systematic review and meta-analysis. *Molecular Psychiatry, 26*(8), 4106–4116. https://doi.org/10.1038/s41380-019-0622-y.

Hildebrand, J. D., & Soriano, P. (1999). Shroom, a PDZ Domain–containing actin-binding protein, is required for neural tube morphogenesis in mice. *Cell, 99*(5), 485–497. https://doi.org/10.1016/S0092-8674(00)81537-8.

Huen, M. S. Y., Huang, J., Leung, J. W. C., Sy, S. M.-H., Leung, K. M., Ching, Y.-P., ... Chen, J. (2010). Regulation of chromatin architecture by the PWWP domain-containing DNA damage-responsive factor EXPAND1/MUM1. *Molecular Cell, 37*(6), 854–864. https://doi.org/10.1016/j.molcel.2009.12.040.

Illingworth, R. S., Gruenewald-Schneider, U., Webb, S., Kerr, A. R. W., James, K. D., Turner, D. J., ... Bird, A. P. (2010). Orphan CpG Islands identify numerous conserved promoters in the mammalian genome. *PLoS Genetics, 6*(9), e1001134. https://doi.org/10.1371/journal.pgen.1001134.

Jahnke, J. R., Terán, E., Murgueitio, F., Cabrera, H., & Thompson, A. L. (2021). Maternal stress, placental 11β-hydroxysteroid dehydrogenase type 2, and infant HPA axis development in humans: Psychosocial and physiological pathways. *Placenta, 104*, 179–187. https://doi.org/10.1016/j.placenta.2020.12.008.

King, L., Robins, S., Chen, G., Yerko, V., Zhou, Y., Nagy, C., ... Zelkowitz, P. (2017). Perinatal depression and DNA methylation of oxytocin-related genes: A study of mothers and their children. *Hormones and Behavior, 96*, 84–94. https://doi.org/10.1016/j.yhbeh.2017.09.006.

Koleva, H., Stuart, S., O'Hara, M. W., & Bowman-Reif, J. (2011). Risk factors for depressive symptoms during pregnancy. *Archives of Women's Mental Health, 14*(2), 99–105. https://doi.org/10.1007/s00737-010-0184-0.

Konjevod, M., Gredicak, M., Vuic, B., Tudor, L., Nikolac Perkovic, M., Milos, T., ... Nedic Erjavec, G. (2023). Overview of metabolomic aspects in postpartum depression. *Progress in Neuro-Psychopharmacology and Biological Psychiatry, 127*, 110836. https://doi.org/10.1016/j.pnpbp.2023.110836.

Kroh, S. J., & Lim, G. (2021). Perinatal depression. *International Anesthesiology Clinics, 59*(3), 45–51. https://doi.org/10.1097/AIA.0000000000000325.

Kuster, J. E., Guarnieri, M. H., Ault, J. G., Flaherty, L., & Swiatek, P. J. (1997). IAP insertion in the murine LamB3 gene results in junctional epidermolysis bullosa. *Mammalian Genome, 8*(9), 673–681. https://doi.org/10.1007/s003359900535.

Lancaster, C. A., Gold, K. J., Flynn, H. A., Yoo, H., Marcus, S. M., & Davis, M. M. (2010). Risk factors for depressive symptoms during pregnancy: A systematic review. *American Journal of Obstetrics and Gynecology, 202*(1), 5–14. https://doi.org/10.1016/j.ajog.2009.09.007.

Landan, G., Cohen, N. M., Mukamel, Z., Bar, A., Molchadsky, A., Brosh, R., ... Tanay, A. (2012). Epigenetic polymorphism and the stochastic formation of differentially methylated regions in normal and cancerous tissues. *Nature Genetics, 44*(11), 1207–1214. https://doi.org/10.1038/ng.2442.

Lund, R. J., Kyläniemi, M., Pettersson, N., Kaukonen, R., Konki, M., Scheinin, N. M., ... Ekholm, E. (2021). Placental DNA methylation marks are associated with maternal depressive symptoms during early pregnancy. *Neurobiology of Stress, 15*, 100374. https://doi.org/10.1016/j.ynstr.2021.100374.

Ma, X., Chen, X., Mu, X., Cao, M., & Zhang, Y. (2025). Epigenetics of maternal-fetal interface immune microenvironment and placental related pregnancy complications. *Frontiers in Immunology, 16*. https://doi.org/10.3389/fimmu.2025.1549839.

McIntosh, A. M., Sullivan, P. F., & Lewis, C. M. (2019). Uncovering the genetic architecture of major depression. *Neuron, 102*(1), 91–103. https://doi.org/10.1016/j.neuron.2019.03.022.

Mentch, S. J., & Locasale, J. W. (2016). One-carbon metabolism and epigenetics: Understanding the specificity. *Annals of the New York Academy of Sciences, 1363*(1), 91–98. https://doi.org/10.1111/nyas.12956.

Michaud, E. J., Vugt, M. J. van, Bultman, S. J., Sweet, H. O., Davisson, M. T., & Woychik, R. P. (1994). Differential expression of a new dominant agouti allele (Aiapy) is correlated with methylation state and is influenced by parental lineage. *Genes & Development, 8*(12), 1463–1472. https://doi.org/10.1101/gad.8.12.1463.

Mikkelsen, T. S., Ku, M., Jaffe, D. B., Issac, B., Lieberman, E., Giannoukos, G., ... Bernstein, B. E. (2007). Genome-wide maps of chromatin state in pluripotent and lineage-committed cells. *Nature, 448*(7153), 553–560. https://doi.org/10.1038/nature06008.

Mill, J., & Heijmans, B. T. (2013). From promises to practical strategies in epigenetic epidemiology. *Nature Reviews. Genetics, 14*(8), 585–594. https://doi.org/10.1038/nrg3405.

Mitchell, C., Notterman, D., Brooks-Gunn, J., Hobcraft, J., Garfinkel, I., Jaeger, K., ... McLanahan, S. (2011). Role of mother's genes and environment in postpartum depression. *Proceedings of the National Academy of Sciences, 108*(20), 8189–8193. https://doi.org/10.1073/pnas.1014129108.

Mohn, F., Weber, M., Rebhan, M., Roloff, T. C., Richter, J., Stadler, M. B., ... Schübeler, D. (2008). Lineage-specific polycomb targets and De Novo DNA methylation define restriction and potential of neuronal progenitors. *Molecular Cell, 30*(6), 755–766. https://doi.org/10.1016/j.molcel.2008.05.007.

Moore, L. D., Le, T., & Fan, G. (2013). DNA methylation and its basic function. *Neuropsychopharmacology: Official Publication of the American College of Neuropsychopharmacology, 38*(1), 23–38. https://doi.org/10.1038/npp.2012.112.

Nagy, C., & Turecki, G. (2012). Sensitive periods in epigenetics: Bringing us closer to complex behavioral phenotypes. *Epigenomics, 4*(4), 445–457. https://doi.org/10.2217/epi.12.37.

Nazzari, S., Grumi, S., Mambretti, F., Villa, M., Giorda, R., & Provenzi, L. (2022). Maternal and infant NR3C1 and SLC6A4 epigenetic signatures of the COVID-19 pandemic lockdown: when timing matters. *Translational Psychiatry, 12*(1), 386. https://doi.org/10.1038/s41398-022-02160-0.

Nemoda, Z., & Szyf, M. (2017). Epigenetic alterations and prenatal maternal depression. *Birth Defects Research, 109*(12), 888–897. https://doi.org/10.1002/bdr2.1081.

Oberlander, T. F., Weinberg, J., Papsdorf, M., Grunau, R., Misri, S., & Devlin, A. M. (2008). Prenatal exposure to maternal depression, neonatal methylation of human glucocorticoid receptor gene (NR3C1) and infant cortisol stress responses. *Epigenetics: Official Journal of the DNA Methylation Society, 3*(2), 97–106. https://doi.org/10.4161/epi.3.2.6034.

Ohi, K., Fujikane, D., Takai, K., Kuramitsu, A., Muto, Y., Sugiyama, S., & Shioiri, T. (2025). Methylation risk scores in psychiatric disorders: Advancing epigenetic research in mental health. *JMA Journal, 8*(2), 363–370. https://doi.org/10.31662/jmaj.2024-0329.

Osborne, L., Clive, M., Kimmel, M., Gispen, F., Guintivano, J., Brown, T., ... Kaminsky, Z. (2016). Replication of epigenetic postpartum depression biomarkers and variation with hormone levels. *Neuropsychopharmacology: Official Publication of the American College of Neuropsychopharmacology, 41*(6), 1648–1658. https://doi.org/10.1038/npp.2015.333.

Payne, J. L., Osborne, L. M., Cox, O., Kelly, J., Meilman, S., Jones, I., ... Kaminsky, Z. A. (2020). DNA methylation biomarkers prospectively predict both antenatal and postpartum depression. *Psychiatry Research, 285*, 112711. https://doi.org/10.1016/j.psychres.2019.112711.

Plank, A.-C., Frey, S., Basedow, L. A., Solati, J., Canneva, F., von Hörsten, S., ... Golub, Y. (2021). Prenatally traumatized mice reveal hippocampal methylation and expression changes of the stress-related genes Crhr1 and Fkbp5. *Translational Psychiatry, 11*, 183. https://doi.org/10.1038/s41398-021-01293-y.

Raghavan, V., Khan, H. A., Seshu, U., Rai, S. P., Durairaj, J., Aarthi, G., ... Thara, R. (2021). Prevalence and risk factors of perinatal depression among women in rural Bihar: A community-based cross-sectional study. *Asian Journal of Psychiatry, 56*, 102552. https://doi.org/10.1016/j.ajp.2021.102552.

Rodríguez, M. A., Valentine, J., Ahmed, S. R., Eisenman, D. P., Sumner, L. A., Heilemann, M. V., & Liu, H. (2010). Intimate partner violence and maternal depression during the perinatal period: A longitudinal investigation of latinas. *Violence Against Women, 16*(5), 543–559. https://doi.org/10.1177/1077801210366959.

Sandoval, J., Heyn, H., Moran, S., Serra-Musach, J., Pujana, M. A., Bibikova, M., & Esteller, M. (2011). Validation of a DNA methylation microarray for 450,000 CpG sites in the human genome. *Epigenetics: Official Journal of the DNA Methylation Society, 6*(6), 692–702. https://doi.org/10.4161/epi.6.6.16196.

Saxonov, S., Berg, P., & Brutlag, D. L. (2006). A genome-wide analysis of CpG dinucleotides in the human genome distinguishes two distinct classes of promoters. *Proceedings of the National Academy of Sciences, 103*(5), 1412–1417. https://doi.org/10.1073/pnas.0510310103.

Schulz, W. A., Steinhoff, C., & Florl, A. R. (2006). Methylation of endogenous human retroelements in health and disease. In W. Doerfler, & P. Böhm (Eds.). *DNA methylation: Development, genetic disease and cancer* (pp. 211–250). Springer. https://doi.org/10.1007/3-540-31181-5_11.

Sullivan, P. F., Eaves, L. J., Kendler, K. S., & Neale, M. C. (2001). Genetic case-control association studies in neuropsychiatry. *Archives of General Psychiatry, 58*(11), 1015–1024. https://doi.org/10.1001/archpsyc.58.11.1015.

Sun, Y. P., Deng, K. J., Wang, F., Zhang, J., Huang, X., Qiao, S., & Zhao, S. (2004). Two novel isoforms of *Adam*23 expressed in the developmental process of mouse and human brains. *Gene, 325*, 171–178. https://doi.org/10.1016/j.gene.2003.10.012.

Tazi, J., & Bird, A. (1990). Alternative chromatin structure at CpG islands. *Cell, 60*(6), 909–920. https://doi.org/10.1016/0092-8674(90)90339-G.

Tesfaye, M., Chatterjee, S., Zeng, X., Joseph, P., & Tekola-Ayele, F. (2021). Impact of depression and stress on placental DNA methylation in ethnically diverse pregnant women. *Epigenomics, 13*(18), 1485–1496. https://doi.org/10.2217/epi-2021-0192.

Thirumoorthy, C., Sharma, K. K., Deepa, M., Yogaprabhu, S., Sneha, J., Rekha, R. P., ... Gokulakrishnan, K. (2025). Epigenome-wide association study identifies a specific panel of DNA methylation signatures for antenatal and postpartum depressive symptoms. *Journal of Affective Disorders, 383*, 195–206. https://doi.org/10.1016/j.jad.2025.04.083.

Thompson, M., Hill, B. L., Rakocz, N., Chiang, J. N., Geschwind, D., Sankararaman, S., ... Halperin, E. (2022). Methylation risk scores are associated with a collection of phenotypes within electronic health record systems. *Npj Genomic Medicine*, 7(1), 50. https://doi.org/10.1038/s41525-022-00320-1.

Ukai, H., Ishii-Oba, H., Ukai-Tadenuma, M., Ogiu, T., & Tsuji, H. (2003). Formation of an active form of the interleukin-2/15 receptor β-chain by insertion of the intracisternal a particle in a radiation-induced mouse thymic lymphoma and its role in tumorigenesis. *Molecular Carcinogenesis, 37*(2), 110–119. https://doi.org/10.1002/mc.10128.

Unternaehrer, E., Bolten, M., Nast, I., Staehli, S., Meyer, A. H., Dempster, E., ... Meinlschmidt, G. (2016). Maternal adversities during pregnancy and cord blood oxytocin receptor (OXTR) DNA methylation. *Social Cognitive and Affective Neuroscience, 11*(9), 1460–1470. https://doi.org/10.1093/scan/nsw051.

Upadhyay, R. P., Chowdhury, R., Aslyeh Salehi, Sarkar, K., Singh, S. K., Sinha, B., ... Kumar, A. (2017). Postpartum depression in India: A systematic review and meta-analysis. *Bulletin of the World Health Organization, 95*(10), 706–717C. https://doi.org/10.2471/BLT.17.192237.

Vaiserman, A. (2015). Epidemiologic evidence for association between adverse environmental exposures in early life and epigenetic variation: A potential link to disease susceptibility? *Clinical Epigenetics,* 7(1), 96. https://doi.org/10.1186/s13148-015-0130-0.

Viktorin, A., Meltzer-Brody, S., Kuja-Halkola, R., Sullivan, P. F., Landén, M., Lichtenstein, P., & Magnusson, P. K. E. (2016). Heritability of perinatal depression and genetic overlap with nonperinatal depression. *American Journal of Psychiatry, 173,* 158–165. https://doi.org/10.1176/appi.ajp.2015.15010085.

Wenzel, E. S., Frye, R., Roberson-Nay, R., & Payne, J. L. (2025). The neurobiology of postpartum depression. *Trends in Neurosciences, 48*(7), 469–482. https://doi.org/10.1016/j.tins.2025.05.005.

Wu, M., Rinchik, E. M., Wilkinson, E., & Johnson, D. K. (1997). Inherited somatic mosaicism caused by an intracisternal a particle insertion in the mouse tyrosinase gene. *Proceedings of the National Academy of Sciences, 94*(3), 890–894. https://doi.org/10.1073/pnas.94.3.890.

Wutz, A., Smrzka, O. W., Schweifer, N., Schellander, K., Wagner, E. F., & Barlow, D. P. (1997). Imprinted expression of the Igf2r gene depends on an intronic CpG island. *Nature, 389*(6652), 745–749. https://doi.org/10.1038/39631.

Zheng, Y., Fan, W., Zhang, X., & Dong, E. (2016). Gestational stress induces depressive-like and anxiety-like phenotypes through epigenetic regulation of BDNF expression in offspring hippocampus. *Epigenetics: Official Journal of the DNA Methylation Society, 11*(2), 150–162. https://doi.org/10.1080/15592294.2016.1146850.

Zucchi, F. C. R., Yao, Y., Ward, I. D., Ilnytskyy, Y., Olson, D. M., Benzies, K., ... Metz, G. A. S. (2013). Maternal stress induces epigenetic signatures of psychiatric and neurological diseases in the offspring. *PLoS One, 8*(2), e56967. https://doi.org/10.1371/journal.pone.0056967.

Zwart, R., Sleutels, F., Wutz, A., Schinkel, A. H., & Barlow, D. P. (2001). Bidirectional action of the Igf2r imprint control element on upstream and downstream imprinted genes. *Genes & Development, 15*(18), 2361–2366. https://doi.org/10.1101/gad.206201.

CHAPTER TWELVE

A short guide on *in vivo* transfection of flag-HDAC4 in dentate gyrus for study of adult neurogenesis in repetitive mild traumatic brain injury

Singh Ankit Satyaprakash[a,1], Nitesh Kumar Gupta[a,1], Humaira Farooq[b,1], Surya Pratap Singh[a], and Rajaneesh Kumar Gupta[b,*]

[a]Department of Biochemistry, Institute of Science, Banaras Hindu University, Varanasi, UP, India
[b]Department of Molecular & Human Genetics, Institute of Science, Banaras Hindu University, Varanasi, UP, India
*Corresponding author. e-mail address: guptarajaneesh@gmail.com; guptarajaneesh@bhu.ac.in

Contents

[1] Equal contribution.

Methods in Enzymology, Volume 734
ISSN 0076-6879, https://doi.org/10.1016/bs.mie.2026.05.058

Abstract

HDAC4, a class II epigenetic regulator, plays a critical role in adult hippocampal neurogenesis and determines the fate of newly proliferating and/or differentiating neural stem cells (NSCs). However, repetitive mild traumatic brain injury (rmTBI) impairs neurogenesis, contributing to the pathophysiology of cognitive deficits observed in neurodegenerative and neuropsychiatric disorders. Here, we present a protocol: (1) to induce rmTBI in mice using a closed head injury (CHI) device and to validate the model by assessing cognitive/behavioral performances and (2) to perform intracranial injection of FLAG-HDAC4 into the dentate gyrus of adult mice using a stereotaxic apparatus, in order to investigate the role of HDAC4 in impaired adult neurogenesis. This is achieved by analyzing $BrdU^+$ and $Nestin^+$ cells (markers of NSC proliferation), DCX^+ and $ND1^+$ cells (markers of neuronal differentiation), along with immunohistochemistry, fluorescence microscopy, and image analysis.

1. Introduction

Traumatic Brain Injury (TBI) is a multifactorial clinical condition involving a cascade of acute and chronic reactions as a secondary response, triggered by single or multiple biomechanical injuries. Major prevalent consequence of which is neurodegeneration, the deteriorating state of neurons eventually leading to neuronal death and cognitive impairment (DeKosky & Asken, 2017; Denniss & Barker, 2023; Dodd, Smith & Chen, 2022). Owing to its heterogeneity, TBI can be classified on the basis of severity, mechanism of injury, frequency, and pathophysiology. This particular protocol constitutes the induction of rmTBI, a class of TBI based on the frequency of injuries (Wright et al., 2016). Constitutive multiple mild brain injuries over time lead to rmTBI, ultimately leading to neurodegeneration and cognitive decline. The pathology exhibits neuroinflammation, oxidative stress, synaptic dysfunction, impaired neurogenesis (especially in hippocampus), and axonal damage (Campbell et al., 2024; Gonçalves et al., 2016; Jimenez-Cyrus et al., 2024; Saha et al., 2019; Simon et al., 2017).

Here, we use a Closed Head Injury device (CHI device) to induce rmTBI. This device enables reproducible injury to the intact skull, eliminating the need for a craniotomy-based approach and associated procedural complexity (Bolton Hall et al., 2016). This enables high-throughput and consistent experimental design, compatible with downstream analyses such as behavioral testing, and immunohistochemistry (IHC). Therefore, behavioral assays such as Barnes Maze Test and Y-Maze test are further used to validate the neurodegeneration caused by the induction of rmTBI (Mouzon et al., 2018) (Fig. 1).

Further, with the objective of counteracting this neurodegeneration by promoting the restoration of neurogenesis, a particular Flag-tagged epigenetic regulator is introduced. Adult hippocampal neurogenesis is a multi-step process in which new neurons are formed by the NSCs. These precursor cells are abundant in the subgranular zone (SGZ) of the Dentate Gyrus of the Hippocampus, where they proliferate and differentiate into immature neurons (Kempermann et al., 2015). These immature neurons further transform into mature neurons in the granule cell layer and integrate into the existing hippocampal neural circuit, contributing to cognitive functions such as learning, memory, and pattern separation (Toni et al., 2007). During neuronal differentiation, epigenetic regulators such as class II histone deacetylases (HDACs), particularly HDAC4, serve a significant role in the transcriptional control of the neuronal differentiation process. Moreover, HDAC4 also

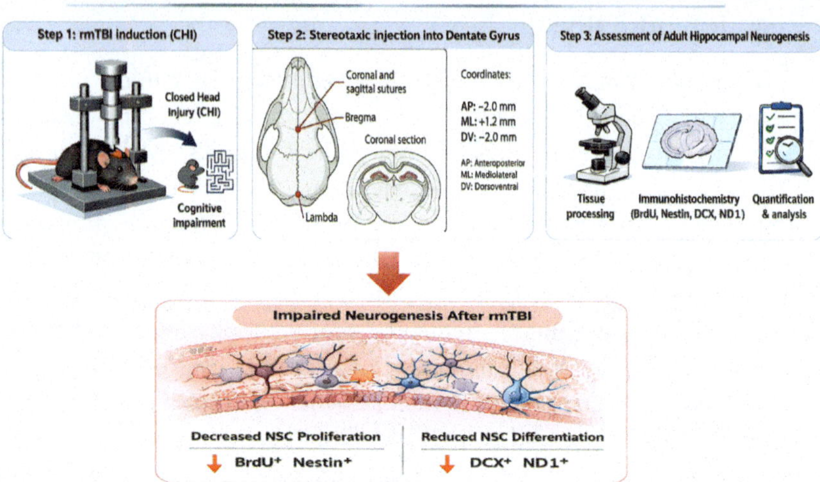

Fig. 1 Workflow for rmTBI induction, HDAC4 stereotaxic injection, and neurogenesis analysis.

binds to the MEF2 gene, a transcription factor, cumulatively modulating the CREB pathway (Shalizi & Bonni, 2018). This further impacts neuronal differentiation and synaptic plasticity. Injecting Flag-tagged HDAC4 into the dentate gyrus of mice enables tracking and localized manipulation of gene expression in the mice, which serves as an important part in the study of neurogenesis (Saha et al., 2019; Urban & Guillemot, 2021; Zhao et al., 2020). This protocol details the precise intracranial delivery of a Flag-tagged HDAC4 expression vector (pcDNA3.1-Flag-HDAC4) into the dentate gyrus (DG) of adult Swiss Webster mice via stereotaxic surgery.

2. Experimental prerequisites

Timing: 1–2 weeks.

Stereotaxic injection is a neurosurgical technique, used to deliver plasmid DNA, viral vector, drugs and peptides in specific regions of the brain. The surgery primarily requires a stereotaxic frame and a brain atlas to assess three-dimensional coordinates of the region of interest (Cetin et al., 2006). To investigate whether administration of HDAC4 into the dentate gyrus can rescue an impaired neurogenesis, we delivered plasmid DNA expressing Flag-tagged HDAC4 into the dentate gyrus of the hippocampus by injecting intracranially using stereotaxic surgery. Preparation of animals, reagents, and surgical instruments is crucial prior to the procedure (Saha et al., 2019). Proper calibration of the injury device and stereotaxic apparatus is required to ensure reproducible results (Dash et al., 2001). The following preparatory steps should be completed before starting the experimental protocol (Erginousakis & Papatheodoropoulos, 2025).

2.1 Preparation of animals and ethical approval

a) Adult mice (8–12 weeks old) should be acclimatized to the animal facility for at least one week before experimentation to minimize stress-induced variability (National Research Council, 2011).

b) Animals should be housed under controlled environmental conditions, including a 12-hour light/dark cycle, stable temperature, and unrestricted access to food and water.

2.1.1 Institutional permissions

All experimental procedures in this protocol must be carried out in accordance with the National Institutes of Health Guide for the Care and Use of Laboratory Animals and must be approved by the Institutional Animal Care and Use Committee (IACUC) (Jobling, 2012). Table 1.

Note: This protocol is standardized according to 8–12 weeks old adult Swiss Webster mice (Jackson Laboratory), gender unspecific. Table 2.

2.2 Sterilization

Ensure all the equipment is sterilized, and if possible, autoclaved. Table 3.

2.3 Calibration of closed head injury device for rmTBI induction

a) The Closed Head Injury Device should be calibrated prior to each experiment to ensure consistent impact force (Bodnar et al., 2019).

b) Verify alignment of the impactor tip with the skull surface to ensure reproducible injury parameters.

c) Test the device using dummy runs before initiating animal experiments.

2.4 Preparation for behavioral assessment of cognitive function

Behavioral tests such as Y-maze and Barnes maze should be planned to evaluate cognitive deficits following rmTBI.

Table 1 Key resources table.

Reagent or resource	Source	Identifier
Antibodies		
Anti-BrdU	Santa Cruz	Cat No. sc-32323
Anti-Nestin	Santa Cruz	Cat No. sc-33677
Anti-Doublecortin (DCX)	Abcam	Cat No. Ab18723
Anti-NeuroD1	Abcam	Cat No. EPR17084
Goat anti-Mouse IgG (H+L) cross absorbed Alexa Fluor 647	Abcam	Cat No. A150115
Goat anti-Rabbit IgG (H+L) cross absorbed Alexa Fluor 488	Abcam	Cat No. A150077
Biological Samples		
Mouse brain tissue (hippocampus)	Experimental samples	Dentate gyrus region
Recombinant DNA		
Flag-HDAC4 plasmid (pcDNA3.1, CMV promoter)	Addgene	Cat No. 13821
In vivo-jetPEI transfection reagent	PolyPlus Transfection	Cat No. 201−10G
Experimental Models: Organisms/Strains		
Swiss Webster Mice (Male, 8–10 weeks)	The Jackson Laboratory	Stock No. 000689
Chemicals, Peptides, and Recombinant Proteins		
Ketamine	Sigma-Aldrich	Cat No. K2753
Xylazine	Sigma-Aldrich	Cat No. X1251
Phosphate buffered saline (PBS)	Thermo Fisher Scientific	Cat No. 10010023
Povidone-iodine solution	Sigma-Aldrich	Cat No. 109036
Sterile 5 % Glucose solution	Sigma-Aldrich	N/A

Meloxicam (analgesic)	Sigma-Aldrich	Cat No. M3935
BrdU (Bromodeoxyuridine)	Sigma-Aldrich	Cat No. B5002
DAPI (4′,6-diamidino-2-phenylindole, dihydrochloride)	Thermo Fisher Scientific	Cat No. D1306
Paraformaldehyde	Thermo Fisher Scientific	Cat No. 28908
0.9 % Normal saline solution	Procell	Cat No. PB180353
Triton X-100	Beyotime	Cat No. ST79
DAPI for nucleic acid staining	Sigma-Aldrich	Cat No. D9542
Critical Commercial Assays		
QIA prep Spin Miniprep Kit	QIAGEN	Cat No. 27104
Software and Algorithms		
ImageJ/Fiji	NIH	N/A
Ethovision XT	Noldus	Version 17
GraphPad Prism	GraphPad	Version 10
Other		
Fluorescence Microscope	Olympus, Japan	Cat No. IX83
Closed Head Injury Device	Laboratory equipment	N/A
Stereotaxic apparatus	Kopf Instruments	Model 940

2.5 Preparation of flag-HDAC4 plasmid

Flag-HDAC4 plasmid is a flag tagged HDAC4 insert in pcDNA3.1 vector having a CMV promoter (Addgene, n.d.). Purify large amounts of high-quality endotoxin-free Flag-HDAC4 plasmid DNA using a commercially available kit, for instance, Qiagen EndoFree Plasmid Maxi Kit. Dilute the plasmid in sterile 5 % (w/v) glucose to a final concentration of 1.0–2.0 μg/μL, i.e., dilute 2.5 μg Flag-HDAC4 plasmid DNA in 2.5 μl 5 % sterile glucose (w/v). Store at −80 °C for long-term use.

Table 2 List of primary antibodies and their applications in neurogenesis analysis.

Target	Primary antibody	Host species	Dilution	Purpose
BrdU	Anti-BrdU	Mouse	1:200	Labels proliferating cells by incorporating into newly synthesized DNA during S-phase
Nestin	Anti-Nestin	Mouse	1:200	Marker of neural stem/ progenitor cells (NSPCs) in the early stage of neurogenesis
DCX	Anti-Doublecortin (DCX)	Rabbit	1:200	Labels immature neurons/ neuroblasts, indicating ongoing neurogenesis
NeuroD1	Anti-NeuroD1	Rabbit	1:200	Marks neuronal differentiation, involved in maturation of progenitors into neurons

Table 3 List of secondary antibodies and their applications in neurogenesis analysis.

Secondary antibody	Target	Primary host	Fluorophore	Dilution
Goat Anti-Mouse IgG (H+L), cross-adsorbed	BrdU	Mouse	Alexa Fluor 647	1:500
Goat Anti-Mouse IgG (H+L), cross-adsorbed	Nestin	Mouse	Alexa Fluor 647	1:500
Goat Anti-Rabbit IgG (H+L), cross-adsorbed	DCX	Rabbit	Alexa Fluor 488	1:500
Goat Anti-Rabbit IgG (H+L), cross-adsorbed	NeuroD1	Rabbit	Alexa Fluor 488	1:500

2.6 Preparation of surgical equipment

a) Sterilize surgical instruments including forceps, scissors, and scalpels using appropriate sterilization procedures.

b) Ensure that the stereotaxic apparatus is properly calibrated to allow accurate targeting of the dentate gyrus. (c) Prepare micro syringes for intracranial plasmid delivery.

2.7 Preparation for immunohistochemistry and imaging

a) Calibrate the cryostat for brain sectioning and ensure the temperature is properly stabilized.

b) Prepare primary and secondary antibodies for BrdU, Nestin, Doublecortin (DCX), and NeuroD1 (ND1) according to the manufacturer's recommended dilution.

c) Prepare a fluorescence microscope or confocal microscope for imaging of stained sections (Schindelin et al., 2012).

d) Install and test image analysis software such as ImageJ for quantitative analysis of neurogenesis markers.

2.8 Preparation of various reagents

2.8.1 Anesthesia cocktail

For anesthetizing, prepare a fresh cocktail of Ketamine (100 mg/ml) and Xylazine (10 mg/ml) in sterile saline.

2.8.2 BrdU

Dissolve BrdU in 0.9 % NaCl and 0.007 N NaOH, to obtain a final concentration of 10 mg/ml.

2.8.3 Buffers

2.8.3.1 Buffer for tissue fixation

Dissolve 4 % Paraformaldehyde (PFA) in 0.1 M PBS.

2.8.3.2 Blocking buffer for immunohistochemistry

Dilute 5 % (v/v) Normal Donkey Serum and 0.3 % (v/v) Triton-X 100 in 1x PBS.

Note: Within each step of the protocol, if the recommended reagent/equipment is not available, you may try a product with similar functionality.

Critical: It is highly recommended that people learn about vernier scales and safe mouse positioning before performing stereotactic injections.

3. Materials, equipment, and experimental resources

3.1 Key resources table

Alternatives: Alternative imaging systems such as spinning disk confocal microscopes or epifluorescence microscopes may be used depending on availability. Similarly, alternative image analysis software capable of cell counting and fluorescence intensity quantification can be used instead of ImageJ/Fiji.

4. Step-by-step method details

4.1 Induction of repetitive mild traumatic brain injury (rmTBI) using closed head injury device and its validation

4.1.1 Equipment

Closed Head injury (CHI) device, syringe with 27-gauge needle, heating pad, animal weighing balance, recovery cage, Barnes maze apparatus, Y-maze apparatus, spatial cues, and a video tracking system.

4.1.2 Reagents

Ketamine, xylazine, and ethanol.

4.1.3 Procedure

4.1.3.1 Induction of repetitive mild traumatic brain injury (rmTBI)

Timing: 30–40 min per animal.

1. Weigh the mouse and calculate the anesthetic dosage accordingly.
2. Handle the mouse securely and disinfect the abdominal site with ethanol.
3. Insert a syringe with 27-gauge needle into the lower abdominal quadrant of the mouse (angling 30° relative to the body surface),
4. Gradually inject the anesthesia cocktail into the peritoneal cavity. Eject carefully and observe.
5. Estimate the depth of anesthesia by the pedal reflex.
6. Place the anesthetized mouse on a heating pad to maintain body temperature during the procedure.
7. Position the head beneath the closed head injury (CHI) device (Fig. 2).
8. Align the impactor tip of the CHI device perpendicular to the skull surface.
9. Deliver a controlled mechanical impact to the skull using predetermined parameters suitable for induction of mild traumatic brain injury.

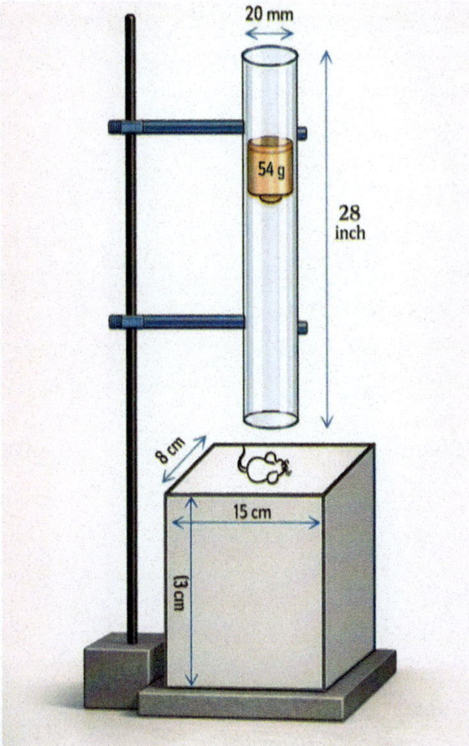

Fig. 2 Closed head injury (CHI) device for induction of rmTBI.

  a) Adjust the impact force according to the desired severity of injury.
  b) Ensure that the impact is delivered consistently across all experimental animals.

10. Following the impact, carefully remove the animal from the injury device and place it in a recovery cage.

11. Monitor the animal until full recovery from anesthesia is observed

4.1.3.2 Validation of rmTBI induction using behavioral assays
Timing: 6–7 days.

4.1.3.2.1 Spatial memory assessment via barnes Maze Test

1. The Barnes maze (Fig. 3A) consists of a circular platform (90–120 cm diameter) with multiple holes (typically 16–20), with one escape box located beneath a target hole.

2. Train mice for 3–5 consecutive days with 2–3 trials per day.

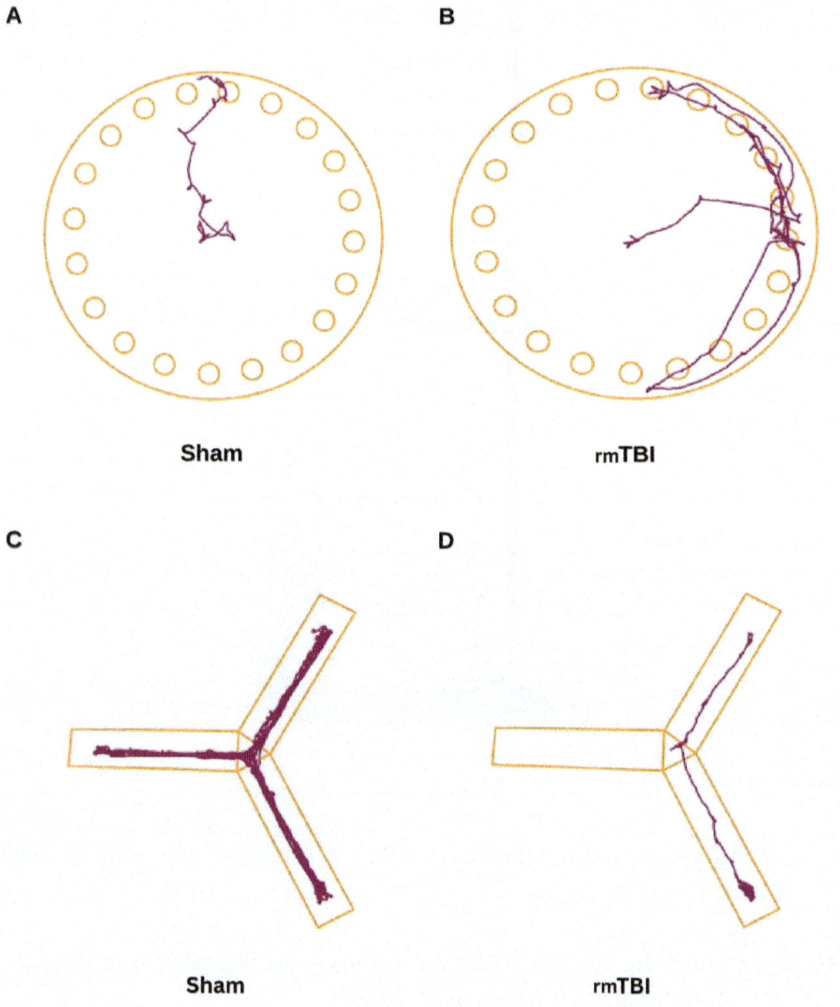

Fig. 3 Diagrammatic overview of behavioral assays performed in the study to validate the induced rmTBI by using CHI device. (A). Schematic diagram for Barnes Maze test on Sham. (B). Schematic diagram for Barnes Maze test on rmTBI Model. (C). Schematic diagram for Y Maze test on Sham. (D). Schematic diagram for Y Maze test on rmTBI Model.

3. Place the mouse at the center of the platform and allow it to locate the escape box using spatial cues.
4. Record latency to find the escape box, number of errors, and path length (Fig. 3B).

5. On the probe trial (without an escape box), assess memory retention by measuring time spent in the target quadrant.
6. Increased latency and errors indicate cognitive deficits associated with rmTBI.

4.1.3.2.2 Working memory assessment via Y-Maze Test

1. The Y-maze apparatus consists of three arms (typically 30–40 cm long, 8–10 cm wide) positioned at 120° angles (Fig. 3C).
2. Place the mouse at the center of the maze and allow free exploration for 5–8 min.
3. Record the sequence and number of arm entries (Fig. 3D) using a video tracking system or manual scoring.
4. Calculate spontaneous alternation percentage using the formula:
   Alternation (%) = [(Number of alternations)/(Total arm entries − 2)] × 100
5. Reduced alternation behavior indicates impairment in spatial working memory following rmTBI.

4.1.4 Notes

1. Before injecting the anesthesia complex, aspirate slightly to confirm that no fluid is drawn.
2. Ensure the head of the animal is positioned securely under the CHI device while positioning.
3. The closed head injury model allows induction of mild traumatic brain injury without performing craniotomy (Saha et al., 2019). Proper alignment of the impactor tip and consistent injury parameters are essential for achieving reproducible rmTBI induction.

4.2 Introduction of formulated Flag-HDAC4 into the dentate gyrus of rmTBI mice

4.2.1 Equipment

Micropipettes and tips, microcentrifuge tubes, vortex mixer, heating pad, stereotaxic apparatus, ear bars and nose clamp, surgical instruments (scalpel, forceps, scissors), surgical drill, Hamilton microsyringe, mouse brain atlas, and recovery cage with warming support.

4.2.2 Reagents

jetPEI (polyethylenimine transfection reagent), sterile 5% glucose solution, Flag-HDAC4 plasmid, anesthesia cocktail (ketamine and xylazine), 70% ethanol, povidone-iodine solution, ophthalmic ointment, antiseptic solution, meloxicam, and sterile saline.

4.2.3 Procedure

4.2.3.1 Formulating DNA-polyethyleneimine transfection complexes

Timing: 30–40 min.

1. Dilute 0.35 μL of jetPEI (Polyethylenimine (PEI)–based transfection reagent) into 2.15 μL of sterile 5 % glucose (N/P ratio of 7).
2. Combine diluted jetPEI with diluted Flag-HDAC4 plasmid (Fig. 4) and mix by vortexing.
3. Incubate for 15–20 min at Room Temperature.

4.2.3.2 Injecting anesthesia intraperitoneally

Timing: 20–30 min.

1. Anesthetize the mouse using an appropriate anesthetic regimen and confirm the absence of reflex responses before initiating the procedure.
2. Place the anesthetized mouse on a heating pad.
3. Shave the scalp region.
4. Use 70 % ethanol and iodine-povidone to sterilize the shaved region.

4.2.3.3 Adjusting to the stereotactic frame and exposing the skull

Timing: 20–30 min.

1. Position the mouse as per the stereotaxic frame.

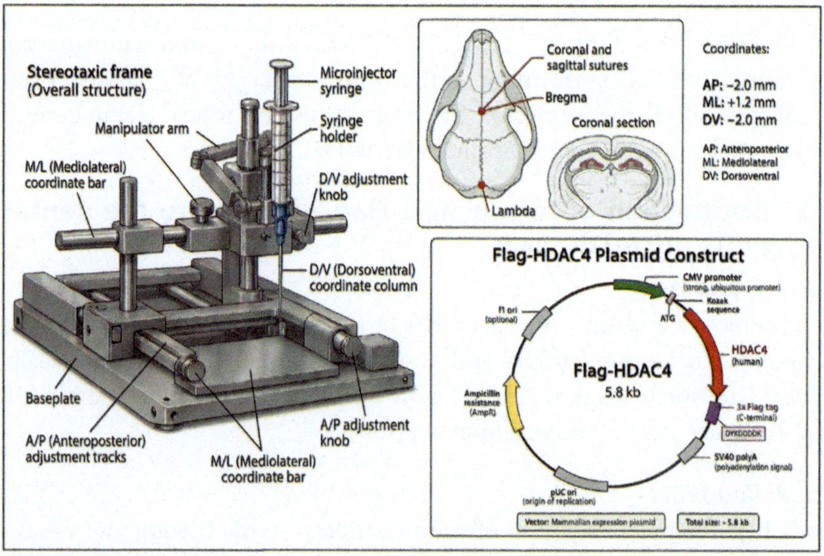

Fig. 4 Stereotaxic Injection of the prepared Flag HDAC4 plasmid into the dentate gyrus of mice.

2. Stabilize the skull using ear bars and a nose clamp.
3. Make a small midline incision on the scalp to expose the skull.

Pause Point: If necessary, the procedure may be paused after stereotaxic positioning of the animal before plasmid injection. Animals should remain under stable anesthesia during this period.

4.2.3.4 Surgery and intracranial injection of prepared flag HDAC4 plasmid

Timing: 30–40 min.

1. Identify bregma and lambda as stereotaxic reference points and find coordinates corresponding to the dentate gyrus using a mouse brain atlas (Paxinos & Franklin, 2001) ensuring they are aligned on the same horizontal plane. (Typical coordinates relative to bregma: AP: −0.2 mm, ML: −0.13 mm, & DV: 0.2 mm.) (Fig. 4).
2. For improved visualization of the injection site, a small burr hole (0.5 mm) can be drilled at the stereotaxic coordinates before needle insertion.
3. Load the formulated transfection complex into a Hamilton micro syringe.
4. Slowly lower the needle into the dentate gyrus region according to the stereotaxic coordinates.
5. Gradually inject 5 μL transfection complex at a rate of approximately 0.5 μL/min.
6. Carefully withdraw the needle.

4.2.3.5 Post-operative care and recovery

Timing: 30–45 min.

1. Suture the scalp.
2. Apply antiseptic solution.
3. Administer analgesic (meloxicam, 5 mg/kg).
4. Introduce warm sterile saline 0.5 ml to avoid dehydration.
5. Shift the mouse to a separate recovery cage having a warming pad.
6. Keep under observation.

Pause Point: Wait for at least 24 h or more, until the mouse regains normal mobility.

4.2.4 Notes

1. Apply ophthalmic ointment to both eyes to prevent drying during surgery.
2. Accurate stereotaxic targeting is essential for successful delivery of plasmid DNA into the dentate gyrus. Incorrect positioning may lead to injection outside the hippocampal region.

3. Ensure that the bregma and lambda are aligned on the same horizontal plane.
4. Use a fine surgical drill to create an opening in the skull.
5. Avoid excessive pressure while drilling to avoid damage to the underlying brain tissue.
6. Hold the syringe for 5−10 min post-infusion in order to avoid backflow.

4.3 BrdU administration

4.3.1 Equipment

Syringes suitable for intraperitoneal injection.

4.3.2 Reagents

BrdU (Bromodeoxyuridine) solution.

4.3.3 Procedure

Timing: 3–5 days.

1. Following HDAC4 injection, administer prepared BrdU (Bromodeoxyuridine) at a dose of 50 mg/kg of body weight. (Yu et al., 2008)
2. Inject intraperitoneally once regularly for consecutive 3−5 days.

4.3.4 Notes

1. Administer at the same time daily to avoid circadian differences.
2. For evaluation of proliferating cells, sacrifice mouse after 24 h of last injection, while for the evaluation of differentiating cells, allow mouse to survive for 1−3 weeks before assessment.

4.4 Specimen collection

4.4.1 Equipment

Perfusion setup, dissection instruments, cryostat, embedding molds, and microscope slides.

4.4.2 Reagents

1 × PBS (Phosphate Buffered Saline), 4% paraformaldehyde (PFA) in 0.1 M PBS, 30% sucrose solution, and OCT (Optimal Cutting Temperature) compound.

4.4.3 Procedure

4.4.3.1 Transcardial brain perfusion and tissue fixation

Timing: 2–3 h (excluding fixation time).

1. At the desired experimental time point following the BrdU administration, deeply anesthetize the mouse using ketamine-xylazine cocktail.

2. Carefully expose the heart by opening the thoracic cavity.
3. Insert the perfusion needle into the left ventricle and cut the right atrium.
4. Perfuse with 50 ml of cold 1x PBS (Phosphate Buffer Saline) and 50 ml of 4 % PFA (Paraformaldehyde).
5. Ensure proper fixation using the prepared fixation buffer (4 % PFA in 0.1 M PBS), overnight at 4 °C.
6. Cryoprotect by transferring to 30 % sucrose solution until the brain sinks.

4.4.3.2 Preparation of coronal brain sections by cryosectioning

Timing: 2–3 h.

1. Embed the brain tissue in OCT (Optimal Cutting Temperature) for cryosectioning.
2. Prepare coronal brain sections (20–30 μm thickness) using a cryostat.
3. Collect sections containing the hippocampus and dentate gyrus region onto microscope slides.

4.4.4 Notes

1. Proper fixation and cryoprotection of brain tissue are essential for maintaining structural integrity during cryosectioning and subsequent immunohistochemical analysis.
2. Brain sections can be stored at −20 °C or in cryoprotectant solution at −80 °C until immunohistochemistry is performed.

4.5 Immunohistochemistry for detection of neurogenesis markers

4.5.1 Equipment

Humid chamber, microscope slides, coverslips, rocking platform, incubator (4 °C), and fluorescence microscope.

4.5.2 Reagents

Primary antibodies: Anti-BrdU, Anti-Nestin, Anti-Doublecortin (DCX), and Anti-NeuroD1 (ND1), fluorophore-conjugated secondary antibodies, 1 × PBS, PBS with 0.05 % Tween-20, 0.3 % Triton X-100 in PBS, blocking buffer, 2 N HCl, 0.1 M borate buffer (pH 8.5), DAPI, and ProLong Gold antifade mounting medium.

4.5.3 Procedure

Timing: 3 days.

- BrdU antibody to detect proliferating neural stem cells (Kuhn et al., 1996).
- Nestin antibody to identify neural progenitor populations (Saha et al., 2019).

- Doublecortin (DCX) antibody to label immature neurons (Bond, Ming & Song, 2015).
- NeuroD1 (ND1) antibody to detect neuronal differentiation (Kempermann, Song & Gage, 2015).

1. Bring brain sections to room temperature
2. Wash thrice for 5 min using 1xPBS.
3. Denature DNA using 2 N HCl at 37 °C for 1 h.
4. Add 0.1 M Borate Buffer (pH 8.5) for neutralization and incubate for 10 min at room temperature.
5. Wash with 0.3 % Triton X-100 in 1x PBS at room temperature, thrice for 10 min.
6. Wash three times for 5 min each using 1xPBS.
7. Transfer to the blocking buffer and incubate for 1 h at room temperature.
8. Stain with Primary antibodies, incubate overnight at 4 °C on a rocking incubator.
9. Wash thrice with 1x PBS (or PBS + 0.05 % Tween-20), 5 min each to remove unbound primary antibodies.
10. Stain with an appropriate fluorophore-tagged secondary antibody for 2 h at room temperature in the dark, and counterstain with DAPI.
11. Wash sections again with PBS to remove excess secondary antibodies.
12. Counterstain nuclei using DAPI.
13. Mount the sections using ProLong Gold Antifade mounting media and seal coverslips.

4.5.4 Notes

1. For BrdU antibody, perform steps 3 and 4, since BrdU is incorporated inside the DNA double helix which requires denaturation for accessibility. For staining with other antibodies, skip steps 3 and 4 of the procedure.
2. Primary antibody concentrations should be optimized to achieve strong signal with minimal background staining.
3. Air dry slides overnight after mounting for better results.
4. Fluorescent samples should be protected from light during staining and storage to prevent photobleaching.

4.6 Fluorescence imaging and quantitative analysis of neurogenesis markers

4.6.1 Equipment

Fluorescence or confocal microscope equipped with appropriate filter sets, imaging workstation, and computer with ImageJ/Fiji software.

4.6.2 Reagents

Mounting medium–preserved stained brain sections.

4.6.3 Procedure

Timing: 3–4 h.

1. Acquire fluorescence images using a fluorescence or confocal microscope equipped with appropriate filter sets.
2. Identify the dentate gyrus region of the hippocampus in each brain section.
   - 20× magnification → region overview (dentate gyrus localization)
   - 40× /63× oil immersion→ cellular morphology and marker expression
3. Capture z-stack images of the stained sections for each neurogenesis marker.
4. Import acquired images into ImageJ/Fiji software for quantitative analysis (Schindelin et al., 2012).
5. Define the dentate gyrus as the region of interest (ROI) for analysis.
6. Count marker-positive cells manually or using automated threshold-based detection.
7. Record the number of BrdU+, Nestin+, DCX+, and ND1+ cells for each section.
8. Calculate average cell counts for each experimental group.
9. Compile data as Mean ± SEM (Standard Error of the Mean) and perform statistical analysis using appropriate tests.

4.6.4 Notes

1. Ensure that imaging parameters such as exposure time and detector gain remain constant across experimental groups.
2. Inconsistent imaging settings may introduce bias in fluorescence intensity measurements and cell counting.

5. Expected outcomes

If this protocol works, it should be possible to repeatedly cause mild traumatic brain injury (rmTBI) in mice and specifically express Flag-HDAC4 in the dentate gyrus of the hippocampus. A proper stereotaxic injection and then an immunohistochemical analysis should show exactly where neurogenesis markers are in the subgranular zone (SGZ) of the dentate gyrus.

After processing the tissue and staining it with immunofluorescence, it is possible to use specific markers to tell different groups of neural stem cells, neuronal progenitors, and immature neurons apart. BrdU incorporation is anticipated to mark proliferating neural stem cells and progenitor cells within the SGZ (Gao et al., 2009). Nestin staining should help find neural progenitor cells and early populations of neural stem cells that are located along the dentate gyrus (Saha et al., 2019). Doublecortin (DCX) staining should show immature neurons with long cell bodies and dendritic processes that reach out to the molecular layer (Bond, Ming & Song, 2015). NeuroD1 (ND1) staining should show that newly formed neurons are becoming more specialized and mature.

Fluorescence microscopy should make it easy to see and label these markers in sections of the hippocampus. BrdU-positive cells usually look like small dots of nuclear staining inside cells that are growing. There are Nestin-positive cells along the SGZ that have unique staining patterns in their cytoplasm. DCX-positive immature neurons are typically arranged along the inner granule cell layer and exhibit dendritic projections that extend into the molecular layer of the dentate gyrus. ND1-positive cells are committed to becoming neurons, and they may be found in the same groups as DCX-positive cells during the early stages of neuronal differentiation (Fig. 5).

Animals that have been exposed to rmTBI may show changes in the number and placement of these marker-positive cells (Sun, 2014). Depending on the severity of the injury and the timeline of the experiment, rmTBI may cause either short-term or long-term problems with adult neurogenesis. By counting the number of marker-positive cells in the dentate gyrus, we can compare the control, rmTBI, and HDAC4-manipulated groups.

After stereotaxic delivery of the Flag-HDAC4 plasmid, alterations in the prevalence of BrdU+, Nestin+, DCX+, and ND1+ cells may indicate the regulatory function of HDAC4 in neural stem cell proliferation and neuronal differentiation (Saha et al., 2019). So, fluorescence imaging and quantitative image analysis should work together to give us data that shows changes in neurogenesis in the hippocampus that can be repeated.

Moreover, the Barnes–Maze and the Y-Maze tests are expected to reveal deficits in spatial learning, memory retention, and working memory, evidenced by increased latency to locate the escape hole in the Barnes-Maze and reduced spontaneous alternation behavior in the Y-Maze.

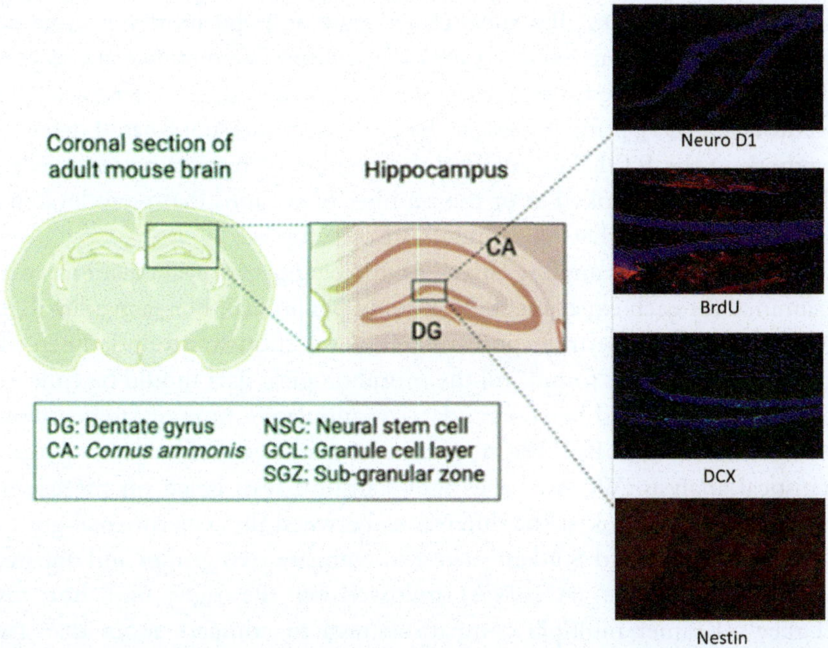

Fig. 5 Coronal brain section showing dentate gyrus and expression of neurogenesis markers (NeuroD1, BrdU, DCX, Nestin).

6. Quantification and statistical analysis

Quantitative analysis of adult neurogenesis was conducted utilizing fluorescence microscopy images derived from immuno-stained hippocampal sections. For each experimental group, coronal brain sections with the dentate gyrus were imaged using the same acquisition settings so that the samples could be compared. To reduce technical differences during image acquisition, the imaging settings, such as exposure time, gain, and laser intensity, were kept the same for all samples.

Behavioral data were expressed as mean ± SEM and analyzed using appropriate statistical tests. Repeated measures ANOVA was used for training sessions in the Barnes maze, followed by post hoc analysis. Differences between experimental groups were considered statistically significant at $p < 0.05$.

We used ImageJ/Fiji image analysis software to count the number of marker-positive cells (Schindelin et al., 2012). The dentate gyrus part of the hippocampus was chosen as the area of interest (ROI) for the study. Using anatomical landmarks, the subgranular zone and the nearby granule cell layer were carefully marked out for each section. We counted or measured BrdU-positive, Nestin-positive, Doublecortin (DCX)-positive, and NeuroD1 (ND1)-positive cells by hand or with threshold-based detection methods in the ROI. We adjusted the number of marker-positive cells to the area that was looked at or the number of sections that were looked at for each animal so that we could compare samples.

To minimize sampling bias, multiple hippocampal sections were examined for each experimental condition per animal. For each animal, the average number of marker-positive cells per section was found. Then, the group averages were found. All the morphological data should be collected as Mean $\pm$ SEM, and behavioral data as cumulative time or latency plots.

We used GraphPad Prism or similar statistical software to perform statistical analysis. We used the right statistical tests based on the experimental design to look at the differences between the experimental groups. We used an unpaired Student's *t*-test to compare two groups and one-way analysis of variance (ANOVA) followed by the right post hoc tests (Tukey's–Kramer multiple comparison test) to compare more than two groups. A p-value of less than 0.05 was taken statistically significant.

To avoid overestimating or misinterpreting neurogenesis markers, it is important to carefully read fluorescence data. For instance, DCX staining should be looked at along with other signs of neuronal differentiation to correctly find immature neurons. BrdU incorporation also shows cells that are dividing, and it should be looked at in the context of the subgranular zone anatomy. To make sure that the experimental results can be repeated and are reliable, it is important to use the same imaging parameters, standardized quantification methods, and correct statistical analysis.

7. Advantages

The protocol delineated herein establishes a consistent experimental framework for investigating adult hippocampal neurogenesis within the context of repetitive mild traumatic brain injury (rmTBI) (Gao et al., 2009). One of the best aspects of this method is that it uses a closed head injury (CHI) device, which allows induction of mild traumatic brain injury without obligating craniotomy. This model closely resembles mild concussive injuries seen in humans and minimizes surgical complications linked to more invasive brain injury models.

Another benefit involves allowance of targeted manipulation of gene expression in the dentate gyrus by delivering Flag-HDAC4 plasmid DNA directly into the brain. This method helps us focus directly on how HDAC4 works at the molecular level to control the growth of neural stem cells and the differentiation of neurons in the adult hippocampus. Localized plasmid delivery is more precise in space and less likely to cause unwanted effects in other parts of the brain than systemic methods.

The protocol also includes several other markers of adult neurogenesis, such as BrdU, Nestin, Doublecortin (DCX), and NeuroD1 (ND1). Using these markers together lets you look at all the different stages of neurogenesis, such as the growth of neural stem cells, the growth of progenitor cell populations, and the formation of immature neurons (Saha et al., 2019). This multi-marker approach gives us a better idea of how traumatic brain injury affects neurogenesis in the dentate gyrus.

Moreover, the integration of fluorescence microscopy with image analysis software facilitates the quantitative evaluation of neurogenesis markers in specific anatomical regions of the hippocampus. Standardized methods for imaging and measuring make it easier to repeat experiments and compare groups of experiments.

In general, this protocol provides a flexible and repeatable way to study molecular mechanisms that control neurogenesis in the adult hippocampus after a traumatic brain injury. The integration of regulated injury induction, precise gene delivery, and quantitative immunohistochemical analysis constitutes a robust methodology for investigating the cellular responses associated with brain injury and neuronal regeneration.

8. Limitations

A major limitation of this protocol corresponds to the requirement of surgical skills with a significant level of expertise. Minor deviation in skull placement on the stereotaxic frame, or inaccurate coordinate selection can lead to injection of the transfection complex outside of the dentate gyrus region (Lowery & Majewska, 2010). Minor changes in the placement of needle could cause the plasmid to be delivered to the wrong part of the hippocampus, which could change how consistently genes are expressed in different experimental animals. Such inaccuracies compromise the specificity of gene manipulation within the Hippocampus. Secondly, different cell types require different optimization levels with respect to large injection volume and the vector. The expression rate may also vary depending on the vector used, the titer, and the transfection efficacy.

Variability is another massive limitation in this protocol. Even when using a calibrated CHI device, small changes in the angle of impact, the force of the impact, or the position of the animal can induce variation in the severity of the injury among different animals. To reduce variability, it is important to carefully standardize injury parameters and experimental conditions.

Moreover, BrdU, Nestin, Doublecortin (DCX), and NeuroD1 (ND1) are markers that show specific stages of neuronal development, but they fail to distinguish cell types. For instance, BrdU incorporation marks proliferating cells, but it fails to differentiate between neuronal and non-neuronal cells. Similarly, DCX expression identifies developing neurons; however, it does not provide information on their long-term survival or their functional integration into hippocampal neural circuits. Consequently, the interpretation of results must take into account the integrated expression patterns of various markers.

For obtaining statistically robust data, a large dataset is required which requires large sample size and longitudinal study. Moreover, behavioral assessment gets majorly affected by environmental factors such as light, noise, temperature, and handling conditions, thereby requiring multiple levels of optimization. Therefore, strict environmental control and experimental standardization are essential to ensure reliable and reproducible behavioral data.

Lastly, the rmTBI model used in mice may not fully capture the complexity of traumatic brain injury seen in humans (Xiong et al., 2013). Variations in brain anatomy, injury mechanisms, and recovery processes may hinder the direct application of findings to clinical settings. Despite these constraints, the protocol continues to be an essential instrument for examining the molecular mechanisms governing adult hippocampal neurogenesis subsequent to brain injury.

9. Optimization and troubleshooting

9.1 Problem: Inefficient drilling

Damage to the hippocampus during drilling.

9.2 Potential solution to optimize the procedure

Ensure that the brain coordinates identified through the bregma point are accurate. Avoid intensified drilling and penetration. Use a needle of an appropriate gauge to avoid injury to the brain tissue.

9.3 Problem: Incorrect targeting during stereotaxic injection

Inaccurate placement of the injection needle during stereotaxic surgery may lead to plasmid delivery outside the dentate gyrus region of the hippocampus. This can result in inconsistent expression of Flag-HDAC4 and variability in experimental outcomes.

9.4 Potential solution to optimize the procedure

Careful calibration of the stereotaxic apparatus and verification of stereotaxic coordinates using a mouse brain atlas are essential for accurate targeting of the dentate gyrus. Performing practice injections and confirming needle placement using dye injections or test runs can improve surgical accuracy. Maintaining stable positioning of the animal during surgery also helps prevent targeting errors.

9.5 Problem: Weak or inconsistent immunofluorescence staining of neurogenesis markers

Weak staining of BrdU, Nestin, Doublecortin (DCX), or NeuroD1 (ND1) may occur due to insufficient antibody concentration, inadequate permeabilization of tissue sections, or improper fixation of brain tissue. Poor fixation can reduce antigen preservation, while inadequate blocking conditions may lead to low signal intensity or uneven staining across sections.

9.6 Potential solution to optimize the procedure

Optimization of antibody concentrations and incubation conditions is recommended to improve staining quality. Primary antibodies should be tested at different dilutions to determine the optimal working concentration. Adequate permeabilization using Triton X-100 and proper blocking using serum or bovine serum albumin can reduce background staining while improving signal detection. Ensuring freshly prepared paraformaldehyde fixation and proper tissue cryoprotection can also improve antigen preservation.

9.7 Problem: High background fluorescence in stained sections

High background fluorescence may occur due to nonspecific antibody binding, inadequate washing steps, or excessive antibody concentration. Background fluorescence can interfere with accurate identification of marker-positive cells during image analysis.

9.8 Potential solution to optimize the procedure

Increasing the duration and number of washing steps between antibody incubations can reduce nonspecific fluorescence. Using appropriate blocking buffers and optimizing antibody concentrations can further minimize background staining. Protecting samples from light and using high-quality mounting media can also improve fluorescence signal clarity.

9.9 Problem: Improper orientation

Disordered histological features.

9.10 Potential solution to optimize the procedure

Ensure proper tissue orientation during embedding and sectioning. Incorrect alignment of the tissue can lead to distorted or disorganized histological structures.

9.11 Problem: Variability in rmTBI injury severity between animals

Variability in injury severity may occur if the closed head injury device is not properly calibrated or if animal positioning during impact delivery is inconsistent. Differences in injury parameters may influence neurogenesis outcomes and reduce experimental reproducibility.

9.12 Potential solution to optimize the procedure

The rmTBI device should be calibrated before each experimental session to ensure consistent impact parameters. Animals should be positioned carefully under the impactor tip to maintain uniform injury conditions. Performing pilot experiments to standardize injury parameters can help improve reproducibility across experimental groups.

9.13 Problem: Stress

Stress can hamper the clear differentiation in behavioral tests.

9.14 Potential solution to optimize the procedure

In order to decrease stress, ensure that the animals are handled for several days prior to testing. Set a standard testing time. Use automated behavioral tracking systems to minimize observer bias and improve data reliability.

9.15 Problem: Photobleaching during prolonged imaging

Loss of fluorescence due to prolonged exposure to excitation light.

9.16 Potential solution to optimize the procedure

Minimize exposure time, use appropriate filter sets, and acquire images sequentially for each channel (Lichtman and Majewska, 2010).

9.17 Problem: Spectral overlap between fluorescence channels

Spectral overlap between fluorescence channels can lead to signal bleed-through, compromising the specificity and accuracy of multichannel imaging.

9.18 Potential solution to optimize the procedure

Capture each channel sequentially using appropriate filter sets (DAPI, FITC, TRITC, Cy5) to avoid bleed-through.

9.19 Problem: Loss of image quality during storage

Loss of image quality during storage can compromise data integrity and downstream analysis.

9.20 Potential solution to optimize the procedure

Save all Z-stack images in a lossless format (e.g., TIFF) and avoid compression.

10. Safety considerations and standards

The protocol's experimental procedures must be carried out in compliance with institutional, national, and international regulations pertaining to the utilization of laboratory animals in research. Before the study can begin, the Institutional Animal Ethics Committee (IAEC) or another regulatory body must review and approve the experimental protocols. It is important to follow the rules for caring for laboratory animals when housing and handling them. This includes providing the right housing conditions, enriching the environment, and keeping an eye on the animals' health throughout the experiment.

All surgical procedures, such as inducing repetitive mild traumatic brain injury and stereotaxic intracranial injections, must be conducted under suitable anesthesia to reduce pain and distress in the animals. During recovery from anesthesia, animals need to be closely watched, and any signs of distress or strange behavior should be dealt with right away. To make sure that animals are safe and that the experiment is reliable, it is important to take care of them and keep an eye on them after surgery.

When working with chemical reagents like paraformaldehyde and Triton X-100, you should follow the safety rules for the lab. Paraformaldehyde is a dangerous chemical that can give off harmful formaldehyde vapors. Because of this, you should make and handle paraformaldehyde solutions in a chemical fume hood while wearing the right personal protective equipment (PPE), like gloves, lab coats, and safety goggles. You should get rid of waste solutions that have paraformaldehyde or other dangerous chemicals according to the rules for biosafety and chemical waste disposal at your institution.

Standard biosafety practices should be followed for all procedures that involve biological samples and potentially dangerous reagents. Before doing the procedures in this protocol, lab workers should get the right training in how to handle animals, how to do surgery, and how to be safe with chemicals. Following institutional biosafety standards and laboratory safety rules will help make sure that the experimental protocol is carried out safely and ethically.

11. Alternative methods / procedures

The protocol in this chapter uses a closed head injury (CHI) device to cause repetitive mild traumatic brain injury (rmTBI). However, other experimental models of traumatic brain injury may also be used, depending on what equipment is available and what the goals of the experiment are. The controlled cortical impact (CCI) model and the fluid percussion injury (FPI) model are two examples of models that are often used. These models enable meticulous regulation of injury severity and have been extensively utilized to investigate cellular and molecular responses subsequent to brain trauma. These methods usually involve craniotomy and may result in more severe injury than closed head injury models.

You might also want to think about other ways to get genes into the dentate gyrus. Adeno-associated virus (AAV) or lentiviral vectors can be used instead of plasmid-based stereotaxic injection to deliver genes to hippocampal neurons in a stable and long-lasting way. Viral vectors frequently exhibit superior transfection efficiency and more reliable expression in comparison to plasmid DNA injection. But viral methods need special biosafety procedures and may also have to follow extra rules.

You can also use different markers of adult neurogenesis based on how you set up your experiment. Thymidine analogs like EdU can also be used to find proliferating cells along with BrdU labeling. These analogs have benefits in staining methods that don't need DNA denaturation. You could also add other markers, like Ki67 for cells that are dividing or NeuN for mature neurons, to get more information about the different stages of neuronal development.

Finally, neurogenesis markers can be looked at in different ways using imaging and quantification. You can use confocal microscopy to get pictures with more in-depth details and to make three-dimensional models of hippocampal sections. Automated cell counting software or machine learning–based image analysis tools may also enhance precision and efficiency in quantifying marker-positive cells relative to manual counting techniques. Researchers studying adult hippocampal neurogenesis can choose from these different methods based on the resources they have, their experimental goals, and their level of technical skill.

12. Conclusion

This chapter presents a standardized and reproducible protocol for evaluating the role of HDAC4 in repetitive mild traumatic brain injury (rmTBI). The approach integrates a controlled closed head injury model with stereotaxic delivery of Flag-HDAC4 into the dentate gyrus of mice, followed by behavioral and immunohistochemical analyses. Careful adherence to this protocol enables detailed functional assessment of HDAC4 expression in the context of rmTBI.

The workflow yields high-quality immunofluorescence data and facilitates accurate quantification of neural stem cell proliferation and differentiation in the dentate gyrus using established markers. Collectively, this methodological framework provides a robust platform for investigating the molecular mechanisms underlying adult hippocampal neurogenesis following repetitive mild traumatic brain injury and supports future studies aimed at identifying therapeutic targets.

Acknowledgements

This work was supported by the Banaras Hindu University (No. R/Dev/D/IoE/Seed Grant (5th Call)/2024–25/81862) and Ramalingaswami Re-entry Fellowship (D.O. NO. BT/HRD/35/02/2006) to RKG.

References

Addgene. (n.d.). Flag-HDAC4 plasmid (pcDNA3.1, CMV promoter) [Plasmid]. Addgene.

Bodnar, C. N., Roberts, K. N., Higgins, E. K., & Bachstetter, A. D. (2019). A systematic review of closed head injury models of mild traumatic brain injury in mice and rats. *Journal of Neurotrauma, 36*(11), 1683–1706.

Bond, A. M., Ming, G.-L., & Song, H. (2015). Adult mammalian neural stem cells and neurogenesis: Five decades later. *Cell Stem Cell, 17*(4), 385–395.

Bolton Hall, A. N., Joseph, B., Brelsfoard, J. M., & Saatman, K. E. (2016). Mild closed head injury in mice induces long-term cognitive deficits and neuroinflammation. *Journal of Neurotrauma, 33*(20), 1881–1891.

Campbell, A., Lai, T., Wahba, A. E., Boison, D., & Gebril, H. M. (2024). Enhancing neurogenesis after traumatic brain injury: The role of adenosine kinase inhibition in promoting neuronal survival and differentiation. *Experimental Neurology, 381*, 114930.

Cetin, A., Komai, S., Eliava, M., Seeburg, P. H., & Osten, P. (2006). Stereotaxic gene delivery in the rodent brain. *Nature Protocols, 1*(6), 3166–3173.

Dash, P. K., Mach, S. A., & Moore, A. N. (2001). Enhanced neurogenesis in the rodent hippocampus following traumatic brain injury. *Journal of Neuroscience Research, 63*(4), 313–319.

DeKosky, S. T., & Asken, B. M. (2017). Injury cascades in TBI-related neurodegeneration. *Brain Injury, 31*(9), 1177–1182.

Denniss, R. J., & Barker, L. A. (2023). Brain trauma and the secondary cascade in humans: Review of biochemical mechanisms. *Behavioral Sciences, 13*(5), 388.

Dodd, W. S., Smith, D. H., & Chen, Y. (2022). Traumatic brain injury and secondary neurodegenerative disease. *Trauma Care, 2*(4), 510–522.

Erginousakis, I., & Papatheodoropoulos, C. (2025). Modifying factors of adult hippocampal neurogenesis: A dorsoventral perspective in health and disease. *Cells, 15*(1).

Gao, Z., Ure, K., Ables, J. L., Lagace, D. C., Nave, K.-A., Goebbels, S., et al. (2009). Neurod1 is essential for the survival and maturation of adult-born neurons. *Nature Neuroscience, 12*(9), 1090–1092.

Gonçalves, J. T., Schafer, S. T., & Gage, F. H. (2016). Adult neurogenesis in the hippocampus: From stem cells to behavior. *Cell, 167*(4), 897–914.

Jimenez-Cyrus, D., Adusumilli, V. S., Stempel, M. H., Maday, S., Ming, G.-L., Song, H., et al. (2024). Molecular cascade reveals sequential milestones underlying hippocampal neural stem cell development into an adult state. *Cell Reports, 43*(6).

Jobling, M. (2012). *National research council (NRC): Nutrient requirements of fish and shrimp.* Springer.

Kempermann, G., Song, H., & Gage, F. H. (2015). Neurogenesis in the adult hippocampus. *Cold Spring Harbor Perspectives in Biology*, 7(9), a018812.

Kuhn, H. G., Dickinson-Anson, H., & Gage, F. H. (1996). Neurogenesis in the dentate gyrus of the adult rat: Age-related decrease of neuronal progenitor proliferation. *Journal of Neuroscience, 16*(6), 2027–2033.

Lowery, R. L., & Majewska, A. K. (2010). Intracranial injection of adeno-associated viral vectors in the mouse brain using stereotaxic coordinates. *Journal of Visualized Experiments, 45*, e2140.

Mouzon, B. C., Bachmeier, C., Ojo, J. O., Acker, C. M., Ferguson, S., Paris, D., ... Crawford, F. (2018). Lifelong behavioral and neuropathological consequences of repetitive mild traumatic brain injury. *Journal of Neurotrauma, 29*(14), 2497–2510.

National Research Council. (2011). *Guide for the care and use of laboratory animals*. National Academies Press.

Paxinos, G., & Franklin, K. B. J. (2001). *The mouse brain in stereotaxic coordinates* (2nd ed.). Academic Press.

Saha, P., Gupta, R., Sen, T., & Sen, N. (2019). Histone deacetylase 4 downregulation elicits post-traumatic psychiatric disorders through impairment of neurogenesis. *Journal of Neurotrauma, 36*(23), 3284–3296.

Schindelin, J., Arganda-Carreras, I., Frise, E., Kaynig, V., Longair, M., Pietzsch, T., ... Cardona, A. (2012). Fiji: An open-source platform for biological-image analysis. *Nature Methods, 9*(7), 676–682.

Shalizi, A., & Bonni, A. (2018). HDAC4 and MEF2 in neuronal survival and differentiation pathways. *Current Opinion in Neurobiology, 48*, 70–76.

Simon, D. W., McGeachy, M. J., Bayır, H., Clark, R. S. B., Loane, D. J., & Kochanek, P. M. (2017). The far-reaching scope of neuroinflammation after traumatic brain injury. *Nature Reviews Neurology, 13*(3), 171–191.

Sun, D. (2014). The potential of endogenous neurogenesis for brain repair and regeneration following traumatic brain injury. *Neural Regeneration Research, 9*(7), 688–692.

Toni, N., Teng, E. M., Bushong, E. A., Aimone, J. B., Zhao, C., Consiglio, A., et al. (2007). Synapse formation on neurons born in the adult hippocampus. *Nature Neuroscience, 10*(6), 727–734.

Urban, N., & Guillemot, F. (2021). Neurogenesis in the adult brain: Epigenetic regulation and function. *Nature Reviews. Neuroscience, 22*, 607–623.

Wright, D. K., Brady, D. R., Kamnaksh, A., Trezise, J., Sun, M., McDonald, S. J., ... Shultz, S. R. (2016). Repeated mild traumatic brain injuries induce persistent changes in plasma protein and magnetic resonance imaging biomarkers in the rat. *Journal of Neurotrauma, 33*(13), 1213–1222.

Xiong, Y., Mahmood, A., & Chopp, M. (2013). Animal models of traumatic brain injury. *Nature Reviews. Neuroscience, 14*(2), 128–142.

Yu, T. S., Zhang, G., Liebl, D. J., & Kernie, S. G. (2008). Traumatic brain injury-induced hippocampal neurogenesis requires activation of early nestin-expressing progenitors. *Journal of Neuroscience, 28*(48), 12901–12912.

Zhao, Z., Smith, J. R., Patel, K., Lee, M. H., Chen, Y., & Brown, T. J. (2020). Stereotaxic delivery of plasmid DNA and viral vectors into mouse hippocampus for gene manipulation. *Journal of Visualized Experiments (JoVE).*

CHAPTER THIRTEEN

The extracellular Zinc-finger ubiquitin-binding domain of histone Deacetylase 6 enhances podosome formation in neuronal cells

Tazeen Qureshi[a,b], Abhishek Ankur Balmik[c], and Subashchandrabose Chinnathambi[a,b,*]

[a]Neurobiology Group, Division of Biochemical Sciences, CSIR-National Chemical Laboratory, Pune, India
[b]Department of Neurochemistry, National Institute of Mental Health and Neuro Sciences Hospital (NIMHANS), Institute of National Importance, Bangalore, Karnataka, India
[c]Department of Pharmacology and Chemical Biology, Emory University School of Medicine, 5109 Rollins Research Center, Atlanta, GA, United States
*Corresponding author. e-mail address: subashneuro@nimhans.ac.in

Contents

Abstract

Neurodegenerative disorders are characterized by progressive synaptic failure, neuronal loss, and the accumulation of pathological protein aggregates. A critical but often overlooked driver of this decline is cytoskeletal dysregulation, which compromises essential cellular functions ranging from intracellular transport to morphological stability. Histone Deacetylase 6 (HDAC6) is a central regulator of these dynamics, yet its role in neurodegeneration remains controversial: while its deacetylase activity is often linked to microtubule instability and toxicity, its ubiquitin-binding functions are essential for

Methods in Enzymology, Volume 734
ISSN 0076-6879, https://doi.org/10.1016/bs.mie.2026.05.060

aggregate clearance. We have previously demonstrated that the ZnF-UBP domain acts as a direct modulator of cytoskeletal architecture, enhancing the formation of actin-rich migratory structures—such as podosomes and lamellipodia—and promoting neuritic outgrowth. It does this by inducing increased localization of actin remodelling proteins to the podosomes, ultimately conferring enhanced migration potential to cells. This chapter highlights the protocols essential for understanding the therapeutic potential of the HDAC6 Zinc Finger Ubiquitin-Binding Protein (ZnF-UBP) domain, in the context of actin remodelling through podosome structures.

1. Introduction

Neurodegenerative conditions such as Alzheimer's disease (AD) are clinically defined by progressive memory loss and cognitive decline, driven by the degeneration of neurons, synaptic connections and a protein buildup of extracellular amyloid-beta plaques and intracellular neurofibrillary tangles. However, the precise link between the molecular pathology and clinical symptoms remains unclear. Emerging evidence suggests cytoskeletal dysregulation as the critical bridge (Bamburg & Bloom, 2009; Muñoz-Lasso et al., 2020; Vickers et al., 2016). Synaptic loss involves cytoskeletal breakdown (Chen et al., 2019; Penzes & VanLeeuwen, 2011; Tracy & Gan, 2018), which disrupts intracellular trafficking.

Actin remodeling is central to this machinery, driving functions such as migration and trafficking through the formation of complex actin structures like podosomes, lamellipodia, filopodia (Fig. 1). Among these are podosomes—actin-rich structures found in migratory cells (Das & Chinnathambi, 2021a; Linder & Aepfelbacher, 2003; Vincent et al., 2012; Zambonin-Zallone et al., 1988) that degrade the extracellular matrix to facilitate movement. Podosomes are stabilized by the scaffold protein TKS5 and adhesion proteins like vinculin, which mediate interactions with the cellular environment (Fig. 2) (Courtneidge et al., 2005; Iizuka et al., 2016; Linder & Aepfelbacher, 2003; Murphy & Courtneidge, 2011; Pęziński et al., 2021; Vincent et al., 2012; Zambonin-Zallone et al., 1989). Histone Deacetylase 6 (HDAC6) is a key cytoplasmic regulator of these dynamics (Balmik & Chinnathambi, 2022; Boyault et al., 2007; Kaluza et al., 2011; Kim et al., 2020; Kozyreva et al., 2014; Pham et al., 2021). Unlike other HDACs, HDAC6 contains two catalytic deacetylase domains and a unique Zinc Finger Ubiquitin-Binding Protein (ZnF-UBP) domain (Balmik et al., 2021; Bertos et al., 2004; Qureshi & Chinnathambi, 2022; Simões-Pires et al., 2013; Verdin et al., 2003; Yao & Yang, 2011). This ZnF-UBP domain

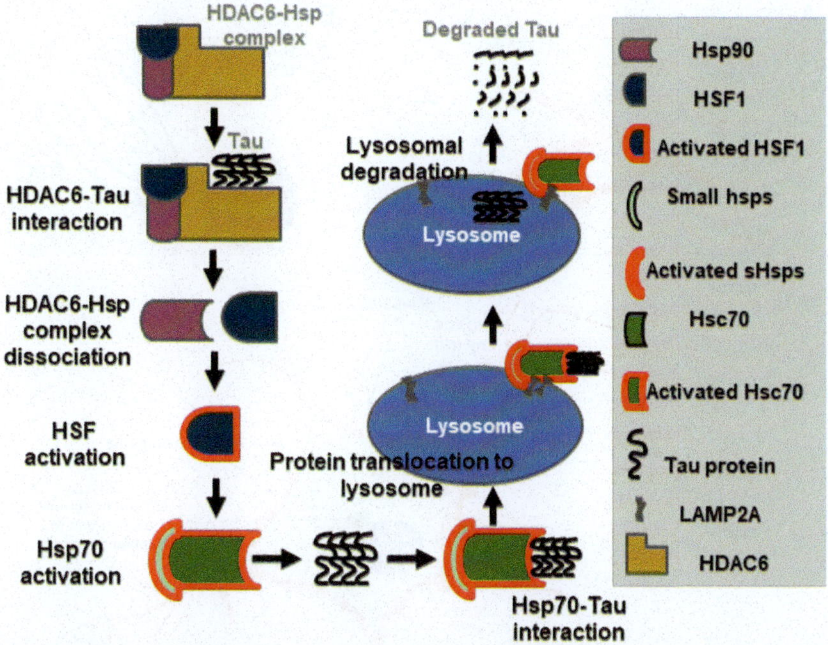

Fig. 1 Histone Deacetylase-6 protein domain. HDAC6 is bound to Heat Shock Factor (HSF1) and (Heat Shock Protein) Hsp90 complex. When HDAC6 binds to misfolded or aggregated proteins, it dissociates from the complex and the HSF1 gets activated. Activated HSF1 further activates Hsp70 and small Hsps. Activated Hsp70-small Hsps complex binds to Tau aggregates and delivers them to the lysosome through Lysosome-Associated Membrane Protein (LAMP-2A).

allows HDAC6 to recruit protein aggregates to the microtubule-organizing center (MTOC) for clearance, distinguishing it from the protein's epigenetic functions. The role of HDAC6 in AD has been controversial. Its deacetylase activity can destabilize microtubules, impairing transport and promoting neurodegeneration. Conversely, it has been shown to support actin assembly and axonal growth (Janke & Chloë Bulinski, 2011; Takemura et al., 1992; Zhang et al., 2012; Zilberman et al., 2009). Recent research suggests the non-catalytic ZnF-UBP domain acts as the key to resolving this paradox, separating the protein's protective functions from its potential toxicity. (Qureshi & Chinnathambi, 2022). Our findings (Balmik et al., 2021) indicate that the HDAC6 ZnF-UBP domain directly modulates cytoskeletal dynamics and Tau pathology, independent of deacetylase activity (Fig. 3). Neurons treated with the ZnF-UBP domain exhibited increased formation of migratory structures (podosomes, lamellipodia, filopodia) and enhanced

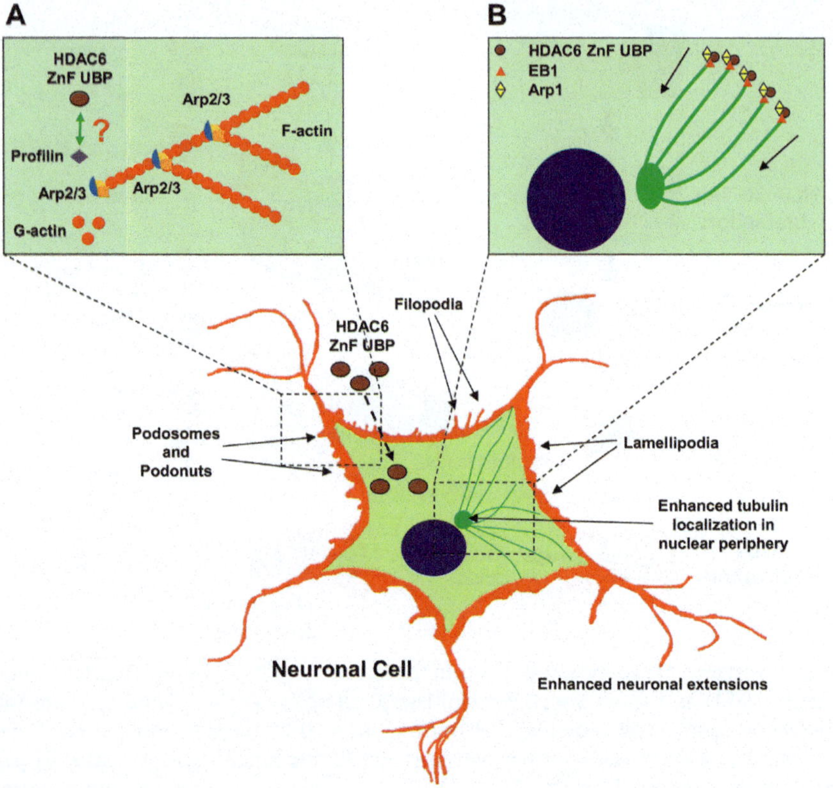

Fig. 2 Effect of HDAC6 ZnF UBP on cytoskeletal organization. A. HDAC6 ZnF UBP modulated the Actin polymerization and tubulin organization in neuro2a cells. B. It associates with profilin and Arp2/3, one of the actin polymerization factor to promote F-actin formation resulting in enhanced neuronal extensions and actin rich structure *viz.* podosomes, podonuts, lamellipodia and filopodia.

neuritic outgrowth. (Balmik et al., 2021; Qureshi, Desale, Sonawane, et al., 2024) HDAC6 ZnF UBP also induced increased localization of the actin remodelling proteins Arp2, WASP and TKS5 to the podosomes. Additionally, exposure to HDAC6 ZnF UBP domain led to increased migration capacity in neuro2a cells (Qureshi, Desale, Sonawane, et al., 2024).

2. Rationale

Podosomes are complex actin-rich cytoskeletal structures, abundant in proteins that interact and degrade the extracellular matrix, enabling cells

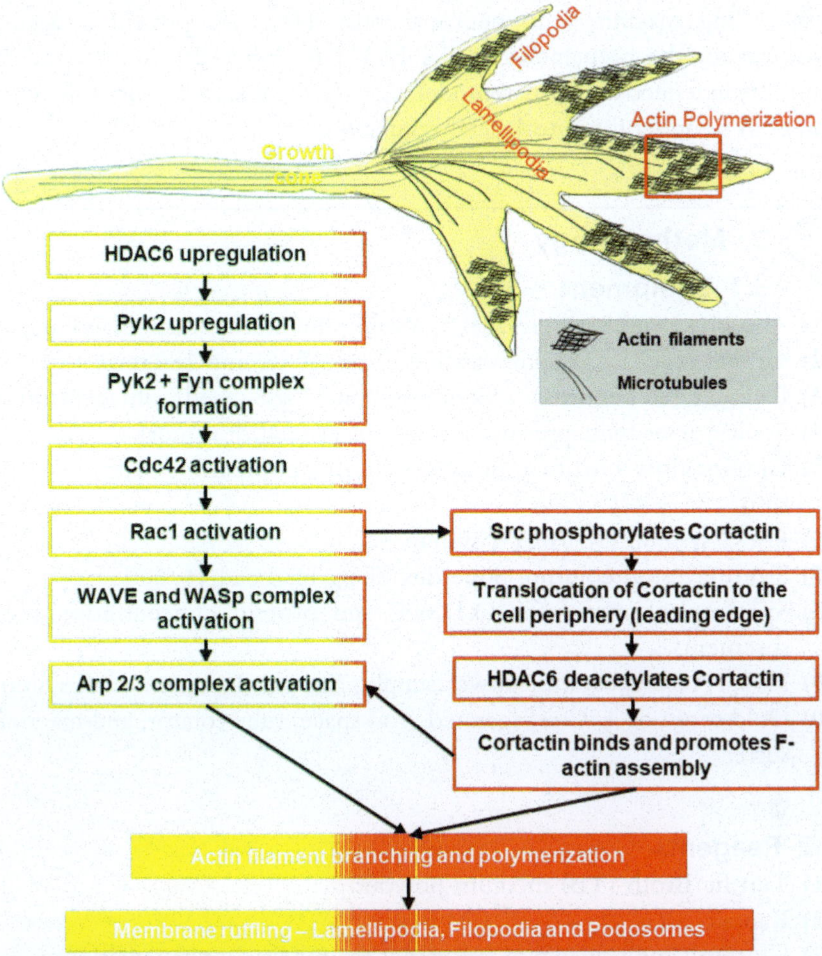

Fig. 3 HDAC6 promotes actin remodelling. Upregulation of HDAC6 levels in AD gives rise to the upregulation of Pyk2, which forms a functional complex with Fyn Kinase. Activated Pyk2-Fyn complex leads to activation of Cdc42 and Rac1. Rac1 activates the Arp2/3 complex via the WAVE complex. On the other side, activated Rac1 promotes phosphorylation of Cortactin through Src. This facilitates its translocation to the cell periphery. HDAC6 localizes at the leading edge and interacts with Cortactin, leading to its deacetylation and activation. Activated Cortactin binds and promotes Factin assembly for actin filament polymerization. Cortactin also directly interacts with and activates the Arp2/3 complex.

to displace and migrate. The formation of podosomes requires extensive actin networks and remodeling. In the recent work from our lab, we have shown the potential of the HDAC6 ZnF UBP domain on the actin remodelling machinery in neuronal cells. Here we present a detailed protocol for the preparation of the HDAC6 ZnF UBP domain and the study of its effect on actin remodelling in neuronal cells, through podosomes as a model for cytoskeletal regulation.

3. Methodology

3.1 Equipment

1) Autoclave and biological safety cabinet for sterile media handling.
2) Orbital incubator shaker capable of 16–37 °C and 180 rpm.
3) Refrigerated centrifuges (bench-top and high-speed), ultracentrifuge.
4) Sonicator or high-pressure homogenizer (cell disruptor).
5) Gravity-flow chromatography columns compatible with Ni-NTA resin.
6) FPLC system: GE AKTA system
7) Size-exclusion column: Superdex 75 pg Hi-Load 16/600
8) Spectrophotometer for O.D.600 and protein concentration measurements.
9) SDS-PAGE apparatus, power supply, and gel documentation system.
10) Cold room or 4 °C refrigerated workspace, tube rotator, and magnetic stirrer

3.2 Reagents

1) Terrific Broth (TB) medium powder.
2) Kanamycin
3) Glycerol (molecular biology grade) for media supplementation.
4) IPTG (Isopropyl β-d-1-thiogalactopyranoside)
5) Ni-NTA agarose beads.
6) Dialysis tubing with appropriate molecular weight cut-off.
7) Centrifugal concentrators (e.g., 3.5 and 5 kDa MWCO) for protein concentration.
8) Reagents for SDS-PAGE (acrylamide, buffers, running buffer, staining solutions).
9) BCA Protein Assay Kit or equivalent and standards (e.g., BSA) for protein quantification.
10) Buffers:

Buffer/ components	Tris	NaCl	PMSF	DTT	Imidazole	PIC	Glycerol
Lysis	50 mM	500 mM	1 mM	—	20 mM	1:1000 (v/v)	—
Wash	50 mM	500 mM	—	—	20 mM	—	—
Elution	50 mM	500 mM	—	—	1 M	—	5 %
Dialysis	500 mM	100 mM	—	2 mM	—	—	5 %
Gel filtration	50 mM	100 mM	—	1 mM	—	1:1000 (v/v)	-

3.3 Procedure

3.3.1 HDAC6 ZnF UBP preparation and purification

(Balmik et al., 2021; Qureshi, Desale, Sonawane, et al., 2024).

3.3.1.1 Transformation into *E.coli* competent cells

Bacterial cells have a tendency to take up free DNA from the environment with a limited efficiency. This property of bacterial cells has been employed in molecular biology to insert foreign DNA in the form of plasmid into bacteria. The bacterial cells must be made competent by chemical or physical methods to take up the DNA. Upon transformation of foreign DNA into bacteria, it can be utilized for maintenance of genetic material or can be induced to express a recombinant protein of interest.

3.3.2 Procedure

1. Competent cells (*DH5α, BL21** or *BL21* Codon plus RIL) were removed from −80 °C and kept on ice bath for 20 min.
2. Keeping the cells on ice bath, 1−2 μl of plasmid DNA is added.
3. Cells were incubated in ice bath for 30 min for binding of plasmid DNA.
4. Heat shock is given to the cells by keeping in water bath at 42 °C for 90 s to transiently disrupt the cell membrane and plasmid DNA intake.
5. The cells are immediately kept on ice bath for 5 min.
6. 750 μl of SOC media is added to the cells, mixed and kept in incubation for 1 h in a shaker incubator.
7. 50−100 μl of media containing transformed cells were taken and spread on an antibiotic containing LB agar culture plate.
8. The plates were kept in an inverted position at 37 °C for 8−10 h.

The bacterial cells transformed with plasmid DNA gives pin-point colonies at the end of 8–10 h which are picked for further scale up.

3.3.2.1 Miniprep for plasmid isolation and purification

After the transformation of desired plasmid vector containing the gene of interest (Tau or HDAC6), isolated colonies of *DH5α* strain were used for maintenance of plasmid. *DH5α* has high transformation efficiency and contains recA mutation which eliminates homologous recombination, endA1 mutation to inactivate intracellular endonuclease which degrades plasmid DNA. Thus, it is considered a suitable strain of E.coli for maintenance of plasmid DNA due to high insert stability. Δ (lacZ) M15 allele, which enables blue-white screening for lacZ based vectors.

Purification of plasmid DNA was carried out according to the manufacturer's protocol (Machery Nagel). The technique for plasmid DNA isolation is based on the alkaline lysis method. The bacterial cells are lysed using a lysis buffer containing SDS and sodium hydroxide (NaOH). SDS dissolves the bacterial cell membrane while NaOH helps in disruption of cell wall as well as denaturation of cellular proteins and DNA. Neutralization buffer is added to the lysates, which allows renaturation of genomic and plasmid DNA. Plasmid DNA, being smaller in size reanneals completely resulting in its solubilization while genomic DNA does not get renatured properly due to large size and stay precipitated. The cellular debris, precipitated proteins and DNA can now be easily separated from plasmid DNA by centrifugation. Plasmid DNA is captured on a silica membrane by ethanolic precipitation and can be eluted by using buffer of low ionic strength in slightly alkaline conditions.

3.3.3 Procedure

1. Single isolated pin-point colonies obtained after transformation are selected for inoculation into 5 ml of LB media and kept for incubation at 37 °C overnight in a shaker incubator. At least four isolated colonies were inoculated in four tubes.
2. Bacterial cells obtained after overnight culture were harvested by centrifugation at 11,000 × *g* for 30 s at room temperature.
3. 250 μl of Buffer A1 was added in each tube and mixed thoroughly by vortexing until the cells were completely lysed.
4. 250 μl of Buffer A2 was added and mixed by inverting the tubes. Tubes were incubated for 5 min at room temperature.
5. 300 μl of Buffer A3 was added and mixed again by inverting the tubes.

6. Cell lysates were clarified by centrifugation at 11,000 × *g* for 30 s at room temperature.
7. Clarified lysates were passed through columns with silica membrane, placed over collection tubes, by centrifugation at 11,000 × *g* for 30 s at room temperature. Flow-through was discarded.
8. Silica membranes were washed with 600 μl Buffer A4 supplemented with ethanol by centrifugation at 11,000 × *g* for 1 min at room temperature. Flow-through was discarded and columns were placed in an empty collection tube.
9. Silica membranes were dried by centrifugation at 11,000 × *g* for 2 min at room temperature. Collection tubes were discarded.
10. 50 μl of Buffer AE was added to each tube and incubated for 1 min at room temperature.
11. Plasmid DNA was eluted by centrifugation at 11,000 × *g* for 1 min at room temperature.
12. The respective concentration of plasmid DNA obtained from each tube was measured in Nanodrop (Simplinano).

3.3.3.1 Protein expression and culture scale-up for protein preparation

To check the expression level of protein of interest, *BL21** or *BL21* Codon plus RIL strains of *E.coli* were used. *BL21** strain contains DE3 lysogen carrying the gene for *T7 RNA polymerase* under the control of lacUV5 promoter. It is widely used for over-expression of recombinant protein cloned in T7 expression vectors by using IPTG as an inducer. *BL21* Codon plus RIL has same elements for protein expression as *BL21** but also contains extra copies of argU, ileY and leuW tRNA genes to resolve the condition of codon bias in AT rich genes.

The expression of recombinant protein was checked by culturing the single isolated colonies after transformation of *BL21** or *BL21* Codon plus RIL strains containing plasmids for Tau (pT7C) and HDAC6 ZnF UBP (pET28-LIC) respectively. The plasmids confer antibiotic resistance to the transformed cells inhibiting the growth of non-transformed cells or contaminants. The gene for protein of interest is under the regulation of lac operator and the *T7 RNA polymerase* can bind to the operator and start transcription under inducible conditions *i.e.* when an inducer is present in the cell. Isopropyl β-D-1-thiogalactopyranoside (IPTG) is a molecular mimic of allolactose, which is the metabolite of lactose responsible for triggering lac operon system. Expression of protein of interest was carried out in the presence of IPTG as an inducer and the levels of protein expression were visualized by comparing with uninduced bacterial culture by using SDS-PAGE. Further, the selected colony with optimum protein expression was scaled-up using large volumes of culture.

3.3.4 Media preparation and Bacterial expression

1) Weigh 70.5 g terrific broth (TB) for each 1.5 l culture and 4.7 g for the 100 ml (for primary culture).
2) Add 6 ml Glycerol to each 1.5 l culture and 0.4 ml to the 100 ml culture. Make up the volume with R.O. water and autoclave.
3) **Primary inoculation:** Add 50 ug/ml Kanamycin followed by 1% of inoculum stock (HDAC6 ZnF UBP in pET28a-LIC plasmid, transformed and expressed in BL21 Codon plus RIL cells) and mix. Incubate the culture at 37 °C and constant shaking at 180 rpm, overnight.
4) **Secondary inoculation:** Add 50 ug/ml Kanamycin to the media, mix well and add 1 % primary inoculum. Incubate (37 °C, 180 rpm) till the solution appears turbid. Check the O.D. at 600 nm; it should be around 0.6 (*See note i*). Keep the flasks in cold room for half an hour (to stop any further growth).
5) **Induction:** After the culture cools down, add 200 uM IPTG to the flasks and incubate at 16 °C, 180 rpm, overnight for expression.
6) **Buffer preparation:** Prepare all buffers (Lysis, Wash, Elution, Dialysis, Gel Filtration) and adjust pH to 8.0.

3.3.5 Cell harvesting and lysis

1) Centrifuge the culture to collect the pellet at 4500 rpm, 15 mins. Decant the supernatant and take out the pellets in 50 ml falcons and centrifuge at 4000 rpm for 10 mins to remove any remaining supernatant. (*See note ii*)
2) Add about equal volume of lysis buffer to the pellets and mix well (vortex) to get rid of any clumps. More buffer can be added and vortexed to attain a smoother consistency which is not too thick. (*See note iii*)
3) Homogenize this pellet suspension in a cell disrupter or homogenizer at 15 Kpsi OR sonicate at Pulse = 10 s, Amplitude = 50 %, Time = 25 mins total/pellet.
4) If the lysate becomes too thick, add a little more lysis buffer and vortex to make a smoother consistency (no clumps). Subject again to sonication till the entire solution becomes homogenous (lysis complete). (*See note iv and v*) Keep checking for lysis for a darker shade of solution, from time to time.
5) Centrifuge the lysate at speed 40,000 rpm, time 1 hr, temperature 4 °C. (*See note vi*) Collect the supernatant in another tube. Keep on ice. (*See note vii*)

3.3.6 Buffers used for HDAC6 ZnF UBP preparation

Extraction/lysis buffer
50 mM Tris-Cl pH 8.0
500 mM NaCl
1 mM PMSF
20 mM Imidazole
Protease Inhibitor Cocktail

3.3.7 Ni-NTA affinity purification

1) Take Ni-NTA beads in 1:50 ratio with the supernatant and spin at 400 rpm for 60 s and discard the supernatant. This is to remove the 20 % ethanol in which the beads are stored. Give 2 more such washes with water. Give one more wash with the wash buffer.
2) Resuspend the washed beads in wash buffer and add to the protein supernatant. Mix well and keep on a tube rotator in the cold room or at 4 °C for 1.5–2 hrs.
3) Add the supernatant with beads to the column burette. The supernatant will flow through, while the column gets packed.
4) Add wash buffer and collect in a separate tube labelled Wash 1. After Wash 1 tube is filled, collect in another tube labelled Wash 2.
5) Put the stopper for the column and add 6 ml elution buffer. Keep 6 1.5 ml tubes ready for collecting 6 elution fractions (F1-F6). Collect 1 ml in each. (*See note viii and ix*)

3.3.8 Ni-NTA affinity chromatography

Wash buffer	Elution buffer
50 mM Tris-Cl pH 8.0	50 mM Tris-Cl pH 8.0
500 mM NaCl	500 mM NaCl
20 mM Imidazole	1 M Imidazole
	5 % Glycerol

3.3.9 Dialysis and concentration

1) Wash a dialysis bag (3 kDa cutoff) with dialysis buffer.

2) Fold the bag from one open side and attach the clamp tightly. Add buffer from the other side and apply pressure on the clamped side to check for leaks.
3) Replace the buffer with the elution fractions. After filling, create enough air gap inside the bag to help it float during dialysis. Ensuring the air gap, fold the other side and attach the 2nd clamp. Apply gentle pressure and check for leaks.
4) Fill a big beaker with dialysis buffer and keep the filled bag inside. Place a magnetic bead in the beaker and keep it on magnetic stirring at 4 °C (Cold room). After 2−3 h, transfer the bag in fresh dialysis buffer. Let it stay overnight in the cold room. (*See note x*)
5) Centrifuge the dialyzed sample in 3500 kDa cutoff centricons at 4000 rpm, temp 4 °C. Check for a reduction in volume after every 20−30 min and mix. Get the final volume to 1 ml.

Dialysis buffer	**Size-exclusion chromatography buffer**
50 mM Tris-Cl pH 8.0	50 mM Tris-Cl pH 8.0
100 mM NaCl	100 mM NaCl
2 mM TCEP/DTT	1 mM TCEP/DTT
5 % Glycerol	

3.3.10 Size-exclusion and final processing

1) Equilibrate the size-exclusion column (Superdex 75 pg Hi-Load 16/600) with Gel Filtration Buffer using an FPLC system, ensuring stable conductivity and baseline.
2) Inject the concentrated sample into the column using the sample loop. (*See note xi, xii*)
3) Identify peak fractions corresponding to the target protein by UV absorbance (214 and 280 nm) and confirm by SDS-PAGE, pooling fractions with a single band at the expected molecular weight. The peak between ~80-~120 ml would correspond to HDAC6 (ZnF UBP domain).
4) Concentrate the pooled fractions as required and determine protein concentration using the BCA assay (*See note xiv*) or UV absorbance at 280 nm (*See note xv and xvi*).
5) Perform a concentration-dependent SDS-PAGE to confirm the estimated concentration. (*See note xvii*)
6) Aliquot the concentrated protein and snap freeze to store at −80 °C.

3.3.11 Bradford assay

Bradford assay is a protein estimation method based on the binding of protein with Coomassie dye under acidic conditions [5]. The Coomassie G-250 dye is doubly protonated in acidic conditions and has red color. Upon binding with protein it gets unprotonated and acquires blue color. The principle of the method is based on transfer of electron from dye to the basic amino acids Arginine, Lysine and Histidine in protein. This leads to the disruption in protein structure and expose hydrophobic patches; to which the dye can bind. The dye bound to protein forms a stable blue colored complex which is proportional to the protein concentration. The intensity of the colored complex can be spectroscopically measured and used to quantify the protein concentration.

3.3.12 Standards, protein dilutions and reagent preparation

1. Working reagent is prepared by mixing 4parts MiliQ Water and 1 Part Bradford Reagent (Bio-Rad).
2. The protein standards are made in a range of 0–250 μg/ml (Stock–1 mg/ml) concentrations in MiliQ water.
3. The protein dilutions are made in MiliQ water in at least three ratios ranging from 1:25 – 1:400.
4. All the standards and protein dilutions are made in duplicates.

3.3.13 Procedure

1. 10 μl of BSA standards and protein dilutions are mixed with 200 μl of working reagent and pipetted into a 96 wells clear bottom plates.
2. Incubate at room temperature for at least 5 min.
3. Absorbance will increase over time; samples should incubate at room temperature for no more than 1 h.
4. The absorption by colored complex is measured at 595 nm in Tecan Infinite 200 PRO spectrophotometer.
5. The protein concentration of the sample is determined by plotting a standard graph using MS Excel.

3.3.14 SDS-PAGE analysis and quantification

SDS-PAGE is an analytical method to separate proteins on the basis of their molecular weight. The proteins migrate under the influence of electric field, through the matrix of poly-acrylamide gel after denaturation by anionic detergent Sodium dodecyl sulfate and reducing agent DTT or β-mercaptoethanol. Upon denaturation, SDS binds to the protein with respect to its relative molecular weight and impart overall negative charge over a wide pH range. Thus, protein migration takes place toward positive electrode (anode) in an electrophoresis system. The smaller proteins migrate fast due to less resistance by gel matrix and *vice versa.*

Discontinuous gel system is used for the separation of proteins which consists of a stacking gel with wells for sample application and a resolving gel. SDS-PAGE running buffer consists of Tris-Cl and Glycine which gets ionized as the electric field is applied. Bromophenol Blue (BPB) is used as the tracking dye. The stacking gel has a lower pH of 6.8 which imparts differential mobility to protein, glycine, chloride and BPB. The ionization potential of glycine is less in stacking gel as its pH is closer to isoelectric pH of glycine. This imparts lesser conductivity to glycine. However, chloride ions, proteins and BPB retains higher conductivity forming a stack of protein between slow migrating glycine ions and faster chloride ions due to local potential difference between them. Upon entering resolving gel at pH 8.8, glycine retains its electrical conductivity but the overall migration rate declines. Each component ion is subjected to constant electric force and protein separation begins as per their molecular weight. A molecular weight marker is loaded along with the protein samples to locate the protein of interest. The electric field is stopped when tracking dye reaches 80–90% of gel length. The gel is removed from the assembly and proceeded further for staining or western blotting.

3.3.15 Buffers and reagents

SDS-PAGE running buffer
25 mM Tris-Cl
250 mM Glycine
10 % SDS

Tracking dye/sample buffer
300 mM Tris-Cl pH 6.8
600 mM DTT
6 % BPB
12 % SDS
60 % Glycerol

3.4 Notes

i. The O.D. for the secondary inoculation takes about 6 hrs to reach O.D. = 0.6, at which point the culture will be just a little less turbid than the primary culture turbidity.

ii. The supernatant can be reused for balancing if any culture is left to be centrifuged.

iii. Avoid adding too little or too much buffer during lysis, to avoid diluting the protein. This can be divided into 2 falcons to avoid too much thickness.

iv. Prepare a beaker with ice to form a sort of ice jacket around the falcon during sonication. This is done to avoid protein degradation.

v. Keep adding lysis buffer in each sonication cycle, till the final total volume equals 5 times the pellet volume. E.g., If the pellet is 20 ml, the total final volume should be equal to 100 ml (25 ml in 4 falcons), each 25 ml falcon subjected to 5 cycles of 5 min sonication.

vi. In the ultracentrifuge, because there are such high speeds, the system needs to be clear of any air and filled with vacuum. Start only once the air in the system is <250 μm. Wait and check for any abnormalities till the high speed is attained. Make sure the tubes are balanced till the 3rd decimal.

vii. Take out 10 ul of supernatant, and pellet (a little scraping is enough) for loading on gel – SDS PAGE for checking protein loss at each stage.

viii. Avoid wasting or touching the top of the tube. Keep the fractions in ice, to avoid degradation.

ix. Prepare tubes for loading on the gel (protein loss checking): Take 30 ul of F1-F6, Wash 1, Wash 2, Flowthrough and 10 ul of beads, after lysis, supernatant and pellet.

x. The dialysis step is for the removal of excess imidazole. However, there have been instances where the dialysis step has led to protein precipitation in which case the dialysate has to be dissolved again in the elution buffer, bringing the imidazole concentration back to being high. Imidazole, being small in size, will be eradicated in the later steps of gel filtration and protein concentration. Thus, the dialysis step can even be skipped in case of instances of protein precipitation.

xi. Let the system undergo washing, to clear the pipes and tubings of air bubbles. Check the conductivity (= zero) to make sure the column is completely washed.

xii. Initially, connect the top loosely, look for the flow from the bottom connector, and then tighten both ends. This will avoid the buildup of pressure, which will avoid column disruption.
xiii. Make sure the fraction collector is clean and completely dry, to avoid protein contamination or degradation.
xiv. The dilutions used for the BCA assay are 1:10, 1:25, 1:50 for the prepared protein.
xv. The blank used for the absorbance-based concentration estimation at this step is the Gel filtration buffer.
xvi. The absorbance-based protein concentration can be checked at various steps in the protein preparation process to estimate the amount of protein at each stage. The blank at each step would be the respective buffer.
xvii. Concentration-dependent SDS-PAGE: For instance, the protein concentration through BCA/Absorbance is 2 mg/ml. Then, the concentrations to be loaded on the gel would be 0.25, 0.5, 1 and 2 mg/ml alongside a previously prepared protein of known concentration. This would help visually compare and evaluate the concentration measured in the protein estimation assays.

3.5 Cell culture and treatment with extracellular HDAC6

(Balmik et al., 2021; Qureshi, Desale, Sonawane, et al., 2024).

3.5.1 3-(4, 5-dimethylthiazol-2-yl)-2, 5-diphenyltetrazolium bromide (MTT) assay

MTT assay is a colorimetric method to measure metabolic functions of cell as an indicator of their viability. 3-(4, 5-dimethylthiazol-2-yl)- 2, 5-diphenyltetrazolium bromide (MTT) is a tetrazolium salt which get reduced to a purple colored MTT-formazan crystals by NADP dependent oxidoreductase enzymes. The crystals formed during the reaction are solubilized and the color intensity is measured at 500–600 nm. The intensity of the colored product is directly proportional to the viability of cells.

3.5.2 Procedure

1. 10^4 neuro2a cells (ATCC CCL-131) were seeded in a 96 well culture plate in DMEM supplemented with 10 % FBS and antibiotic penicillin-streptomycin for 24 h at 37 °C CO_2 incubator.
2. The cells were treated with HDAC6 ZnF UBP (0- 500 nM) in serum-starved media for 24 h.

3. MTT was added to the cells at a concentration of 0.5 mg/ml and incubated for 3 h.
4. The reduction of MTT by cellular enzymes forms purple formazan crystals, which were dissolved by adding DMSO.
5. The developed color was quantified by reading at 570 nm in a Tecan Infinite 200 PRO spectrophotometer.

3.5.3 Lactate dehydrogenase (LDH) assay

Lactate Dehydrogenase (LDH) is a cytoplasmic enzyme involved in cellular respiration and present in all cells. LDH is released from the cells if damage to cell membrane occurs. Thus, LDH release form the cell can be used as an indicator of cyto-toxicity. LDH assay is a colorimetric assay which quantifies the LDH released by cell by using a tetrazolium salt 2-(4-iodophenyl)-3-(4-nitrophenyl)-5-phenyl tetrazolium chloride (INT). INT gets converted into a red formazan product upon reduction by NADH produced as LDH converts lactate to pyruvate. The intensity of red color is proportional to the released LDH and thus the number of damaged cells. We studied the effect of HDAC6 ZnF UBP treatment on cell membrane integrity by LDH assay.

3.5.4 Procedure

1. 10^4 neuro2a cells (ATCC CCL-131) were seeded in a 96 well culture plate in DMEM supplemented with 10% FBS and antibiotic penicillin-streptomycin for 24 h at 37 °C CO_2 incubator.
2. The cells were treated with HDAC6 ZnF UBP (0- 500 nM) in serum-starved media for 24 h.
3. After the treatment with HDAC6 ZnF UBP, supernatant media was separated and used for the assay as per the manufacturer's protocol.
4. In brief, 50 μL of cell supernatant was incubated with 50 μL of the provided reaction mixture for 30 min at room temperature.

50 μL of stop solution was added to each well and the color developed was measured at 490 nm and background subtraction was done at 680 nm in a Tecan Infinite 200 PRO spectrophotometer.

3.5.5 Cell culture

1) Seed 2 × 104 cells on an 18 mm coverslip in a 12-well culture plate.
2) After the cells were attached to the substratum (24 h), replace the complete media with the serum-starved (0.5% FBS) media and add HDAC6 ZnF UBP (50 nM). Incubate at 37 °C, 5 % CO_2 for 24 h.

3.5.6 Immunofluorescence

Immunofluorescence refers to the technique used for visualizing the proteins in cells or tissues by using fluorescent tagged primary or secondary antibodies. This technique provides the cellular or subcellular localization, distribution and levels of protein in question. The immunostained proteins can be visualized by a microscope based on the principle of fluorescence *i.e.* simple fluorescence microscope, confocal microscope *etc* (Fig. 4).

3.5.7 Procedure

1) Neuro2a cells from passage number 10–20 were cultured in advanced DMEM supplemented with penstrep-glutamine, anti-mycotic and 10 % FBS at 37 °C, 5 % CO_2, until the cells were 70–80 % confluent.
2) For immunofluorescence studies, $5*10^4$ cells were seeded on a glass coverslip (Bluestar) in a 12 well cell culture plate and allowed to attach for 24 h.
3) Cells were given the respective treatment in serum starved media (0.5 % FBS) for 24 h.
4) Fix cells with 4 % Paraformaldehyde (PFA) for 20 mins at R.T.
5) Wash with 1 % PBST. Block cells using 5 % blocking buffer (5 % Horse Serum + 0.2 % Triton X-100).
3) Incubate cells with the primary antibody (prepared in the blocking buffer) overnight at 4 °C. Wash thrice.
4) Further incubate with the secondary antibody for 1 h at R.T., followed by washing unbound antibodies.
5) Incubate the cells with nuclear counterstain DAPI for 5 mins and mount on a glass slide using a mounting medium.
6) Capture images at 63X or 100X magnification in an epifluorescence or confocal microscope.
7) Analyze intensities of each protein and their cellular localization through the quantification analysis of the images in the microscope software.

3.6 Podosome analysis

(Qureshi, Desale, Chidambaram, et al., 2024; Qureshi, Desale, Sonawane, et al., 2024).

3.6.1 Confirmation of podosomes—TKS5 immunomapping

1) Podosome structures can be confirmed in the cells though immunomapping of the protein TKS5 in the podosomes (Fig. 4B).

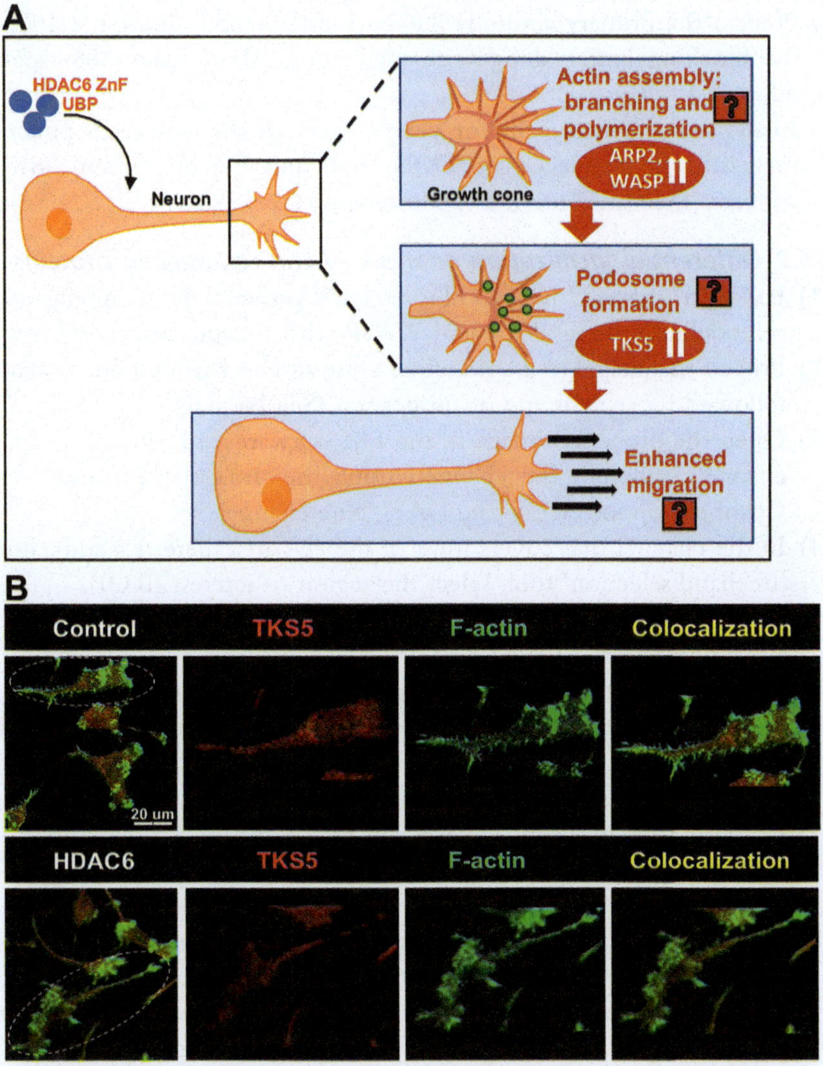

Fig. 4 HDAC6 shows potential to enhance migration through increased podosome formation in neurons. (A) Schematic denoting the hypothesis of the study – HDAC6 ZnF UBP treatment to N2a cells results in: increased actin branching (Arp2 protein), increased podosome formation (TKS5 protein) and increased migration. (B) Confocal images (2D) with emphasis on a single cell from the field, showing localization of F-actin (green), TKS5 (red) and their interaction (colocalization) in the merge (DAPI – blue, Merge - yellow). The zoomed images highlight (white dotted circles) 2 regions - a podosome-specific region and a region with no podosomes.

2) Prepare the primary antibody TKS5 (Host – rabbit; dilution 1:100) in the blocking buffer along with phalloidin (Alexa Fluor 488 tagged; dilution 1:40).
3) Analyze the imaged cells for yellow spots in the actin-rich punctae, suggesting co-incidence of TKS5 (red) and F-actin (green), which confirms their attribution as podosomes.

3.6.2 Differential localization analysis—Actin remodeling proteins

1) Follow the above immunofluorescence protocol for mapping actin remodeling proteins Arp2 and WASP with F-actin, respectively.
2) Process the images to get rid of the noise and background fluorescence, followed by analysis and quantification (See Note 1).
3) Open the processed image in the FIJI software with the merge of all channels and split the channels using the 'Image – Colors – Split Channels' option. Close the DAPI/blue channel.
4) In the channel that covers most of the cell area (here, F-actin), using 'freehand selection' tool, select the region of interest (ROI).
5) Perform the colocalization analysis for the selected ROI using 'Analyze – Colocalization – Coloc 2'.
6) The colocalization parameters for the ROI will be displayed in a separate dialogue box. The 'Pearson's R value (no threshold)' gives the Pearson's value for average colocalization coefficient for all the points in the region selected.
7) Quantify the processed images in FIJI software for colocalization coefficients between Arp2-Factin, WASP-Factin, and TKS5-Factin, respectively.
8) For differential localization analysis, two different kinds of ROIs must be analyzed for each cell – a. Overall colocalization, implies selection of the entire cell and b. Podosome colocalization, where only single podosome punctae or a podosome-rich region would be selected as ROI.
9) Note the values for different ROIs and save in a separate sheet for the analysis of significance between different groups and the trends followed.
10) The difference in the colocalization of the proteins between the two ROIs in the control and treated groups implies the effect of the treatment on the translocation of the protein from or to the podosome regions from other parts of the cell.

11) For example, if the treated group shows higher colocalization between Arp2 and F-actin in the podosomes as compared to the whole cell (overall), while the control group shows relatively lower colocalization in the podosome region, this suggests that the treatment is inducing the translocation or the increased localization of Arp2 towards/within podosomes.
12) Determine the statistical significance of the effect of treatment using unpaired *t*-test at 99% confidence interval, where $p > 0.1$ = non-significant (n.s.), $p < 0.1$ = *, $p < 0.01$ = **, $p < 0.001$ = ***.

3.7 Note

1) Image processing is done in the imaging software (Zen), where the black component of each channel represents the background, while the white component represents the brightness of the color intensity. The black and white components are adjusted to improve the signal to noise ratio for each channel. The processed images are saved in compatible formats like TIFF or JPEG.

3.8 Cell migration

(Das & Chinnathambi 2021b, 2021a; Desale & Chinnathambi, 2020; Qureshi, Desale, Sonawane, et al., 2024).

1) Seed 50×10^4 cells (N2a neuron and N9 microglia CVCL-0452) on 6-well culture plates. Incubate at 37 °C, 5 % CO_2 for 24 h.
2) After the cells settled, create wounds in the wells using a tip, followed by a PBS wash to clear the disturbed cells from the wound.
3) Add 50 nM HDAC6 ZnF UBP in the serum-starved media to the wells. Add 50 μM ATP as the inducer for migration (positive control).
4) 0 and 24 h images of the wounds were captured in a bright field microscope (10X magnification).
5) The wound diameter is noted at three different points across the wound lines.[7,28,29]
6) Wound closure/healing was calculated as:
   % Wound closure = (Wound diameter at 24 h/Wound diameter at 0 h) × 100.

3.8.1 Cell migration (Trans well by Smita)

(Desale & Chinnathambi, 2020; Qureshi, Desale, Sonawane, et al., 2024).

1) Add treatment media to the base of a 24-well transwell plate, and the cell suspension (0.5×10^4) to the insert (0.5 μm pore membrane). Incubate at 37 °C, 5% CO_2 for 24 h (Fig. 4B).

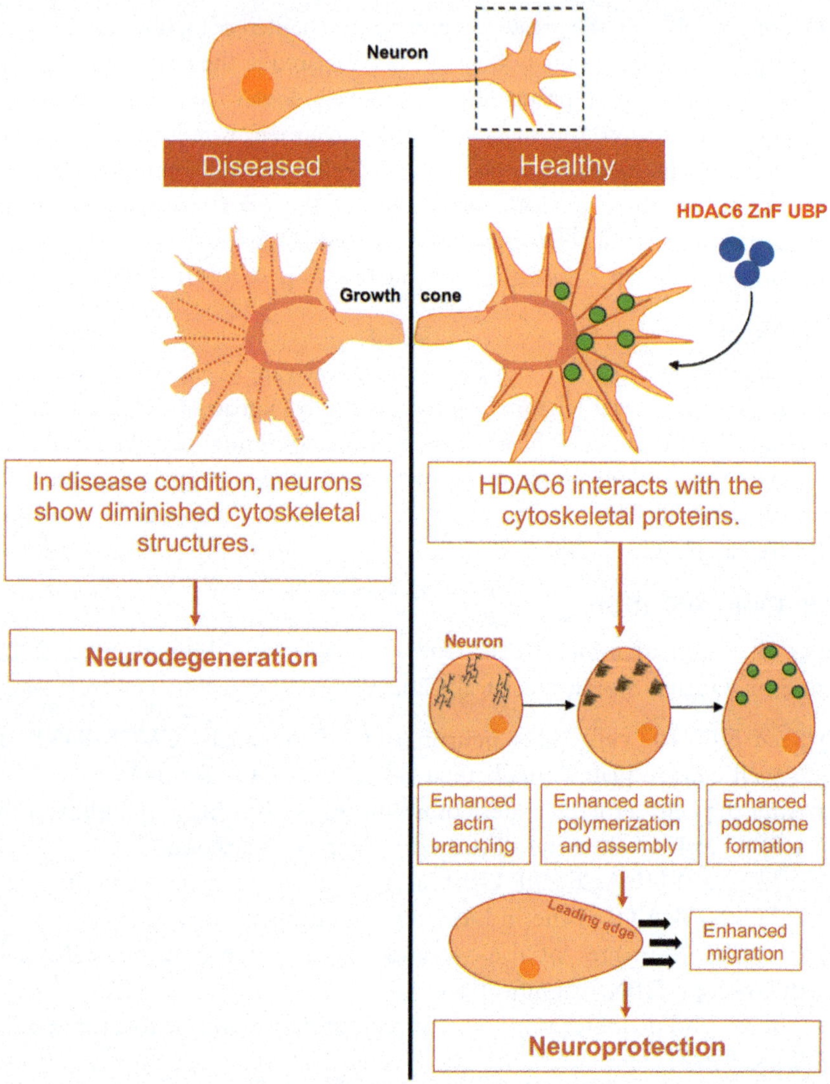

Fig. 5 Therapeutics aspect of HDAC6 in AD. The neuronal loss in Alzheimer's disease involves a myriad of factors, which are either 'cause or effect' of pathological cascade. The therapeutic approach for the prevention and cure of AD would require a multifactorial approach on various levels, which can promote neuronal survival as well as Tau clearance. Augmentation of HDAC6 ZnF UBP function can have a potential to develop a beneficial therapeutic advancement.

2) Replace the media with 4 % PFA for cell fixation. Replace PFA with PBS for washing.
3) Add 0.2 % crystal violet (or any other primary stain) in the wells to stain the cells underneath the insert membrane.
4) Swab the upper/inner side of the insert, leaving only the cells that have migrated to the other side.
5) Capture bright-field images (10X and 20X magnification).
6) Count the number of cells, using particle counter plugin in FIJI (Fiji Is Just ImageJ).

4. Concluding remarks

Alzheimer's disease (AD) is the most progressive neurodegenerative disease, characterized by the amyloid-beta plaques on the extracellular side of the neuron and the deposits of misfolded Tau protein within the neurons, making AD one of the most important Tauopathies. The development of neuronal extensions and connections helps in navigation in the CNS (central nervous system), efficient transfer and information processing as well as for the overall maintenance of neuronal homeostasis. Growth cones are specialized structures in neurons, responsible for the formation of such extensions. These neuritic extensions are composed of protrusions of filopodia (filamentous actin) alternating with fan-shaped lamellipodia (branched actin). These structures harbour dynamic coordination of the various cytoskeletal proteins such as actin, tubulin and their associated proteins in a process called 'actin remodelling'. We have previously reported that the treatment of neuro2a cells with HDAC6 ZnF UBP demonstrates enhancement in the formation of neurites and migratory structures such as the lamellipodia, filopodia and podosome-like structures, potentially regulating actin dynamics in the neuro2a cells (Fig. 5). To summarize our findings, HDAC6 ZnF UBP treatment to Neuro2a cells results in better actin branching and polymerization, which enhances the formation of podosomes and other migratory structures (filopodia), thereby changing the morphology of the cells and ultimately conferring neuro2a cells with an enhanced ability to migrate. Our observations, along with other reports, form a foundation for the field of neuronal podosomes as key players in neuronal health and disease.

References

Balmik, A. A., & Chinnathambi, S. (2022). Inter-relationship of histone Deacetylase-6 with Tau-cytoskeletal organization and remodeling. *European Journal of Cell Biology, 101*(2), 151202. https://doi.org/10.1016/j.ejcb.2022.151202.

Balmik, A. A., Sonawane, S. K., & Chinnathambi, S. (2021). The extracellular HDAC6 ZnF UBP domain modulates the actin network and post-translational modifications of Tau. *Cell Communication and Signaling, 19*(1), 49. https://doi.org/10.1186/s12964-021-00736-9.

Bamburg, J. R., & Bloom, G. S. (2009). Cytoskeletal pathologies of Alzheimer disease. *Cell Motility and the Cytoskeleton, 66*(8), 635–649. https://doi.org/10.1002/cm.20388.

Bertos, N. R., Gilquin, B., Chan, G. K. T., Yen, T. J., Khochbin, S., & Yang, X.-J. (2004). Role of the tetradecapeptide repeat domain of human histone deacetylase 6 in cytoplasmic retention. *The Journal of Biological Chemistry, 279*(46), 48246–48254. https://doi.org/10.1074/jbc.M408583200.

Boyault, C., Sadoul, K., Pabion, M., & Khochbin, S. (2007). HDAC6, at the crossroads between cytoskeleton and cell signaling by acetylation and ubiquitination. *Oncogene, 26*(37), 5468–5476. https://doi.org/10.1038/sj.onc.1210614.

Chen, Y., Fu, A. K. Y., & Ip, N. Y. (2019). Synaptic dysfunction in Alzheimer's disease: Mechanisms and therapeutic strategies. *Pharmacology & Therapeutics, 195*, 186–198. https://doi.org/10.1016/j.pharmthera.2018.11.006.

Courtneidge, S. A., Azucena, E. F., Pass, I., Seals, D. F., & Tesfay, L. (2005). The SRC substrate Tks5, podosomes (invadopodia), and cancer cell invasion. *Cold Spring Harbor Symposia on Quantitative Biology, 70*, 167–171. https://doi.org/10.1101/sqb.2005.70.014.

Das, R., & Chinnathambi, S. (2021a). Microglia degrade extracellular tau oligomers deposits via purinergic P2Y12-driven podosomes, filopodia formation and induce chemotaxis (SSRN Scholarly Paper No. ID 3919950). Social Science Research Network. https://doi.org/10.2139/ssrn.3919950.

Das, R., & Chinnathambi, S. (2021b). Microglial remodeling of actin network by Tau oligomers, via G protein-coupled purinergic receptor, P2Y12R-driven chemotaxis. *Traffic (Copenhagen, Denmark), 22*(5), 153–170. https://doi.org/10.1111/tra.12784.

Desale, S.E., & Chinnathambi, S. (2020). α-Linolenic acid modulates phagocytosis of extracellular Tau and induces microglial migration (p. 2020.04.15.042143). *bioRxiv*. https://doi.org/10.1101/2020.04.15.042143.

Iizuka, S., Abdullah, C., Buschman, M. D., Diaz, B., & Courtneidge, S. A. (2016). The role of Tks adaptor proteins in invadopodia formation, growth and metastasis of melanoma. *Oncotarget*, 7(48), 78473–78486. https://doi.org/10.18632/oncotarget.12954.

Janke, C., & Chloë Bulinski, J. (2011). Post-translational regulation of the microtubule cytoskeleton: Mechanisms and functions. *Nature Reviews. Molecular Cell Biology, 12*(12), 773–786. https://doi.org/10.1038/nrm3227.

Kaluza, D., Kroll, J., Gesierich, S., Yao, T.-P., Boon, R. A., Hergenreider, E., ... Urbich, C. (2011). Class IIb HDAC6 regulates endothelial cell migration and angiogenesis by deacetylation of cortactin: HDAC6 and angiogenesis. *The EMBO Journal, 30*(20), 4142–4156. https://doi.org/10.1038/emboj.2011.298.

Kim, J., Hwang, H.-G., Lee, J.-Y., Kim, M., & Kim, J.-Y. (2020). Cortactin deacetylation by HDAC6 and SIRT2 regulates neuronal migration and dendrite morphogenesis during cerebral cortex development. *Molecular Brain, 13*(1), 105. https://doi.org/10.1186/s13041-020-00644-y.

Kozyreva, V. K., McLaughlin, S. L., Livengood, R. H., Calkins, R. A., Kelley, L. C., Rajulapati, A., ... Pugacheva, E. N. (2014). NEDD9 regulates actin dynamics through cortactin deacetylation in an AURKA/HDAC6–dependent manner. *Molecular Cancer Research, 12*(5), 681–693. https://doi.org/10.1158/1541-7786.MCR-13-0654.

Linder, S., & Aepfelbacher, M. (2003). Podosomes: Adhesion hot-spots of invasive cells. *Trends in Cell Biology, 13*(7), 376–385. https://doi.org/10.1016/S0962-8924(03)00128-4.

Muñoz-Lasso, D. C., Romá-Mateo, C., Pallardó, F. V., & Gonzalez-Cabo, P. (2020). Much more than a scaffold: Cytoskeletal proteins in neurological disorders. *Cells, 9*(2), 358. https://doi.org/10.3390/cells9020358.

Murphy, D. A., & Courtneidge, S. A. (2011). The 'ins' and 'outs' of podosomes and invadopodia: Characteristics, formation and function. *Nature Reviews. Molecular Cell Biology, 12*(7), https://doi.org/10.1038/nrm3141 Article 7.

Penzes, P., & VanLeeuwen, J.-E. (2011). Impaired regulation of synaptic actin cytoskeleton in Alzheimer's disease. *Brain Research Reviews, 67*(1–2), 184–192. https://doi.org/10.1016/j.brainresrev.2011.01.003.

Pęziński, M., Maliszewska-Olejniczak, K., Daszczuk, P., Mazurek, P., Niewiadomski, P., & Rędowicz, M. J. (2021). Tks5 regulates synaptic podosome formation and stabilization of the postsynaptic machinery at the neuromuscular junction. *International Journal of Molecular Sciences, 22*(21), 12051. https://doi.org/10.3390/ijms222112051.

Pham, T. Q., Robinson, K., Xu, L., Pavlova, M. N., Skapek, S. X., & Chen, E. Y. (2021). HDAC6 promotes growth, migration/invasion, and self-renewal of rhabdomyosarcoma. *Oncogene, 40*(3), 578–591. https://doi.org/10.1038/s41388-020-01550-2.

Qureshi, T., & Chinnathambi, S. (2022). Histone deacetylase-6 modulates Tau function in Alzheimer's disease. *Biochimica et Biophysica Acta (BBA) - Molecular Cell Research, 1869*(8), 119275. https://doi.org/10.1016/j.bbamcr.2022.119275.

Qureshi, T., Desale, S. E., Chidambaram, H., & Chinnathambi, S. (2024). Understanding actin remodeling in neuronal cells through podosomes. In S. K. Ray (Ed.). *Neuroprotection: Method and protocols* (pp. 257–266). Springer US. https://doi.org/10.1007/978-1-0716-3662-6_18.

Qureshi, T., Desale, S.E., Sonawane, S.K., Balmik, A.A., Chidambaram, H., & Chinnathambi, S. (2024). Extracellular HDAC6 ZnF UBP domain enhances podosome-mediated neuronal migration (p. 2024.12.04.626746). *bioRxiv*. https://doi.org/10.1101/2024.12.04.626746.

Simões-Pires, C., Zwick, V., Nurisso, A., Schenker, E., Carrupt, P.-A., & Cuendet, M. (2013). HDAC6 as a target for neurodegenerative diseases: What makes it different from the other HDACs? *Molecular Neurodegeneration, 8*(1), 7. https://doi.org/10.1186/1750-1326-8-7.

Takemura, R., Okabe, S., Umeyama, T., Kanai, Y., Cowan, N. J., & Hirokawa, N. (1992). Increased microtubule stability and alpha tubulin acetylation in cells transfected with microtubule-associated proteins MAP1B, MAP2 or tau. *Journal of Cell Science, 103*(Pt 4), 953–964.

Tracy, T. E., & Gan, L. (2018). Tau-mediated synaptic and neuronal dysfunction in neurodegenerative disease. *Current Opinion in Neurobiology, 51*, 134–138. https://doi.org/10.1016/j.conb.2018.04.027.

Verdin, E., Dequiedt, F., & Kasler, H. G. (2003). Class II histone deacetylases: Versatile regulators. *Trends in Genetics, 19*(5), 286–293. https://doi.org/10.1016/S0168-9525(03)00073-8.

Vickers, J. C., Kirkcaldie, M. T., Phipps, A., & King, A. E. (2016). Alterations in neurofilaments and the transformation of the cytoskeleton in axons may provide insight into the aberrant neuronal changes of Alzheimer's disease. *Brain Research Bulletin, 126*, 324–333. https://doi.org/10.1016/j.brainresbull.2016.07.012.

Vincent, C., Siddiqui, T. A., & Schlichter, L. C. (2012). Podosomes in migrating microglia: Components and matrix degradation. *Journal of Neuroinflammation, 9*(1), 190. https://doi.org/10.1186/1742-2094-9-190.

Yao, Y.-L., & Yang, W.-M. (2011). Beyond histone and Deacetylase: An overview of cytoplasmic histone deacetylases and their nonhistone substrates. *Journal of Biomedicine & Biotechnology, 2011*, 146493. https://doi.org/10.1155/2011/146493.

Zambonin-Zallone, A., Teti, A., Carano, A., & Marchisio, P. C. (1988). The distribution of podosomes in osteoclasts cultured on bone laminae: Effect of retinol. *Journal of Bone and Mineral Research: The Official Journal of the American Society for Bone and Mineral Research, 3*(5), 517–523. https://doi.org/10.1002/jbmr.5650030507.

Zambonin-Zallone, A., Teti, A., Grano, M., Rubinacci, A., Abbadini, M., Gaboli, M., & Marchisio, P. C. (1989). Immunocytochemical distribution of extracellular matrix receptors in human osteoclasts: A β3 integrin is colocalized with vinculin and talin in the podosomes of osteoclastoma giant cells. *Experimental Cell Research, 182*(2), 645–652. https://doi.org/10.1016/0014-4827(89)90266-8.

Zhang, L., Sheng, S., & Qin, C. (2012). The role of HDAC6 in Alzheimer's disease. *Journal of Alzheimer's Disease, 33*(2), 283–295. https://doi.org/10.3233/JAD-2012-120727.

Zilberman, Y., Ballestrem, C., Carramusa, L., Mazitschek, R., Khochbin, S., & Bershadsky, A. (2009). Regulation of microtubule dynamics by inhibition of the tubulin deacetylase HDAC6. *Journal of Cell Science, 122*(19), 3531–3541. https://doi.org/10.1242/jcs.046813.

CHAPTER FOURTEEN

Histone deacetylases: From acetylation homeostasis to oncogenic and neurodegenerative disorders

Niharika[a,b] and Samir Kumar Patra[a,b,*]
[a]Biochemistry and Molecular Biology Group, Epigenetics and Cancer Research Laboratory, Department of Life Science, National Institute of Technology Rourkela, Sundergarh, Odisha, India
[b]Department of Life Science, ICMR Collaborating Centre of Excellence (ICMR-CCoE), National Institute of Technology Rourkela, Sundergarh, Odisha, India
*Corresponding author. e-mail address: samirp@nitrkl.ac.in; skpatra99@yahoo.com

Contents

Abstract

Histone deacetylases (HDACs) are central regulators of acetylation homeostasis, governing chromatin architecture, transcriptional dynamics, and diverse cellular processes through reversible lysine deacetylation. Dysregulation of HDAC activity disrupts epigenetic balance and is strongly implicated in oncogenic transformation and the progression of neurodegenerative disorders. This chapter provides a comprehensive

Methods in Enzymology, Volume 734
ISSN 0076-6879, https://doi.org/10.1016/bs.mie.2026.05.057

overview of HDAC biology with a particular emphasis on experimental and analytical methodologies used to investigate their function. We describe the structural and functional diversity of HDAC classes and their roles in multiprotein complexes that regulate gene expression and cellular signaling. A major focus is placed on screening-compatible and mechanistic assays, including fluorometric, colorimetric, radiometric, fluorescence polarization, TR-FRET, AlphaScreen/AlphaLISA, and differential scanning fluorimetry approaches for quantitative measurement of enzymatic activity and inhibitor profiling. In addition, advanced methodologies such as mass spectrometry–based acetylome analysis, chromatin immunoprecipitation sequencing (ChIP-seq), recombinant enzyme assays, and cell-based reporter systems are discussed in the context of functional genomics and drug discovery. The integration of high-throughput screening, structural biology, and multi-omics strategies is highlighted as essential for dissecting HDAC-mediated regulatory networks. Collectively, this chapter serves as a methodological framework for studying HDAC function and developing targeted epigenetic therapies in cancer and neurodegenerative diseases.

1. Introduction

Histone acetylation represents a fundamental epigenetic modification that regulates chromatin structure, gene expression, and cellular homeostasis across eukaryotic organisms. The dynamic balance between histone acetylation and deacetylation controls chromatin accessibility and transcriptional activity by modulating the interaction between DNA and histone proteins (Bannister & Kouzarides, 2011; Kouzarides, 2007). Histone acetyltransferases (HATs) catalyze the addition of acetyl groups to lysine residues on histone tails, leading to chromatin relaxation and transcriptional activation, whereas histone deacetylases (HDACs) remove these acetyl groups, resulting in chromatin condensation and transcriptional repression (Allis & Jenuwein, 2016; Seto & Yoshida, 2014) in association with DNA methylation (Kar et al., 2014; Niharika et al., 2025; Parbin et al., 2014; Parbin et al., 2016; Patra et al., 2001). Recently we have shown that inhibition of DNMT1 by curcumin provoke HDAC1 overexpression and repress DNMT1 gene (Niharika et al., 2025). This reversible acetylation–deacetylation equilibrium is essential for maintaining genome stability, regulating cell cycle progression, and coordinating developmental and stress-responsive gene expression programs (Shahbazian & Grunstein, 2007). Dysregulation of this balance has been strongly associated with pathological conditions, particularly cancer and neurodegenerative disorders, highlighting the central role of HDACs in disease biology (Dawson & Kouzarides, 2012; Haberland et al., 2009).

Histone deacetylases constitute a highly conserved family of enzymes that regulate both histone and non-histone protein acetylation, thereby influencing diverse cellular processes, including transcriptional regulation, DNA damage repair, apoptosis, and signal transduction (Choudhary et al., 2009; Verdin & Ott, 2015). In mammals, HDACs are categorized into four major classes (Class I, II, III, and IV) based on sequence homology, catalytic mechanism, and subcellular localization (Gregoretti et al., 2004). Class I, II, and IV HDACs are Zn^{2+}-dependent enzymes, whereas Class III HDACs, known as sirtuins, require NAD^{+} as a cofactor (Haigis & Sinclair, 2010). This structural and functional diversity allows HDACs to participate in multiprotein complexes such as Sin3, NuRD, and CoREST, which provide target specificity and regulate epigenetic gene silencing (Millard et al., 2016; You et al., 2001).

Accumulating evidence indicates that aberrant HDAC activity disrupts acetylation homeostasis and contributes to oncogenic transformation and neurodegenerative disease progression. In cancer, overexpression or hyperactivation of HDACs results in transcriptional repression of tumor suppressor genes, chromatin remodeling, and activation of proliferative and survival pathways (Li & Seto, 2016; West & Johnstone, 2014). HDAC-mediated deacetylation of non-histone proteins such as p53, NF-κB, STAT3, and α-tubulin further promotes tumor progression, metastasis, and therapeutic resistance (Glozak & Seto, 2007; Xu et al., 2007). Similarly, in neurodegenerative disorders including Alzheimer's disease, Parkinson's disease, and Huntington's disease, dysregulated HDAC activity alters neuronal gene expression, synaptic plasticity, and protein aggregation pathways, ultimately leading to neuronal dysfunction and cell death (Gräff & Tsai, 2013; Kazantsev & Thompson, 2008). These findings emphasize the importance of understanding HDAC biology through robust experimental and analytical methodologies. Potential role of HDACs are depicted in Fig. 1.

The development of reliable and reproducible experimental methods to study HDAC function has significantly advanced the field of epigenetics. Biochemical approaches such as fluorometric and radiometric HDAC activity assays enable quantitative measurement of enzymatic activity and inhibitor screening (Wegener et al., 2003a). Chromatin immunoprecipitation (ChIP) and ChIP-sequencing have facilitated genome-wide mapping of HDAC binding and histone acetylation patterns (Park, 2009). Mass spectrometry-based acetylome profiling allows identification of HDAC-regulated substrates and post-translational modifications (Nalawansha et al., 2018).

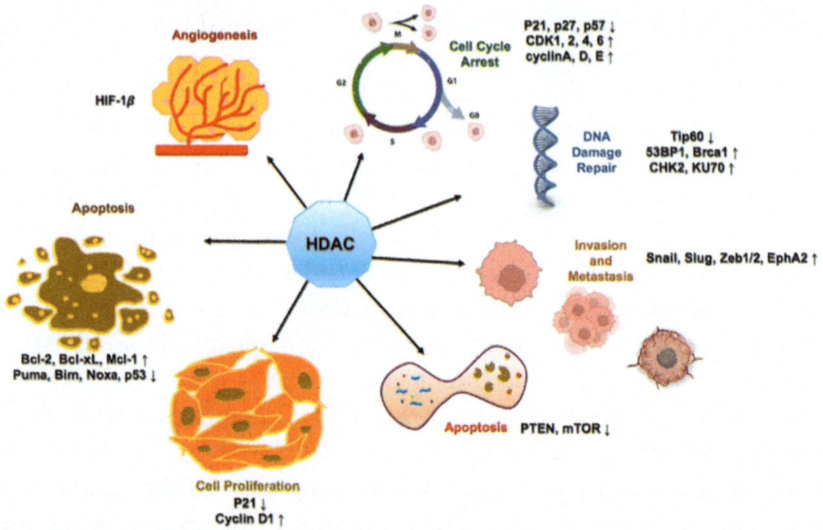

Fig. 1 HDACs as central regulators of cancer-associated cellular processes. HDAC activity influences angiogenesis via HIF-1β, promotes cell cycle progression through modulation of cyclins and CDKs, and regulates DNA damage repair pathways by affecting proteins such as 53BP1, BRCA1, CHK2, and KU70. HDACs suppress apoptosis by upregulating anti-apoptotic factors (Bcl-2, Bcl-xL, Mcl-1) and downregulating pro-apoptotic proteins (Puma, Bim, Noxa, p53), while also enhancing cell proliferation through cyclin D1 induction and p21 repression. Additionally, HDACs contribute to invasion and metastasis by regulating EMT-associated markers including Snail, Slug, ZEB1/2, and EphA2. Collectively, these pathways highlight the multifaceted role of HDACs in tumor progression and survival.

In addition, CRISPR/Cas9-mediated genome editing and RNA sequencing approaches have enabled functional characterization of HDAC-dependent regulatory networks in both cancer and neurodegenerative disease models (Cong et al., 2013; Wang et al., 2014). Structural biology techniques, including X-ray crystallography and cryo-electron microscopy, have further provided detailed insights into HDAC catalytic mechanisms and inhibitor interactions (Vannini et al., 2004).

Modern HDAC research increasingly relies on integrated multi-omics and high-throughput screening strategies to dissect acetylation homeostasis in physiological and pathological contexts. Chemical biology approaches using selective HDAC inhibitors have enabled functional studies and therapeutic targeting in cancer and neurological disorders (Falkenberg & Johnstone, 2014). Single-cell epigenomics and advanced imaging technologies have provided new insights into HDAC dynamics, chromatin organization, and cellular heterogeneity (Harpaz et al., 2022). In cancer research, high-throughput drug screening and functional genomic approaches have identified HDAC-dependent vulnerabilities and therapeutic targets, while in neurodegenerative disease research, epigenomic profiling has revealed HDAC-mediated transcriptional dysregulation in neuronal populations (Beaver et al., 2022; Danks et al., 2025; Wegener et al., 2003a, 2003b).

This chapter provides a comprehensive overview of histone deacetylases and their role in acetylation homeostasis, oncogenic transformation, and neurodegenerative disorders, with a special emphasis on experimental methodologies used to study HDAC biology. We describe biochemical, molecular, genomic, and pharmacological approaches, including HDAC activity assays, chromatin profiling, functional genomics, inhibitor screening, and structural analysis. By integrating mechanistic insights with methodological advances, this chapter aims to serve as a practical guide for researchers investigating HDAC function and developing epigenetic therapeutic strategies in cancer and neurodegenerative diseases.

2. Common screening compatible assays

Amplified Luminescent Proximity Homogeneous (AlphaScreen® and AlphaLISA®) assays represent widely utilized, screening-compatible technologies for investigating the activity of diverse biological targets (Eglen et al., 2008; Roberts & Bradner, 2015; Wang et al., 2017; Yasgar et al., 2016). These proximity-based assays rely on the interaction between two types of beads: donor beads and acceptor beads. Donor beads contain a photosensitizer, typically phthalocyanine, which upon excitation at 680 nm converts ambient molecular oxygen into a highly reactive singlet oxygen species. This singlet oxygen has a limited diffusion range of approximately 200 nm and a short half-life of ~4 μs in solution. When an acceptor bead is positioned within this proximity, energy transfer occurs from the singlet oxygen to thioxene derivatives embedded within the acceptor bead, leading to the emission of a measurable luminescent signal at 520–620 nm (AlphaScreen®) or specifically at 615 nm (AlphaLISA®). A key advantage of these assays is their homogeneous, no-wash format, which simplifies workflow and enhances suitability for high-throughput screening applications compared to conventional ELISA-based methods (Sierecki et al., 2013).

2.1 Colorimetric assays

This assay is based on differences in the electronic absorption spectra between the substrate and product of an enzymatic reaction (Fig. 2A). These assays typically employ chromogenic substrates composed of organic molecules containing conjugated π-electron systems, characterized by alternating single and double bonds that facilitate electron delocalization. Upon absorption of ultraviolet (UV) or visible light (approximately 200–800 nm), electrons within the chromophore are excited from the ground state to higher energy states. The wavelength of absorbed light is largely determined by the extent of conjugation, such that increased conjugation results in absorption at longer wavelengths (Sittampalam et al., 1997).

In certain cases, both substrate and product exhibit overlapping absorbance spectra, necessitating selective monitoring of product formation at wavelengths where substrate interference is minimal. Therefore, optimal detection wavelengths should be determined empirically using purified substrate and product standards (general methodology is represented in Fig. 3). When the native substrate is intrinsically chromogenic, enzymatic activity can be monitored directly without the need for synthetic labels, thereby minimizing steric hindrance and preserving native enzyme–substrate interactions within the catalytic site. This is particularly advantageous in avoiding artefacts that may affect the identification of substrate-competitive inhibitors.

Despite their widespread historical use, colorimetric assays have largely been supplanted by fluorescence-based methods due to inherent limitations in sensitivity and dynamic range (Gribbon & Sewing, 2003). These assays typically require relatively high product concentrations (often in the low micromolar range) for reliable detection. Furthermore, they are highly susceptible to optical interference from coloured compounds commonly present in small-molecule libraries, which can lead to false-positive hits during screening campaigns (Eggeling et al., 2003). Such artefacts necessitate rigorous validation and secondary screening prior to progression in drug discovery pipelines. Although kinetic-mode measurements can help mitigate optical interference, this approach reduces assay throughput and limits its applicability in large-scale screening (Baell, 2016).

It provide a straightforward and cost-effective alternative for measuring enzymatic activity based on chromogenic substrates. These assays typically employ acetylated peptide substrates linked to color-producing groups that, upon HDAC-mediated deacetylation, undergo subsequent enzymatic or chemical processing to yield a detectable colorimetric signal. The intensity of

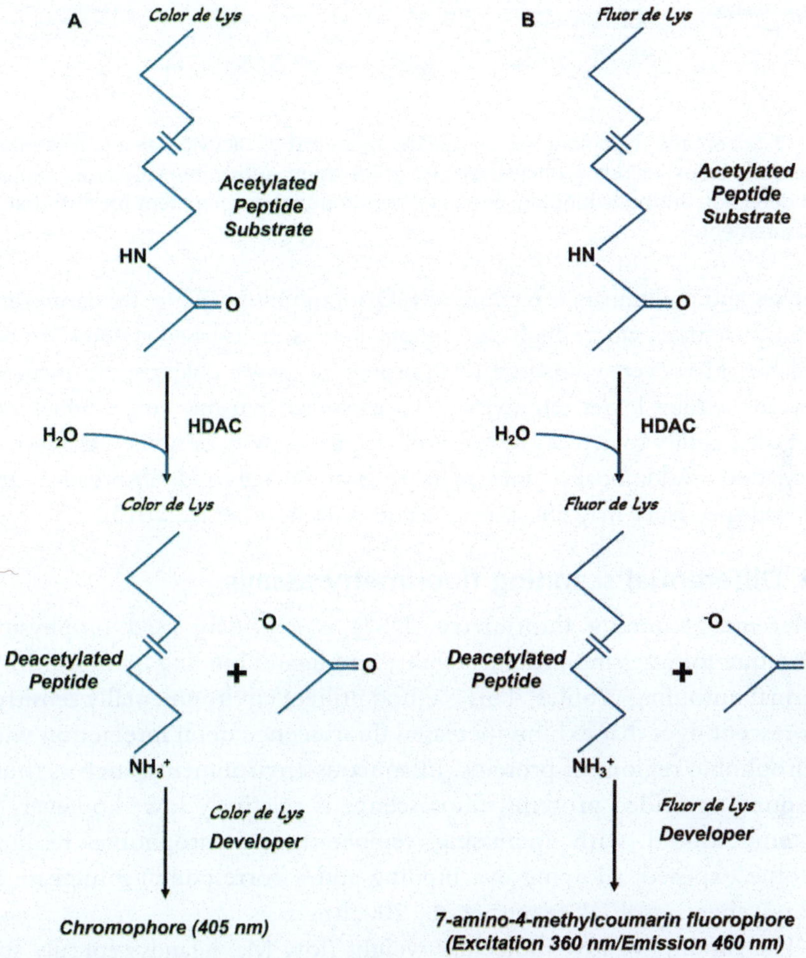

Fig. 2 Principle of colorimetric and fluorometric HDAC activity assays. (A) In the colorimetric assay, histone deacetylase (HDAC) catalyzes the hydrolytic removal of the acetyl group from an acetylated lysine-containing peptide substrate, generating a deacetylated peptide and acetate. The deacetylated product is subsequently processed by a developer reagent to release a chromophore, which is detected spectrophotometrically at ~405 nm. (B) In the fluorometric assay, HDAC-mediated deacetylation of a fluorogenic substrate enables cleavage by a developer to release 7-amino-4-methylcoumarin (AMC), producing a fluorescent signal (excitation ~360 nm, emission ~460 nm) proportional to enzymatic activity.

the color change, measured spectrophotometrically, is proportional to HDAC activity. Owing to their simplicity, minimal equipment requirements, and ease of implementation, colorimetric assays are well suited for routine laboratory

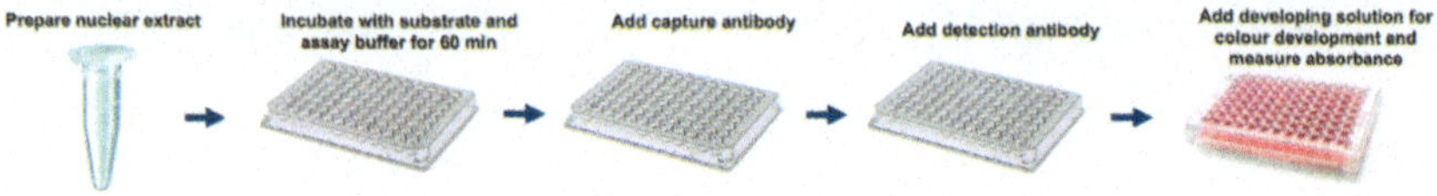

Fig. 3 Schematic workflow of an ELISA-based HDAC activity assay, illustrating sequential steps including nuclear extract preparation, substrate incubation, capture and detection antibody binding, and colorimetric signal development for absorbance measurement.

analyses and preliminary screening of HDAC inhibitors. They are particularly useful in settings where high-throughput fluorescence-based systems are not available. However, compared to fluorometric assays, colorimetric methods generally exhibit lower sensitivity and a narrower dynamic range, which can limit their ability to detect low levels of enzyme activity or subtle variations in acetylation. Additionally, interference from background absorbance and reduced specificity may affect assay accuracy (Sotoud et al., 2013).

2.2 Differential scanning fluorimetry assays

Differential scanning fluorimetry (DSF) is a widely used biophysical technique for assessing protein stability and ligand binding by monitoring thermal unfolding profiles. This method utilizes environmentally sensitive fluorescent dyes that exhibit increased fluorescence upon interaction with hydrophobic regions of proteins. In aqueous environments, such as those surrounding folded proteins, fluorescence is relatively low; however, as proteins unfold with increasing temperature, hydrophobic residues become exposed, allowing dye binding and a corresponding increase in fluorescence signal (Wegener et al., 2003b).

The binding of low molecular weight (low M_r) ligands typically stabilizes the protein structure, resulting in an increase in the thermal denaturation midpoint, or melting temperature (T_m). This shift in T_m can be quantitatively determined from fluorescence–temperature curves and serves as an indicator of ligand binding affinity and stabilizing effects. Consequently, DSF is widely employed in drug discovery and protein characterization to evaluate ligand–protein interactions and screen for potential modulators based on their ability to alter protein thermal stability (Niesen et al., 2007; Simeonov, 2013).

2.3 Enzyme-linked immunosorbent assay (ELISA)

It is a widely employed analytical technique with applications spanning diagnostics, biomolecular detection, and quality control processes. In a typical ELISA format, an antigen is immobilized onto a solid surface, such as a microplate, where it serves as a capture target for a specific antibody or interacting molecule. Following incubation, a series of wash steps is performed to remove non-specifically bound components, thereby enhancing assay specificity. A secondary antibody conjugated to an enzyme is then introduced, which binds to the target complex. Upon addition of an appropriate substrate, the enzyme catalyzes a reaction that generates a measurable signal, commonly colorimetric or fluorometric in nature (Lim et al., 2010; Ma et al., 2016).

Despite its high specificity and sensitivity, a key limitation of ELISA in high-throughput screening applications is its non-homogeneous format, which necessitates multiple washing steps. These additional steps increase assay time, reduce scalability, and introduce variability, thereby limiting its efficiency for large-scale screening workflows. It offer a quantitative approach for measuring HDAC activity using immobilized acetylated substrates and acetylation-specific antibodies. These assays are scalable and reproducible, making them suitable for clinical and translational studies, although their sensitivity and specificity depend on antibody performance and assay optimization (Wang et al., 2010).

2.4 Fluorescence-based assays

This includes fluorescence polarization (FP) and fluorometric HDAC activity assays (Patra et al., 2001; Patra, 2021), represent powerful and widely utilized approaches for studying enzyme activity and molecular interactions due to their high sensitivity, reproducibility, and compatibility with high-throughput screening (HTS) platforms. Fluorescence polarization assays rely on changes in the rotational mobility of a fluorescently labeled ligand upon binding to a larger macromolecule, such as a protein, resulting in an increase in hydrodynamic radius and a corresponding rise in polarization signal (Daśko et al., 2022; Edwards et al., 2004; Lakowicz et al., 2000). In solution, free ligands exhibit rapid rotational diffusion and low polarization, whereas binding to a target protein restricts motion and increases polarization. Displacement of the labeled ligand by a competing molecule leads to a decrease in polarization, enabling quantitative analysis of binding interactions and inhibitor screening (Zhang et al., 2015).

In parallel, fluorometric HDAC activity assays utilize synthetic acetylated lysine substrates conjugated to fluorogenic moieties, such as 7-amino-4-methylcoumarin (AMC), which remain non-fluorescent until enzymatic deacetylation occurs. Following HDAC-mediated deacetylation, the substrate becomes susceptible to proteolytic cleavage, releasing a fluorescent signal proportional to enzymatic activity. This indirect detection method facilitates rapid and quantitative assessment of HDAC kinetics and is extensively applied in drug discovery (Fig. 2B). Despite their advantages, these assays may exhibit substrate-specific biases, limited isoform selectivity, and dependence on secondary enzymatic steps, which can influence assay accuracy (Huang & Aulabaugh, 2016; Janzen, 2014; Ren et al., 2016).

2.5 Time-resolved fluorescence resonance energy transfer (TR-FRET)-based HDAC assays

It represent highly sensitive and homogeneous platforms for monitoring enzyme activity and inhibitor interactions. Commercial systems such as the LANCE® Ultra TR-FRET assay utilize a Europium chelate donor (W1024-Eu) conjugated to acetylation-specific antibodies, such as anti-H3K9(Ac) or anti-H3K27(Ac), in combination with a streptavidin-ULight™ acceptor. These assays employ biotinylated histone H3 peptides as substrates, and a TR-FRET signal is generated when the donor and acceptor are brought into close proximity through antibody-mediated recognition of acetylated substrates. Upon excitation at 340 nm, energy transfer from the Europium donor to the ULight™ acceptor produces a characteristic emission at 665 nm, enabling sensitive detection of acetylation status and HDAC activity. In addition to activity-based measurements, TR-FRET assays can be adapted to evaluate inhibitor binding affinity using a competitive displacement format. In such approaches, an Alexa Fluor® 647-labeled HDAC inhibitor serves as a tracer (acceptor), which binds to GST-tagged HDAC enzymes in the presence of a Europium-labeled anti-GST antibody (donor). Displacement of the tracer by competing compounds results in a reduction in TR-FRET signal, allowing quantitative assessment of inhibitor binding (Ren et al., 2016; Wegener et al., 2003a, 2003b). More broadly, FRET-based assays employ donor–acceptor fluorophore pairs linked to acetylated substrates, where enzymatic deacetylation alters the FRET efficiency. These approaches enable real-time, continuous monitoring of HDAC activity without the need for additional processing steps, making them particularly advantageous for kinetic studies and high-throughput screening applications (Kota et al., 2010) (Fig. 4).

2.6 Radiometric assays

This assay considered a gold standard for measuring HDAC activity due to their high sensitivity and direct quantification of deacetylation. In these

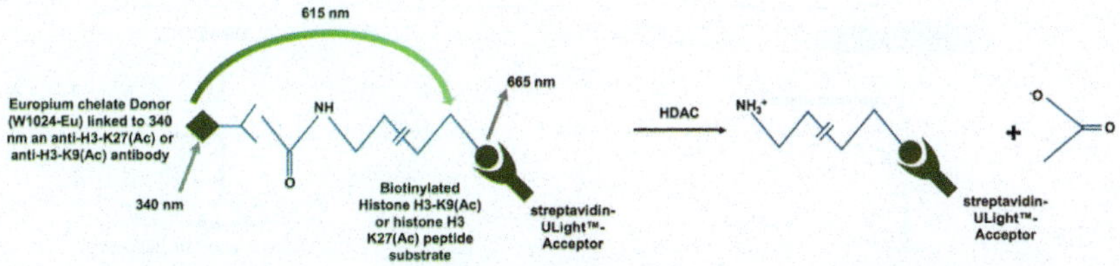

Fig. 4 Time-resolved fluorescence resonance energy transfer (TR-FRET)-based histone deacetylase (HDAC) assay operates through detection of deacetylated peptide substrates via proximity-dependent energy transfer. Following HDAC-mediated deacetylation, the peptide substrate is recognized and bound by a Europium-labeled donor antibody and streptavidin-conjugated ULight™ acceptor, thereby bringing the donor and acceptor fluorophores into close spatial proximity. Upon excitation at 340 nm, energy emitted from the Europium donor is transferred to the ULight™ acceptor through resonance energy transfer, resulting in fluorescence emission at 665 nm, which quantitatively reflects HDAC enzymatic activity.

assays, radiolabeled acetylated substrates, typically incorporating [^{3}H]-acetyl groups, are used to monitor the release of radiolabeled acetate upon enzymatic cleavage. Despite their accuracy and robustness in kinetic studies, the requirement for radioactive materials, specialized handling procedures, and low throughput restricts their routine use in many laboratories (Marks et al., 2011).

2.7 Mass spectrometry

This approaches have emerged as powerful tools for studying HDAC activity at the proteome level. These methods enable precise identification and quantification of acetylated and deacetylated peptides, facilitating global acetylome profiling and the discovery of HDAC substrates. Liquid chromatography coupled with tandem mass spectrometry (LC-MS/MS) allows site-specific resolution of acetylation changes, providing insights into the functional impact of HDAC activity on diverse protein networks. Although highly informative, these approaches require advanced instrumentation, extensive sample preparation, and sophisticated computational analysis (Remans et al., 2025; Verma et al., 2023) (Fig. 5).

2.8 Western blot assay

Western blot assay based detection of histone acetylation levels is a widely used method to indirectly assess HDAC activity in cellular systems. This approach employs antibodies specific to acetylated lysine residues on histones or non-histone proteins, allowing comparative analysis of acetylation

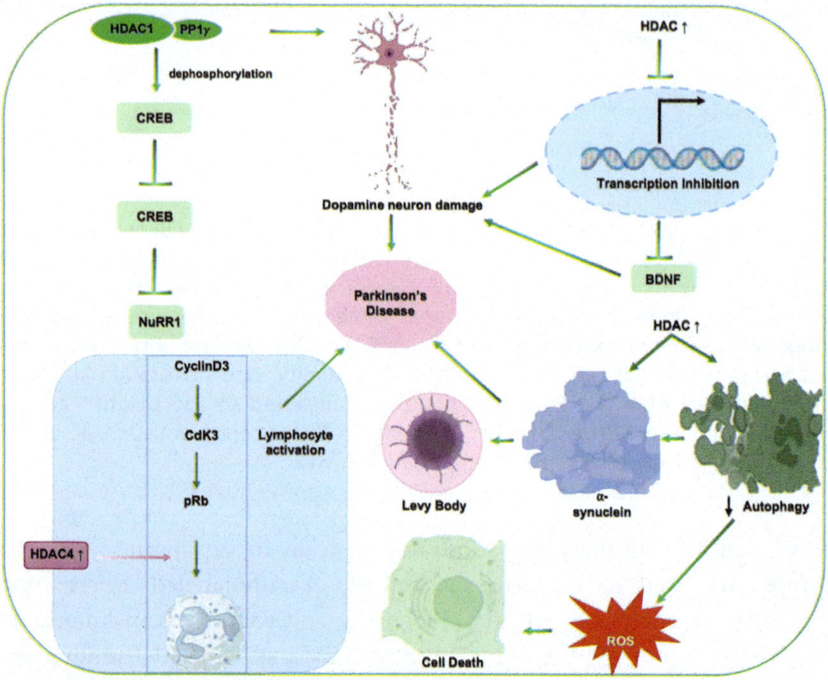

Fig. 5 Workflow for histone post-translational modification (PTM) profiling: Histones are extracted from tissue or cells. Purified histones undergo chemical propionylation both prior to and after tryptic digestion to enhance peptide resolution and PTM specificity. The resulting peptides are subsequently desalted through automated liquid-handling platforms and subjected to mass spectrometric analysis using data-independent acquisition (DIA) mode. PTM identification and quantification are performed using EpiProfile 2.1, while downstream statistical analysis and data visualization are conducted using MetaboAnalyst 6.0.

states under different experimental conditions. While this method is accessible and applicable to a variety of biological samples, it is inherently semi-quantitative and dependent on antibody specificity and signal normalization strategies (Choudhary et al., 2009; Lundby et al., 2012; Remans et al., 2025; Rigby et al., 2012; Verma et al., 2023; Wolfson et al., 2014; Xiao et al., 2023) (Fig. 6).

2.9 Chromatin immunoprecipitation (ChIP) and ChIP-sequencing (ChIP-seq)

It provides powerful approaches for examining HDAC function in the context of chromatin. By immunoprecipitating HDAC-associated chromatin

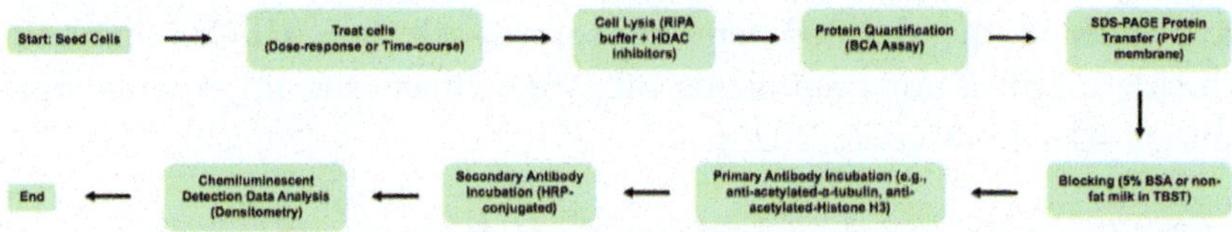

Fig. 6 Workflow of western blot assay based detection of histone acetylation levels.

or acetylated histone marks, these techniques enable genome-wide mapping of HDAC occupancy and histone acetylation patterns. ChIP-seq, in particular, allows high-resolution identification of regulatory regions influenced by HDAC activity, thereby linking enzymatic function to transcriptional regulation. However, these methods require high-quality antibodies, optimized crosslinking conditions, and careful data analysis (Hino et al., 2023; Niharika et al., 2025).

2.10 Biochemical assays

By using recombinant HDAC enzymes offer a controlled system for mechanistic and kinetic studies. Purified HDAC proteins are incubated with defined acetylated substrates to evaluate catalytic activity, substrate specificity, and inhibitor interactions. These assays are particularly useful for structure–function analyses and drug development, although they may not fully recapitulate the complexity of cellular environments where HDACs function within multiprotein complexes (Adams et al., 2019; Kontopidis & Patergiannakis, 2025).

2.11 Cell-based HDAC activity assays

This provides a physiologically relevant context by measuring enzyme activity within living cells. These assays often utilize cell-permeable fluorogenic substrates or reporter constructs that respond to changes in acetylation status. Such approaches are valuable for evaluating the efficacy of HDAC inhibitors and studying regulatory pathways in intact cellular systems. However, they may lack isoform specificity and can be influenced by cellular heterogeneity and off-target effects (Asawa et al., 2020; Hsu et al., 2016).

2.12 HDAC inhibitor screening assays

It is essential for identifying and characterizing small molecules with therapeutic potential. These assays typically combine enzymatic or cell-based readouts with high-throughput screening platforms to evaluate compound libraries for inhibitory activity. Advances in chemical biology have enabled the development of isoform-selective HDAC inhibitors, which are critical for minimizing off-target effects and improving clinical efficacy (Baselious et al., 2025).

2.13 Live-cell imaging and reporter-based assays

It is an advanced assay to study the HDAC activity by enabling spatio-temporal analysis in real time. Fluorescent biosensors and genetically encoded reporters allow visualization of acetylation dynamics within specific cellular compartments, providing insights into HDAC regulation under physiological and pathological conditions. While highly informative, these approaches require careful design and validation of reporter systems (Ni et al., 2018; Zhuang et al., 2026).

Collectively, these methodologies provide complementary strategies for assessing HDAC activity across multiple biological scales, from purified enzymes to complex cellular systems. Integration of biochemical, genomic, and imaging approaches is increasingly essential for dissecting the multifaceted roles of HDACs in acetylation homeostasis and their contribution to oncogenic and neurodegenerative processes.

2.14 HDAC luminescence assays

It represents a commercially available coupled enzymatic platform analogous to colorimetric and fluorometric HDAC assays, but utilize amino-luciferin-conjugated ε-acetylated lysine peptide substrates specifically designed for

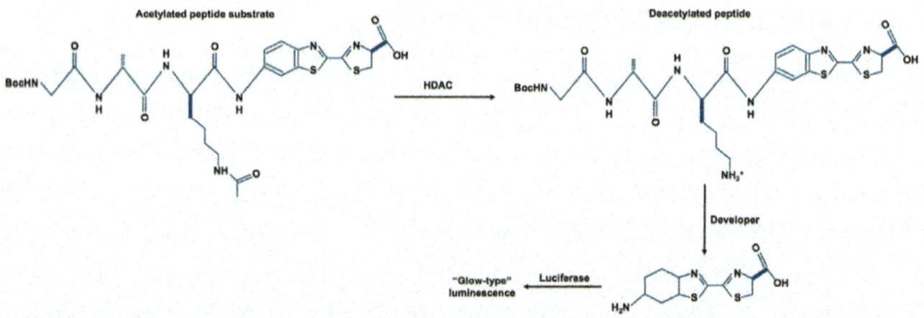

Fig. 7 Luminescence-coupled histone deacetylase (HDAC) assays employ amino-luciferin-conjugated ε-acetylated lysine peptide substrates specifically designed for Class I/II HDAC enzymes. Upon HDAC-mediated deacetylation, the modified substrate becomes accessible to the developer reagent, which catalyzes the release of free amino-luciferin. The liberated amino-luciferin subsequently serves as a substrate for luciferase, also present within the developer reagent, resulting in the generation of a stable glow-type luminescent signal proportional to HDAC enzymatic activity.

Class I/II HDAC enzymes (e.g., Promega HDAC-Glo™ I/II Assays and Screening Systems) (HDAC-Glo™ I/II Assay and Screening System, 2017). In this system, HDAC-mediated deacetylation of the substrate generates a deacetylated product that becomes accessible to a proprietary developer reagent, leading to the release of free amino-luciferin. The liberated amino-luciferin subsequently serves as a substrate for luciferase, which is also included in the developer reagent, producing a stable glow-type luminescent signal proportional to HDAC enzymatic activity. This assay platform has been experimentally validated and successfully applied in high-throughput screening of natural product libraries for HDAC modulators (Halley et al., 2011) (Fig. 7).

3. Conclusion

In conclusion, histone deacetylases (HDACs) represent pivotal regulators of epigenetic homeostasis, orchestrating chromatin organization, transcriptional control, and cellular signaling through dynamic modulation of protein acetylation. Their functional versatility, encompassing both histone and non-histone substrates, underlies their critical involvement in diverse biological processes and disease pathogenesis, particularly in cancer and neurodegenerative disorders. The complexity of HDAC biology necessitates the use of robust, sensitive, and complementary methodological approaches to accurately assess enzymatic activity, substrate specificity, and regulatory mechanisms. This chapter highlights a comprehensive repertoire of experimental strategies, ranging from classical biochemical assays to advanced high-throughput, proteomic, and genomic technologies, each contributing unique insights into HDAC function. The integration of fluorescence-based assays, TR-FRET platforms, mass spectrometry, chromatin profiling, and cell-based systems has significantly enhanced our ability to dissect HDAC-mediated regulatory networks at multiple biological scales. Importantly, these methodological advances have facilitated the identification and optimization of HDAC inhibitors with therapeutic potential. Future progress in the field will depend on the continued refinement of isoform-specific assays, single-cell epigenomic approaches, and real-time imaging technologies. Such innovations will be essential for translating mechanistic insights into precision epigenetic therapies for complex human diseases.

References

Adams, M. K., Banks, C. A. S., Miah, S., Killer, M., & Washburn, M. P. (2019). Purification and enzymatic assay of class I histone deacetylase enzymes. *Methods in Enzymology, 626*, 23–40. https://doi.org/10.1016/bs.mie.2019.07.014.

Allis, C. D., & Jenuwein, T. (2016). The molecular hallmarks of epigenetic control. *Nature Reviews. Genetics, 17*(8), 487–500. https://doi.org/10.1038/nrg.2016.59.

Asawa, R. R., Zakharov, A., Niehoff, T., Chitsaz, A., Jadhav, A., Henderson, M. J., ... Martinez, N. J. (2020). A comparative study of target engagement assays for HDAC1 inhibitor profiling. *SLAS Discovery: Advancing Life Sciences R & D, 25*(3), 253–264. https://doi.org/10.1177/2472555219883625.

Baell, J. B. (2016). Feeling nature's pains: Natural products, natural product drugs, and pan assay interference compounds (PAINS). *Journal of Natural Products, 79*(3), 616–628. https://doi.org/10.1021/acs.jnatprod.5b00947.

Bannister, A. J., & Kouzarides, T. (2011). Regulation of chromatin by histone modifications. *Cell Research, 21*(3), 381–395. https://doi.org/10.1038/cr.2011.22.

Baselious, F., Hilscher, S., Handke, L., Barinka, C., Schutkowski, M., & Sippl, W. (2025). In silico screening of a designed focused chemical space identifies novel alkyl hydrazides as potent HDAC11 inhibitors. *Computers in Biology and Medicine, 196*(Pt A), 110695. https://doi.org/10.1016/j.compbiomed.2025.110695.

Beaver, M., Karisetty, B. C., Zhang, H., Bhatnagar, A., Armour, E., Parmar, V., ... Elefant, F. (2022). Chromatin and transcriptomic profiling uncover dysregulation of the Tip60 HAT/HDAC2 epigenomic landscape in the neurodegenerative brain. *Epigenetics: Official Journal of the DNA Methylation Society, 17*(7), 786–807. https://doi.org/10.1080/15592294.2021.1959742.

Choudhary, C., Kumar, C., Gnad, F., Nielsen, M. L., Rehman, M., Walther, T. C., ... Mann, M. (2009). Lysine acetylation targets protein complexes and co-regulates major cellular functions. *Science (New York, N. Y.), 325*(5942), 834–840. https://doi.org/10.1126/science.1175371.

Cong, L., Ran, F. A., Cox, D., Lin, S., Barretto, R., Habib, N., ... Zhang, F. (2013). Multiplex genome engineering using CRISPR/Cas systems. *Science (New York, N. Y.), 339*(6121), 819–823. https://doi.org/10.1126/science.1231143.

Danks, M. R., Manasterski, P. J., Beetham, H., Dawson, J. C., Elliott, R. J. R., Culley, J., ... Brunton, V. G. (2025). High-throughput screening identifies the activity of histone deacetylase inhibitors in patient-derived models of soft tissue sarcoma. *Cancer Biology & Therapy, 26*(1), 2589666. https://doi.org/10.1080/15384047.2025.2589666.

Daśko, M., de Pascual-Teresa, B., Ortín, I., & Ramos, A. (2022). HDAC inhibitors: Innovative strategies for their design and applications. *Molecules (Basel, Switzerland), 27*(3), 715. https://doi.org/10.3390/molecules27030715.

Dawson, M. A., & Kouzarides, T. (2012). Cancer epigenetics: From mechanism to therapy. *Cell, 150*(1), 12–27. https://doi.org/10.1016/j.cell.2012.06.013.

Edwards, B. S., Oprea, T., Prossnitz, E. R., & Sklar, L. A. (2004). Flow cytometry for high-throughput, high-content screening. *Current Opinion in Chemical Biology, 8*(4), 392–398. https://doi.org/10.1016/j.cbpa.2004.06.007.

Eggeling, C., Brand, L., Ullmann, D., & Jäger, S. (2003). Highly sensitive fluorescence detection technology currently available for HTS. *Drug Discovery Today, 8*(14), 632–641. https://doi.org/10.1016/s1359-6446(03)02752-1.

Eglen, R. M., Reisine, T., Roby, P., Rouleau, N., Illy, C., Bossé, R., & Bielefeld, M. (2008). The use of AlphaScreen technology in HTS: Current status. *Current Chemical Genomics, 1*, 2–10. https://doi.org/10.2174/1875397300801010002.

Falkenberg, K. J., & Johnstone, R. W. (2014). Histone deacetylases and their inhibitors in cancer, neurological diseases and immune disorders. *Nature Reviews. Drug Discovery, 13*(9), 673–691. https://doi.org/10.1038/nrd4360.

Glozak, M. A., & Seto, E. (2007). Histone deacetylases and cancer. *Oncogene, 26*(37), 5420–5432. https://doi.org/10.1038/sj.onc.1210610.

Gräff, J., & Tsai, L. H. (2013). Histone acetylation: Molecular mnemonics on the chromatin. *Nature Reviews. Neuroscience, 14*(2), 97–111. https://doi.org/10.1038/nrn3427.

Gregoretti, I. V., Lee, Y. M., & Goodson, H. V. (2004). Molecular evolution of the histone deacetylase family: Functional implications of phylogenetic analysis. *Journal of Molecular Biology, 338*(1), 17–31. https://doi.org/10.1016/j.jmb.2004.02.006.

Gribbon, P., & Sewing, A. (2003). Fluorescence readouts in HTS: No gain without pain? *Drug Discovery Today, 8*(22), 1035–1043. https://doi.org/10.1016/s1359-6446(03)02895-2.

Haberland, M., Montgomery, R. L., & Olson, E. N. (2009). The many roles of histone deacetylases in development and physiology: Implications for disease and therapy. *Nature Reviews. Genetics, 10*(1), 32–42. https://doi.org/10.1038/nrg2485.

Haigis, M. C., & Sinclair, D. A. (2010). Mammalian sirtuins: Biological insights and disease relevance. *Annual Review of Pathology, 5*, 253–295. https://doi.org/10.1146/annurev.pathol.4.110807.092250.

Halley, F., Reinshagen, J., Ellinger, B., Wolf, M., Niles, A. L., Evans, N. J., ... Gul, S. (2011). A bioluminogenic HDAC activity assay: Validation and screening. *Journal of Biomolecular Screening, 16*(10), 1227–1235. https://doi.org/10.1177/1087057111416004.

Harpaz, N., Mittelman, T., Beresh, O., Griess, O., Furth, N., Salame, T. M., ... Shema, E. (2022). Single-cell epigenetic analysis reveals principles of chromatin states in H3.3-K27M gliomas. *Molecular Cell, 82*(14), 2696–2713.e9. https://doi.org/10.1016/j.molcel.2022.05.023.

HDAC-Glo™ I/II Assay and Screening System. https://worldwide.promega.com/resources/protocols/technical-manuals/101/hdac-glo-i-ii-assay-and-screening-system-protocol/. Accessed 19 Apr 2017.

Hino, S., Sato, T., & Nakao, M. (2023). Chromatin immunoprecipitation sequencing (ChIP-seq) for detecting histone modifications and modifiers. *Methods in Molecular Biology (Clifton, N. J.), 2577*, 55–64. https://doi.org/10.1007/978-1-0716-2724-2_4.

Hsu, C. W., Shou, D., Huang, R., Khuc, T., Dai, S., Zheng, W., ... Xia, M. (2016). Identification of HDAC inhibitors using a cell-based HDAC I/II assay. *Journal of Biomolecular Screening, 21*(6), 643–652. https://doi.org/10.1177/1087057116629381.

Huang, X., & Aulabaugh, A. (2016). Application of fluorescence polarization in HTS assays. *Methods in Molecular Biology (Clifton, N. J.), 1439*, 115–130. https://doi.org/10.1007/978-1-4939-3673-1_7.

Janzen, W. P. (2014). Screening technologies for small molecule discovery: The state of the art. *Chemistry & Biology, 21*(9), 1162–1170. https://doi.org/10.1016/j.chembiol.2014.07.015.

Kar, S., Sengupta, D., Deb, M., Shilpi, A., Parbin, S., Rath, S. K., ... Patra, S. K. (2014). Expression profiling of DNA methylation-mediated epigenetic gene-silencing factors in breast cancer. *Clinical Epigenetics, 6*(1), 20. https://doi.org/10.1186/1868-7083-6-20.

Kazantsev, A. G., & Thompson, L. M. (2008). Therapeutic application of histone deacetylase inhibitors for central nervous system disorders. *Nature Reviews. Drug Discovery*, 7(10), 854–868. https://doi.org/10.1038/nrd2681.

Kontopidis, G., & Patergiannakis, I. S. (2025). Comparative analysis of biochemical and cellular assay conditions and the need for a buffer that mimics cytoplasmic environments. *Molecules (Basel, Switzerland), 30*(17), 3630. https://doi.org/10.3390/molecules30173630.

Kota, S., Scampavia, L., Spicer, T., Beeler, A. B., Takahashi, V., Snyder, J. K., ... Strosberg, A. D. (2010). A time-resolved fluorescence-resonance energy transfer assay for identifying inhibitors of hepatitis C virus core dimerization. *Assay and Drug Development Technologies, 8*(1), 96–105. https://doi.org/10.1089/adt.2009.0217.

Kouzarides, T. (2007). Chromatin modifications and their function. *Cell, 128*(4), 693–705. https://doi.org/10.1016/j.cell.2007.02.005.

Lakowicz, J. R., Gryczynski, I., & Gryczynski, Z. (2000). Novel fluorescence sensing methods for high throughput screening. *Journal of Biomolecular Screening, 5*(3), 123–132. https://doi.org/10.1177/108705710000500304.

Li, Y., & Seto, E. (2016). HDACs and HDAC inhibitors in cancer development and therapy. *Cold Spring Harbor Perspectives in Medicine, 6*(10), a026831. https://doi.org/10.1101/cshperspect.a026831.

Lim, M. J., Foster, G. J., Gite, S., Ostendorff, H. P., Narod, S., & Rothschild, K. J. (2010). An ELISA-based high throughput protein truncation test for inherited breast cancer. *Breast Cancer Research: BCR, 12*(5), R78. https://doi.org/10.1186/bcr2722.

Lundby, A., Lage, K., Weinert, B. T., Bekker-Jensen, D. B., Secher, A., Skovgaard, T., ... Olsen, J. V. (2012). Proteomic analysis of lysine acetylation sites in rat tissues reveals organ specificity and subcellular patterns. *Cell Reports, 2*(2), 419–431. https://doi.org/10.1016/j.celrep.2012.07.006.

Ma, Z., Liu, Z., Jiang, T., Zhang, T., Zhang, H., Du, L., & Li, M. (2016). Discovery of fluorescence polarization probe for the elisa-based antagonist screening of α_1-adrenergic receptors. *ACS Medicinal Chemistry Letters, 7*(10), 967–971. https://doi.org/10.1021/acsmedchemlett.6b00048.

Marks, B. D., Fakhoury, S. A., Frazee, W. J., Eliason, H. C., & Riddle, S. M. (2011). A substrate-independent TR-FRET histone deacetylase inhibitor assay. *Journal of Biomolecular Screening, 16*(10), 1247–1253. https://doi.org/10.1177/1087057111422102 [134].

Millard, C. J., Varma, N., Saleh, A., Morris, K., Watson, P. J., Bottrill, A. R., ... Schwabe, J. W. (2016). The structure of the core NuRD repression complex provides insights into its interaction with chromatin. *eLife, 5*, e13941. https://doi.org/10.7554/eLife.13941.

Nalawansha, D. A., Zhang, Y., Herath, K., & Pflum, M. K. H. (2018). HDAC1 substrate profiling using proteomics-based substrate trapping. *ACS Chemical Biology, 13*(12), 3315–3324. https://doi.org/10.1021/acschembio.8b00737.

Ni, Q., Mehta, S., & Zhang, J. (2018). Live-cell imaging of cell signaling using genetically encoded fluorescent reporters. *The FEBS Journal, 285*(2), 203–219. https://doi.org/10.1111/febs.14134.

Niesen, F. H., Berglund, H., & Vedadi, M. (2007). The use of differential scanning fluorimetry to detect ligand interactions that promote protein stability. *Nature Protocols, 2*(9), 2212–2221. https://doi.org/10.1038/nprot.2007.321.

Niharika, Kumar, M., Verma, A., Dash, P., Roy, A., Manna, S., ... Patra, S. K. (2025). Inhibition of DNMT1 by synergistic application curcumin and 5-AZA-2′-deoxycytidine implicates enhanced therapeutic potential against lung adenocarcinoma. *Computers in Biology and Medicine*, 2025 Jul 15, *196*(Pt B), 110763. https://doi.org/10.1016/j.compbiomed.2025.110763.

Parbin, S., Kar, S., Shilpi, A., Sengupta, D., Deb, M., Rath, S. K., & Patra, S. K. (2014). Histone deacetylases: A saga of perturbed acetylation homeostasis in cancer. *The Journal of Histochemistry and Cytochemistry: Official Journal of the Histochemistry Society, 62*(1), 11–33. https://doi.org/10.1369/0022155413506582.

Parbin, S., Shilpi, A., Kar, S., Pradhan, N., Sengupta, D., Deb, M., ... Patra, S. K. (2016). Insights into the molecular interactions of thymoquinone with histone deacetylase: Evaluation of the therapeutic intervention potential against breast cancer. *Molecular Biosystems, 12*(1), 48–58. https://doi.org/10.1039/c5mb00412h.

Park, P. J. (2009). ChIP-seq: Advantages and challenges of a maturing technology. *Nature Reviews. Genetics, 10*(10), 669–680. https://doi.org/10.1038/nrg2641.

Patra, S. K. (2021). Emerging histone glutamine modifications mediated gene expression in cell differentiation and the VTA reward pathway. *Gene, 768*, 145323. https://doi.org/10.1016/j.gene.2020.145323.

Patra, S. K., Patra, A., & Dahiya, R. (2001). Histone deacetylase and DNA methyltransferase in human prostate cancer. *Biochemical and Biophysical Research Communications, 287*(3), 705–713. https://doi.org/10.1006/bbrc.2001.5639.

Remans, K., Sehr, P., Steimbach, R. R., Gunkel, N., & Miller, A. K. (2025). TR-FRET assay for profiling HDAC10 inhibitors and PROTACs. *Methods in Enzymology, 715*, 41–63. https://doi.org/10.1016/bs.mie.2025.01.061.

Ren, D., Wang, J., Wang, B., & You, Z. (2016). Probes for biomolecules detection based on RET-enhanced fluorescence polarization. *Biosensors & Bioelectronics, 79*, 802–809. https://doi.org/10.1016/j.bios.2016.01.015.

Rigby, L., Muscat, A., Ashley, D., & Algar, E. (2012). Methods for the analysis of histone H3 and H4 acetylation in blood. *Epigenetics: Official Journal of the DNA Methylation Society*, 7(8), 875–882. https://doi.org/10.4161/epi.20983.

Roberts, J. M., & Bradner, J. E. (2015). A bead-based proximity assay for BRD4 ligand discovery. *Current Protocols in Chemical Biology*, 7(4), 263–278. https://doi.org/10.1002/9780470559277.ch150024.

Seto, E., & Yoshida, M. (2014). Erasers of histone acetylation: The histone deacetylase enzymes. *Cold Spring Harbor Perspectives in Biology, 6*(4), a018713. https://doi.org/10.1101/cshperspect.a018713.

Shahbazian, M. D., & Grunstein, M. (2007). Functions of site-specific histone acetylation and deacetylation. *Annual Review of Biochemistry, 76*, 75–100. https://doi.org/10.1146/annurev.biochem.76.052705.162114.

Sierecki, E., Giles, N., Polinkovsky, M., Moustaqil, M., Alexandrov, K., & Gambin, Y. (2013). A cell-free approach to accelerate the study of protein-protein interactions in vitro. *Interface Focus, 3*(5), 20130018. https://doi.org/10.1098/rsfs.2013.0018.

Simeonov, A. (2013). Recent developments in the use of differential scanning fluorometry in protein and small molecule discovery and characterization. *Expert Opinion on Drug Discovery, 8*(9), 1071–1082.

Sittampalam, G. S., Kahl, S. D., & Janzen, W. P. (1997). High-throughput screening: Advances in assay technologies. *Current Opinion in Chemical Biology, 1*(3), 384–391. https://doi.org/10.1016/s1367-5931(97)80078-6.

Sotoud, H., Gribbon, P., Ellinger, B., Reinshagen, J., Boknik, P., Kattner, L., ... Eschenhagen, T. (2013). Development of a colorimetric and a fluorescence phosphatase-inhibitor assay suitable for drug discovery approaches. *Journal of Biomolecular Screening, 18*(8), 899–909. https://doi.org/10.1177/1087057113486000.

Vannini, A., Volpari, C., Filocamo, G., Casavola, E. C., Brunetti, M., Renzoni, D., ... Di Marco, S. (2004). Crystal structure of a eukaryotic zinc-dependent histone deacetylase, human HDAC8, complexed with a hydroxamic acid inhibitor. *Proceedings of the National Academy of Sciences of the United States of America, 101*(42), 15064–15069. https://doi.org/10.1073/pnas.0404603101.

Verdin, E., & Ott, M. (2015). 50 years of protein acetylation: From gene regulation to epigenetics, metabolism and beyond. *Nature Reviews. Molecular Cell Biology, 16*(4), 258–264. https://doi.org/10.1038/nrm3931.

Verma, A. K., Noumani, A., Yadav, A. K., & Solanki, P. R. (2023). FRET based biosensor: Principle applications recent advances and challenges. *Diagnostics (Basel, Switzerland), 13*(8), 1375. https://doi.org/10.3390/diagnostics13081375.

Wang, J., Fang, P., Chase, P., Tshori, S., Razin, E., Spicer, T. P., ... Guo, M. (2017). Development of an HTS-compatible assay for discovery of melanoma-related microphthalmia transcription factor disruptors using AlphaScreen technology. *SLAS Discovery: Advancing Life Sciences R & D, 22*(1), 58–66. https://doi.org/10.1177/1087057116675274.

Wang, L., Bao, Y., Yang, Y., Wu, Y., Chen, X., Si, S., & Hong, B. (2010). Discovery of antagonists for human scavenger receptor CD36 via an ELISA-like high-throughput screening assay. *Journal of Biomolecular Screening, 15*(3), 239–250. https://doi.org/10.1177/1087057109359686.

Wang, T., Wei, J. J., Sabatini, D. M., & Lander, E. S. (2014). Genetic screens in human cells using the CRISPR-Cas9 system. *Science (New York, N. Y.), 343*(6166), 80–84. https://doi.org/10.1126/science.1246981.

Wegener, D., Hildmann, C., Riester, D., & Schwienhorst, A. (2003b). Improved fluorogenic histone deacetylase assay for high-throughput-screening applications. *Analytical Biochemistry, 321*(2), 202–208. https://doi.org/10.1016/s0003-2697(03)00426-3.

Wegener, D., Wirsching, F., Riester, D., & Schwienhorst, A. (2003a). A fluorogenic histone deacetylase assay well suited for high-throughput activity screening. *Chemistry & Biology, 10*(1), 61–68. https://doi.org/10.1016/s1074-5521(02)00305-8.

West, A. C., & Johnstone, R. W. (2014). New and emerging HDAC inhibitors for cancer treatment. *The Journal of Clinical Investigation, 124*(1), 30–39. https://doi.org/10.1172/JCI69738.

Wolfson, N. A., Pitcairn, C. A., Sullivan, E. D., Joseph, C. G., & Fierke, C. A. (2014). An enzyme-coupled assay measuring acetate production for profiling histone deacetylase specificity. *Analytical Biochemistry, 456*, 61–69. https://doi.org/10.1016/j.ab.2014.03.012.

Xiao, Y., Hale, S., Awasthee, N., Meng, C., Zhang, X., Liu, Y., ... Liao, D. (2023). HDAC3 and HDAC8 PROTAC dual degrader reveals roles of histone acetylation in gene regulation. *Cell Chemical Biology, 30*(11), 1421–1435.e12. https://doi.org/10.1016/j.chembiol.2023.07.010.

Xu, W. S., Parmigiani, R. B., & Marks, P. A. (2007). Histone deacetylase inhibitors: Molecular mechanisms of action. *Oncogene, 26*(37), 5541–5552. https://doi.org/10.1038/sj.onc.1210620.

Yasgar, A., Jadhav, A., Simeonov, A., & Coussens, N. P. (2016). AlphaScreen-based assays: Ultra-high-throughput screening for small-molecule inhibitors of challenging enzymes and protein-protein interactions. *Methods in molecular biology (Clifton, N. J.), 1439*, 77–98. https://doi.org/10.1007/978-1-4939-3673-1_5.

You, A., Tong, J. K., Grozinger, C. M., & Schreiber, S. L. (2001). CoREST is an integral component of the CoREST- human histone deacetylase complex. *Proceedings of the National Academy of Sciences of the United States of America, 98*(4), 1454–1458. https://doi.org/10.1073/pnas.98.4.1454.

Zhang, H., Wu, Q., & Berezin, M. Y. (2015). Fluorescence anisotropy (polarization): From drug screening to precision medicine. *Expert Opinion on Drug Discovery, 10*(11), 1145–1161. https://doi.org/10.1517/17460441.2015.1075001.

Zhuang, Z., Feng, Z., Wang, J., Liu, X., Song, L., Sun, C., ... Ren, N. (2026). Advanced imaging for live-cell spatiotemporal monitoring: Technologies and applications. *Research (Washington, D. C.), 9*, 1085. https://doi.org/10.34133/research.1085.

CHAPTER FIFTEEN

Experimental approaches to study histone deacetylation in *Caenorhabditis elegans*

Anam Naseer and Aamir Nazir*
Division of Toxicology and Experimental Medicine, CSIR-Central Drug Research Institute (CDRI), Lucknow, India
*Corresponding author: e-mail address: anazir.cdri@csir.res.in

Contents

Abstract

Histones, being basic in nature, form the core of a nucleosome which governs DNA compaction and regulate the expression of genetic information. These undergo various modifications at their flanking tail region. One such important epigenetic modifications include deacetylation. To study histone deacetylation various experimental models and approaches have been developed till date. However, one of the best in vivo models to study this modification is a soil-dwelling nematode, *Caenorhabditis elegans*. This tiny, transparent worm shares around 70% genetic homology with humans which becomes a key feature in employing it as a model system. This chapter highlights major experimental approaches for studying histone deacetylation using *C. elegans*.

1. Introduction

Every human cell consists of DNA as a genetic material that extends upto 2 meters in length. It stores hereditary information in the form of genes which

Methods in Enzymology, Volume 734
ISSN 0076-6879, https://doi.org/10.1016/bs.mie.2026.05.056

constitute the basis of inheritance and is stored compactly in a cell (Dna & Packaged, 2002). Human genome is comprised of active euchromatin and inactive heterochromatin regions. DNA compaction is governed by several structures such as nucleosomes which are present as 'beads on a string' and linker DNA with the help of activity of proteins such as histones and non-histones (like scaffold proteins) (Naseer & Nazir, 2023). Histones are the basic proteins rich in arginine and lysine residues, which form the major part of the nucleosome. The nucleosome core component if formed by four types of histones such as H2A, H2B, H3 and H4; however, the two nucleosome cores are joined by short segment having H1 protein known as linker histone (Mariño-ramírez et al., 2014; Annunziato et al., 2008). The core histones are arranged as two sets of each of the four histones forming the octamer having the tail region outside the core that undergoes modifications post-translationally. These modifications in turn govern the nature of genes, that is whether it will active or inactive genes or downstream targets associated with them. Major post-translational histone modifications include: methylation, acetylation, phosphorylation, ubiquitinylation, sumoylation (Talbert & Henikoff, 2021).

Histone acetylation or deacetylation form one of the major histone modifications and is a topic of study for many researchers. This includes a reversible epigenetic change which either adds or removes acetyl group at the histone tails majorly on lysine resides, respectively. This phenomenon is regulated by a balanced activity of acetylating and deacetylating enzymes known as histone acetylases (HATs) or acetyl transferases and histone deacetylases (HDACs) which maintains cellular homeostasis. These enzymes aid in shifting the chromatin state from condensed heterochromatin state to relaxed euchromatin state or vice-versa, regulating transcription and gene expression (Naseer & Nazir, 2023).

2. Rationale

Histone deacetylases (HDACs), also known as erasers, work in opposite fashion as of HATs (Seto & Yoshida, no.date). They remove the acetyl group from lysine residues of histone tails, thus compacting the chromatin. Humans have 18 HDACs distributed under different classes as: class-I, II, III, IV. Here, classes I, II and IV HDACs are zinc-dependent while class III HDACs, known as sirtuins, are NAD-dependent enzymes (Fig. 1).

HDACs are crucial for cell cycle maintenance, metabolism and development. Errors in regulation of histone acetylation results in various diseases

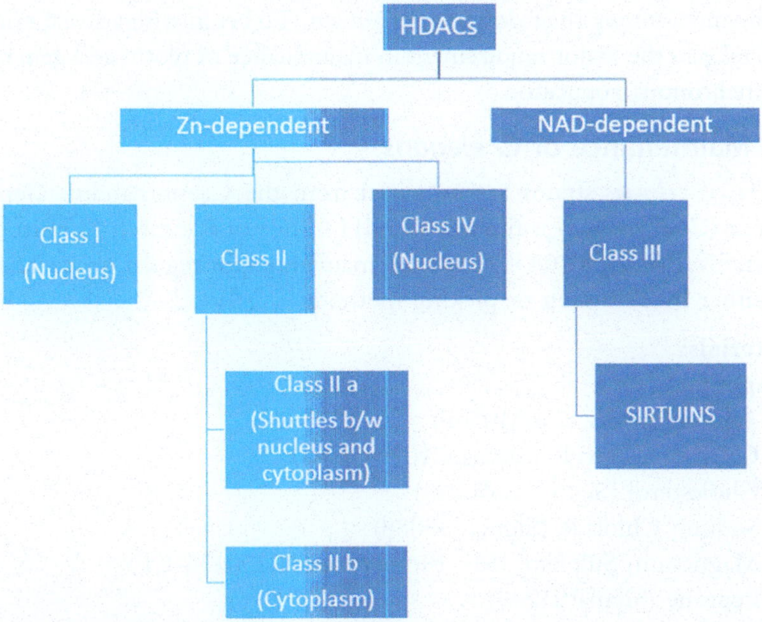

Fig. 1 Classification of HDACs.

and histone deacetylases have extensively been studied for their crucial role in regulating these diseases like neurodegeneration, cancer, metabolic and inflammatory disorders (King et al., 2021). HDAC inhibitors prove to be the promising targets for treating certain types of cancers like haematological malignancies. With the advent of technologies, studying these modifications involve basic as well as high throughput techniques like sequencing. However, by using a very simple model system the effects and pathways regulated by HDACs can easily be comprehended. For this the best model to employ is tiny nematode called *Caenorhabditis elegans*. It is a 1 mm long soil worm which has around 70% genetic homology with humans and has a completely mapped genome (Naseer & Nazir, 2023). Thus, in this chapter we will discuss the methods to study histone deacetylation employing *C. elegans* as model.

3. Methods to study histone deacetylation in *C. elegans*

For assessing histone deacetylation in *C. elegans*, we employ various techniques like silencing or genetic manipulation of genes coding for HDACs, quantifying acetylation levels, screening using fluorescently tagged strains and assessing their downstream effects. To begin with any experiment in *C. elegans* the major requirement in maintenance of worm and generation of synchronous population.

3.1 Maintenance of *C. elegans*

Firstly, *C. elegans* strains are procured from the Caenorhabditis Genetics Center (CGC) which is funded by NIH Office of Research Infrastructure Programs (P40 OD010440) and then maintained for regular propagation in lab either by chunking or picking method.

Materials:

Reagents:

1. Agar (Himedia, GRM026P)
2. Calcium Chloride (Sigma,C1016-500G)
3. Cholesterol (Sigma, SML204)
4. Sodium Chloride (Sigma, S9888)
5. Magnesium Sulphate (MP Biomedicals 02150136-CF)
6. Peptone (Sigma 91249)
7. Potassium Phosphate dibasic (Sigma P3786)
8. Sodium Phosphate Monobasic (Sigma 342483)
9. Uracil (Sigma U0750)
10. Ethanol (Merck 1.11727)
11. Dextrose (SRL, 0449130)
12. Platinum wire picker
13. Bacteria: *E. coli* strain OP50
14. *C. elegans* strains

Equipment:

1. Incubator-shaker (New Brunswick Scientific Innova42, Incubator shaker series)
2. Centrifuge (Eppendorf centrifuge 5702)
3. Bio-Safety Cabinet (Thermo Scientific 1300 series A2)
4. Sterile pipettes, glass slides and tips
5. Precision Weighing Balance (Mettler Toledo JB1603)
6. Autoclave (STERI VERTICAL; 10J-1464)
7. Stereo zoom microscope (Leica S8AP0)
8. B.O.D. Incubators (Labtech LBI500M; Science Tech India)

Protocol:

Step 1: Preparation of NGM (nematode growth medium) for maintaining strains: Nematode growth medium (NGM) is prepared by adding 50 mM NaCl, 2.5 gL-1 Peptone, 17 gL-1 Agar in 1000 ml of distilled water and autoclaving at 121 °C for 45 min at 15 psi. After that the media is cooled down to 50 °C– 60 °C and 5 μgml-1 cholesterol solution (dissolved in ethanol), 1 mM $CaCl_2$, 1 mM $MgSO_4$ and 25 mM KH_2PO_4 is added to it. The contents are mixed and poured onto sterilized petri-plates. The plates are allowed to cool down in the presence of UV light in a Bio-Safety Cabinet and are stored at 4 °C (Naseer et al., 2025).

Step 2: *Escherichia coli (E. coli)* strain OP50 culture preparation: For preparing 100 ml *E. coli* OP50 (uracil auxotroph) culture, add 300 mg KH_2PO_4, 600 mg Na_2HPO_4 and 500 mg NaCl to the distilled water (100 ml) and autoclave at 121 °C for 45 min at 15 psi. Then add 200 μl Uracil (2 mg/ml of MilliQ water), 1 ml Dextrose (200 mg/ml of milliQ water), 100 μl $MgSO_4$, 1 ml NH_4Cl, 500 μl *E. coli* OP50 inoculum and incubate @ 37 °C overnight on incubator-shaker at 180 rpm.

To prepare OP50 lawn om NGM plate, around 300–500 μl of cultured *E. coli* OP50 culture should be seeded on previously prepared NGM plates in a Bio-Safety Cabinet and incubated overnight at 22 °C.

Step 3: Procure the *C. elegans* strains from the Caenorhabditis Genetics Center (CGC) by ordering online from the website (Caenorhabditis Genetics Center CGC – Official Resource for C, no.date).

Step 4: After receiving the strains, cut small pieces of the NGM agar (known as chunk) containing worms and transfer it to the *E. coli* (OP50 strain) seeded NGM plates. This will allow fresh food to the dauer worms and will help them revive.

Alternate to chunking is the picking method. Here, worms are picked one by one using a platinum wire picker and are transferred to the *E. coli* (OP50 strain) seeded NGM plates. This is done in special strains such as rollers, which require transferring of only roller phenotype to get maximum population of roller worms.

Step 5: Close the lid of plate and wrap parafilm around it and maintain it in a B.O.D. incubator, either @ 20 °C or @ 15 °C depending upon the requirement. (Fig. 2).

Notes:

1. Allow the media to cooled down to 50 °C– 60 °C and then only add the remaining mixtures to ensure proper mixing.
2. One should be very cautious while writing the name of the strain to be chunked; there should be no strain-mixing.

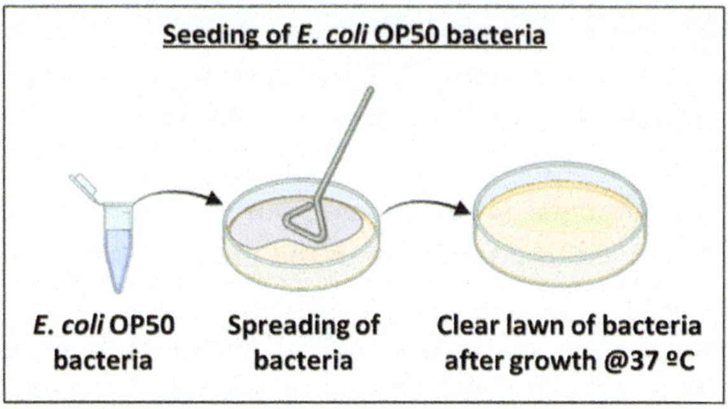

Fig. 2 Seeding of E. coli OP50 bacteria.

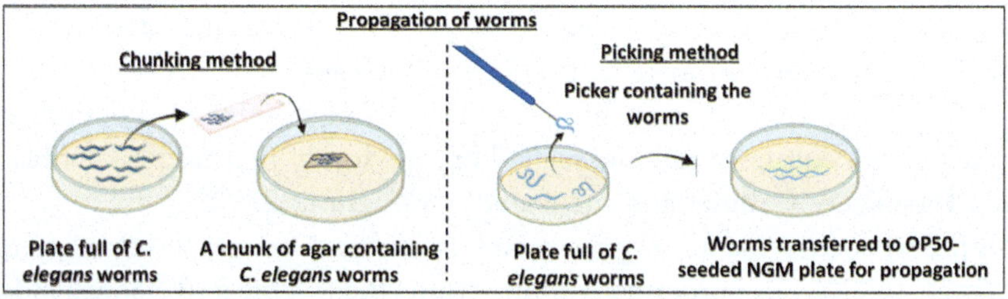

Fig. 3 Propagation of worms by chunking and picking method.

3.2 Embryo isolation to get synchronous population of worms

To study the effect of any gene or treatment, it is generally suggested to have a homogenous population of worms. Naturally, upon chunking or picking worms, the plate contains heterogenous population of worms, that is they have all the stages from eggs <L1 <L2 <L3 <L4 <gravid worms, which shows different response towards different stimuli. Thus, it is necessary to age-synchronise the worms, either by L1 filtration (where heterogenous population of worms are simply filtered out through a 5 μm Nylon net filter giving L1 worms in the filtrate and leaving all the other stages behind) or by embryo isolation (which is a chemical method for age-synchronisation). (Fig. 3).

Materials:

Reagents:

1. M9 buffer
2. NGM plates
3. Axenizing solution (5 ml NaOH + 2 ml Bleach)
4. *C. elegans* worms

Equipment:
1. Centrifuge (Eppendorf centrifuge 5702)
2. B.O.D. Incubator (Labtech LBI500M; Science Tech India)
3. Sterile pipettes and tips

Protocol:
Step 1: Choose a plate with well-fed gravid (egg carrying) worms and collect them in 15 ml centrifuge tubes by washing the plates with M9 buffer by gently swirling the buffer on the plate; avoid rigorous shaking to prevent taking the debris. This step removes the adherent bacteria.
Step 2: Centrifuge @ 2000 rpm for 2 min. Aspirate the supernatant and to the pellet add M9 buffer and repeat the washing step two more times.
Step 3: To the pellet of worms add axenizing solution (5 ml NaOH + 2 ml Bleach) and open the lid of falcon and allow the axenizing solution to dissolve worm body in the presence of air. Avoid exposing the worms to axenizing solution for long duration or else eggs may not hatch.
Step 4: Centrifuge @ 2300 rpm for 1.5 min. Decant the supernatant and to the pellet of eggs add M9 buffer and centrifuge @ 2300 rpm for 2.5 min.
Step 5: Repeat step 4 for two more times, to remove traces of axenizing solution.
Step 6: Seed the isolated eggs (in the pellet) to the experimental plate; air-dry it and incubate @ 22 °C for 48 h to get L4/ young adult (YA) stage. (Fig. 4).

Notes:
1. Washing should be gentle, otherwise there will be agar left and eggs will be trapped in it and they won't hatch.
2. While air-drying the isolated eggs, semi-dry them.

3.3 Genetic manipulation and silencing

3.3.1 RNA interference (RNAi) feeding

RNA interference is a conserved phenomenon of gene silencing that mediates the expression of coding genes post transcriptionally in response to external as well internal stimulus. It was first discovered by Andrew Z. Fire and Craig C. Mello in 1990s in a widely used model system, that is C. elegans. This finding later led to a Nobel Prize in Physiology and Medicine in the year 2006. During this technique, the worms are fed with the HT115 (DE3) bacterial strain of *E. coli* which contains L4440 vector that carries short double-stranded RNA (dsRNA) segments against the gene that needs to be

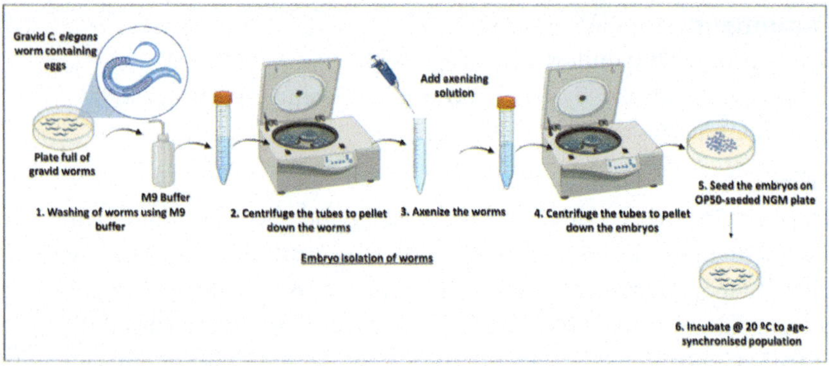

Fig. 4 Embryo isolation of worms.

silenced. This vector contains two T7 promoters in opposite orientation within which the fragment corresponding to the gene of interest is cloned and is transformed into a bacterial strain (HT115(DE3), which lacks double-strand-specific RNase III, enhancing the RNAi efficiency by feeding. This bacterial strain also contains IPTG inducible response of T7 polymerase.

When this vector enters the worm body it gets replicated along with the worm genome and silences the gene of interest. However, this post transcriptional gene silencing occurs naturally in eukaryotes and is important for gene regulation. For studying the histone deacetylation, we employ RNA interference of genes that majorly govern this modification such as, sirtuins (*sir-2.1, sir-2.2, sir-2.3, sir-2.4), hda-1, had-2, hda-3* (Class I HDACs), *had-4, hda-5, hda-6, hda-7* (Class II HDACs).

The detailed protocol for RNA interference-based gene silencing in C. elegans:

Gene silencing is carried out by employing the feeding protocol. HT115 (DE3) bacterial clone producing dsRNA against targeted gene is obtained from Ahringer's RNAi library (purchased from SA Biosciences, Cambridge, UK) and is cultured for 6–8 h in LB containing ampicillin (since, HT115 bacteria has Ampicillin resistance in its plasmid, thus it will ensure the growth of our desired strain of bacteria) (Naseer et al., 2025).

Materials:

Reagents:

1. Bacteria*: E. coli* strain HT115 (RNase III deficient)
2. RNAi plasmid (e.g., L4440 vector containing gene of interest)
3. Ampicillin Sodium Salt (MP Biomedicals, 194526)
4. NGM and NGM-IPTG plates

5. IPTG (MP Biomedicals, 102101)
6. LB broth (HIMEDIA, M1245)
7. Milli-Q water
8. Ahringer's RNAi library

Equipment:
1. B.O.D. Incubator (Labtech LBI500M; Science Tech India) (20 °C for worms, 37 °C
2. for bacteria)
3. Incubator-shaker (New Brunswick Scientific Innova42, Incubator shaker series)
4. Bio-Safety Cabinet (Thermo Scientific 1300 series A2)
5. Sterile pipettes, falcons and tips
6. Precision Weighing Balance (Mettler Toledo JB1603)
7. Autoclave (STERI VERTICAL; 10J-1464)

Protocol:
Step 1: Prepare LB (Luria Bertani) solution by dissolving 25 g LB powder in 1000 ml of RO water and then autoclave it.
Step 2: After cooling down store the solution @ 4 °C.
Step 3: Prepare ampicillin solution by dissolving 5 mg of Ampicillin Sodium Salt in 1 ml of autoclaved Milli-Q water. To make a 1% working concentration, 10 μl ampicillin stock was mixed with 1 ml LB media. Store the solution @ 4 °C.
Step 4: Prior to use, UV treat the LB and Ampicillin solutions and make 1 % Ampicillin solution in a laminar air flow hood with flame on to avoid contamination.
Step 5: Take UV treated falcons (15 ml), and label them as control (having LB+Amp culture only), ev (having LB+Amp solution and empty vector L4440 in HT115 bacteria) and respective culture whose gene silencing is to be done (having LB+Amp solution and L4440 vector containing gene of interest for silencing).
Step 6: To silence the gene of interest from Ahringer's RNAi library, we must first know its sequence and location on the plate, then take 1 μl of the culture and activate it in 50 μl LB + Amp solution to make primary culture and incubate overnight for bacterial growth @ 37 °C. for preparing secondary culture take 20 μl of primary culture and dd it to 100 μl of LB + Amp solution and incubate for bacterial growth @ 37 °C for 5–6 h.

For working 1 ml culture, add 10 µl of activated secondary culture of ev/ gene of interest and incubate for bacterial growth @ 37 °C for 5–6 h.

Step 7: After optimum growth, check for OD and seed around 300 µl-500 µl of culture onto labelled IPTG-NGM plates (that have been prepared at least 48 h prior to use and stored @ 4 °C) and incubate the pates @ 22 °C for overnight. IPTG (Isopropyl-D-1-thiogalactopyranoside) to induce the production of double-stranded RNA (dsRNA) within *E. coli* HT115 (DE3) bacteria. IPTG acts as an inducer for the T7 RNA polymerase, which drives the transcription of the gene-specific dsRNA from a plasmid (usually L4440) in the feeding strain. Embryos are then transferred onto these plates to obtain age-synchronous population. These worms when grown to L4 stage re then observed for phenotypic analysis.

Notes:

1. When using the secondary or tertiary cultures, active the culture @ 3737 °C for 3−4 h.

3.3.2 *Use of genetic mutants*

Genetic mutants consist of a mutated gene of interest inserted into their genome. *C. elegans* offers the best model for creating transgenic strain. These strains are deposited and maintained by the Caenorhabditis Genetics Center (CGC), which is supported by the National Institutes of Health – Office of Research Infrastructure Programs. To study histones, specific strains such as WF1828 [*hda-1*(cw2) II], RB2416 [*hda-10*(ok3311) II], OF1355 [*hda-3*(ix261) I], UL2996 [leEx2996 [*sir-2.1*::mCherry + *rol-6*(su1006)]], VC199 [*sir-2.1*(ok434) IV], DQM1059 [bmdSi245 I; *hda-1*(bmd134[*hda-1*::GFP::loxP]) V] that can easily be procured from CGC and then be cultured by chunking or picking and maintaining inside the B.O.D. There are multiple ways to create transgenics, a few of them include microinjection, crossing of strains.

Protocol for genetic crossing to create genetic mutants in *C. elegans*: To generate a cross between two different strains, firstly males are to be generated of one strain which has a stronger expression which, will then be crossed with hermaphrodites of another strain to get a population having features of both strains. This mutant thus obtained is then stabilised by back crossing.

Materials:

Reagents:

1. *C. elegans* strains
2. NGM plates

Equipment:
1. Sterile pipettes and tips
2. Platinum wire picker
3. B.O.D. Incubator (Labtech LBI500M; Science Tech India)
4. Stereo zoom microscope (Leica S8AP0)
5. Fluorescent microscope (Carl Zeiss Axio Imager.M2)

Protocol:

Step 1: For generation of males, firstly age-synchronised the worms of first strain, let it be strain 'A', by embryo isolation.

Step 2: After 30 h of embryo isolation, observe the worms for post-L3 stage at which they will be ready for embryogenesis. To identify this stage, look for a dot -like structure near the vulva.

Step 3: Transfer around 20 worms to a fresh OP50-seeded NGM plate using a picker.

Step 4: Incubate this plate @ 30 °C for 5–6 h to induce male formation. Shift this plate to 20 °C and incubate for 3 days with continuous monitoring.

Step 5: On 4th day, search for male and transfer then to the hermaphrodite plate.

Step 6: For preparing the hermaphrodite plate, chunk the other strain, let it be strain 'B', while the strain 'A' is being incubated for male generation. When gravid population of strain 'B' is achieved, perform embryo isolation to get age-synchronised population.

Step 7: After 30 h of embryo isolation, observe the worms for post-L3 stage. This step is crucial, since this stage is required to occur synchronously with the 4th day of male generation.

Step 8: Crossing is performed by transferring 4th day of male to the 30 mm OP50-seeded NGM plate containing post-L3 hermaphrodite worms in 1:4 ratio (male: hermaphrodite). There should be multiple plates with this set-up to increase he chances of crossing.

Step 9: Incubate @ 20 °C. On 4th day, remove the parent adults and transfer them to fresh plates and check for recombinant expression.

Step 10: When recombination is achieved then stabilise this strain by back-crossing with N2 strain 6–8 times.

Notes:
1. Before crossing strains, always look for their reporter phenotype and accordingly decide which strains' males are to be taken.
2. Run parallel repeats to ensure timely generation of the genetic cross.

3.4 Monitoring histone deacetylation levels

To assess the levels of histone deacetylation, multiple approaches can be applied either at protein level or at gene level. These include western blotting, immunofluorescence or mRNA expression analysis by qPCR.

3.4.1 Western blotting

Materials:

Reagents:

1. *C. elegans* strains
2. M9 buffer
3. 1X Phosphate buffer saline (PBS)
4. Pierce™ BCA Protein Assay Kit (Thermo Fisher Scientific; 23225).
5. Acrylamide (MP- 193982)
6. 1.5 M Tris-HCl (pH 8.8)
7. 0.5 M Tris-HCl (pH 6.8)
8. Milli-Q Water
9. SDS (Sigma-L3771-500G)
10. APS (MP-190556)
11. TEMED (Sigma 110-18-9)
12. Protein ladder (PUREGENE Genetix; PG-PMT2922)
13. β-mercaptoethanol (Sigma M3148)
14. 5X protein loading dye
15. Methanol (SDFCL, 39192)
16. BSA (bovine serum albumin – MP Biomedicals; BSASG100)
17. Tween-20 (Sigma-Aldrich; P9416-ml)
18. Primary and secondary antibodies

Equipment:

1. Cooling Microfuge (SIGMA 1–14 K; SIGMA 3−30 K)
2. Sterile pipettes and tips
3. Microcentrifuge tubes
4. Sonicator (Haatman; JY92-IIN)
5. Dry bath incubator (Helix DTH-100)
6. Western blotting assembly (Bio-Rad Mini PROTEAN Tetra Cell; 552BR 244213)
7. Semi-dry transfer assembly (Thermo Scientific Pierce G2 Fast Blotter)
8. PVDF membrane (0.2 μm; Parablot; 741260)
9. Chemi-doc (Invitrogen; iBright 1500).

Protocol:

1) Sample preparation in *C. elegans*:
Preparation of sample for western blotting is a crucial step, as it requires disruption of worm's body, release of biomolecules and clearance of debris to contain whole worm protein.
Step 1: Age-synchronised young-adult worms are washed thrice with M9 buffer to remove excess of adherent bacteria and are re-suspended in (1 ×) PBS.
Step 2: The worms are either first frozen with liquid nitrogen, grinded using pestle-mortar and then sonicated or are directly sonicated in (1 ×) PBS using a sonicator at 25 % amplitude for 3 min (pulse time ON/OFF, 30 s).
Step 3: The worm lysate is then centrifuged at 13,000 *g* for 20 min at 4 °C. The supernatant is then transferred to a fresh labelled micro-centrifuge tube containing the total protein, while debris is discarded.
Step 4: Protein concentration is calculated using a Pierce™ BCA Protein Assay Kit.
2) **Western blotting:**
Step 1: To separate protein using SDS-PAGE firstly prepare separating and stacking gels. For smaller molecular weight proteins, separating gel should have more percentage of Acrylamide solution.
- **Separating gel preparation (12 % and 15 %):**
12 % 15 %
Acrylamide solution = 4 ml = 5
1.5 M Tris-HCl (pH 8.8) = 2.5 ml = 2.5
Water = 3.3 ml = 2.3
10 % SDS = 50 μl = 100 μl
10 % APS = 50 μl = 100 μl
TEMED = 6 μl = 8ul
- **5 % stacking gel preparation:**
Acrylamide solution = 650 μl.
0.5 M Tris-HCl (pH 6.8) = 1.25 ml.
Water = 3.05 ml.
10 % SDS = 50 μl.
10 % APS = 50 μl.
TEMED = 4 μl.

Step 2: Protein sample for western blotting is prepared by adding 5X protein loading dye containing reducing agents like 5 % β-mercaptoethanol and heating @ 95 °C for 10 min to ensure denaturation. This will allow the proteins to get denatured, negatively charged, and get separated on the basis of molecular weight.

Step 3: Around 20 μg of protein sample along with protein ladder are loaded into the wells of Bio-Rad; Mini-PROTEAN® System. The gel is ran @ 100 V for 1.5–2 h.

Step 4: When the bands have reached their lowest point on the gel, turn off the power supply and activate the PVDF membrane using methanol (shaking for 1–2 min). The protein from then gel is then transferred to the activated PVDF membrane using semi-dry transfer in a sandwich form, for 30 min–1 h. During this step the gel, PVDF membrane and paper wick are submerged in the transfer buffer and when the set-up is ready, extra transfer buffer is removed using a roller.

Step 5: When the transfer is complete, the non-specific sites are blocked using 5 % BSA (bovine serum albumin) solution and are incubated overnight at 4 °C with primary antibody.

Step 6: Wash the unbound antibody using 0.1 % PBST (0.1 % Tween-20 in PBS) and incubate with secondary antibody for 2 h at RT.

Step 7: Wash the unbound antibody using 0.1 % PBST and develop the blot using chemiluminescence detection in a chemi-doc.

Notes:

1. While sonicating, keep the sample throughout on ice to avoid the heat generated from ultrasonic waves to damage the protein.
2. Use fresh reagents for running western blot and pH should be measured routinely.

3.4.2 Immunofluorescence (IF)

To detect the site-specific localisation of histone acetylation, which indirectly gives the idea of deacetylation as well, the widely used technique involves immunofluorescence (Mareninov et al., 2020). It is a process which allows to detect largely accessible region for a specific modification. In case of studying histone modifications in *C. elegans* via immunofluorescence, first and foremost required is the disruption of worm body via freeze-fracture or fixation to allow the access to nuclear chromatin region so that antibodies can interact with the target histones (Kristy et al., 2013; Wu et al., 2025).

Materials:

Reagents:

1. *C. elegans* strains
2. M9 buffer
3. 1X Phosphate buffer saline (PBS)
4. 4 % paraformaldehyde solution (PFA)
5. Milli-Q Water
6. Liquid nitrogen
7. β-mercaptoethanol (Sigma, M3148)
8. Tris-base (MP Biomedicals, 194557)
9. Triton X-100 (Sigma, X100-ml)
10. DPX (Dibutylphthalate Polystyrene Xylene) (SRL, 042848)

Equipment:

1. Sterile pipettes, tips and glass slides
2. Microcentrifuge tubes (MCT)
3. Cooling Microfuge (SIGMA 1−14 K; SIGMA 3−30 K)
4. Fluorescent microscope (Carl Zeiss Axio Imager.M2)
5. Leica stellaris confocal microscope
6. pH meter (Mettler Tolledo)
7. −30 °C Deep freezer

Protocol:

Step 1: Synchronous young-adult worms were washed three times with M9 buffer to remove excess adherent bacteria and were re-suspended in (1 ×) PBS.

Step 2: Prepare 4 % paraformaldehyde solution (PFA), by weighing 1.6 g PFA and dissolve it in 40 ml (1 ×) PBS; heat @ 60 °C till it dissolves, cool it and add NaOH to maintain the pH of 7.2–7.4. store the solution in dark @ 4 °C. Fix the washed worms in 4% PFA solution (light-sensitive) and incubate @ 4 °C overnight.

Step 3: Wash twice with (1 ×) PBS and once with MQ water. Place around 100 μl of worm pellet on a glass slide and cover it with another glass slide, creating a sandwich.

Step 4: Now pour the liquid nitrogen over this set-up to immediately freeze fracture or allow the sandwich set-up to freeze overnight @ −30 °C.

Step 5: Separate the glass slides using mechanical force to expose the frozen worms.

Step 6: Wash using (1 ×) PBS and collect the worms in micro-centrifuge tubes (MCT) and centrifuge it to pellet down the worms.

Step 7: Add 500 ul of permealization solution and incubate for 48 h @ 37 °C. Permealization solution: 125 mM Tris-base (pH.4) + 1 % Triton X-100 + 5 % β-mercaptoethanol + 10 ml MilliQ water.
Step 8: Wash three times using (1 ×) PBS to remove excess of permealization solution. Block the non-specific sites 5% BSA and incubate for 3 h at room-temperature using end-to-end rotor.
Step 9: Add primary antibody against specific histone acetylation (e.g., H4K5ac or H4K8ac, usually at 1:100 dilution or as recommended by the manufacturer). Incubate it overnight @ 4 °C.
Step 10: Wash thrice with 0.1 % PBST ((1 ×) PBS + Tween-20) for 10 min each and collect the pellet in MCT and incubate in secondary antibody (tagged with fluorescently labelled marker such as Alexa Fluor 647 or 488) for 3 h at RT.
Step 11: Wash thrice with 0.1 % PBST ((1 ×) PBS + Tween-20) for 10 min each and collect the pellet in MCT and mount the pellet of around 20 µl on a glass slide and add 10 µl of DPX (Dibutylphthalate Polystyrene Xylene) and cover using cover slip.
Step 12: Observe the slide using a Leica stellaris confocal microscope.

Notes:
1. Take large number of worms from the start to prevent worm loss while washing and freeze-fracturing.
2. Washing should be gentle to avoid debris.

3.4.3 Real time PCR analysis to measure the mRNA content of HDACs

To measure the histone deacetylation activity, one of the approaches Is to quantify the mRNA content of the gens regulating this modification, such as histone deacetylases. For this, first step includes RNA isolation of the age-synchronised worms followed by cDNA preparation and mRNA quantification using qPCR.

Materials:

Reagents:
1. *C. elegans* strains
2. M9 buffer
3. DEPC-treated water (Sigma-Aldrich; D5758)
4. RNAzol (Molecular Research Center; RN190)
5. Verso cDNA synthesis kit (Thermo Fisher Scientific; AB-1453/B)
6. Takara Master Mix (TB Green® Primer Ex Taq™, RR420A)
7. Chloroform (Merck-60244525001730)
8. Isopropanol (Spectrochem-67-63-0)

Equipment:

1. Nano-drop (Thermo Fisher Scientific; NANO DROP ONE^{C}).
2. Verso cDNA synthesis kit (Thermo Fisher Scientific; AB-1453/B)
3. PCR (Agilent Technologies; Sure Cycler 8800).
4. qPCR machine QuantStudio 5; Applied Biosystems by Thermo Fisher Scientific
5. Cooling Microfuge (SIGMA 1–14 K; SIGMA 3–30 K)
6. Sterile pipettes and tips
7. Microcentrifuge tubes (MCT)

Protocol:

1) RNA isolation:

Step 1: Young-adult worms are harvested in M9 buffer and washed thrice with M9 buffer.

Step 2: Then the worm pellet is washed three times with 0.2 % DEPC-treated water.

Step 3: Around 200 μl RNAzol is added to the pellet and the worms are homogenised using pestles, to rupture the worms without damaging the RNA. This work is performed while placing the MCTs on ice.

Step 4: When 90 % worms are homogenised, around 500 μl RNAzol is again added and vortexed for 20–30 s and incubated for 5 min in ice.

Step 5: Add 200 μl chloroform and vortex for 20–30 s, till it turns milky. Incubate for 10 min on ice. Chloroform is used to facilitate phase-separation method for RNA isolation, here it separates the content of the tube into upper clear aqueous phase containing RNA and lower coloured organic phase having proteins white DNA at the white interphase.

Step 6: Centrifuge at 12,000 × g for 10 min @ 4 °C. collect the upper clear phase (having RNA) in a fresh MCT and add 500 μl isopropanol and vortex for 20 s. Incubate it for 10 min on ice.

Step 7: Centrifuge at 12,000 × g for 10 min @ 4 °C. remove the supernatant.

Step 8: To the RNA pellet, add 5 % ethanol (1 ml) and centrifuge at 7500 × *g* for 5 min @ 4 °C. Repeat this washing step one more time.

Step 9: Air-dry the pellet to remove excess of ethanol. Once the RNA pellet is completely air-dried, the RNA is dissolved in 15 μl DEPC-water (0.1 %0.1Diethylpyrocarbonate) to remove any RNases that degrade RNA in the sample and RNA concentration is taken using a nano-drop.

2) cDNA preparation:

Step 1: After isolating the RNA, cDNA is prepared using Verso cDNA synthesis kit in a PCR machine.

Step 2: For this, the RNA is diluted to achieve 1000 ng/µl concentration and use 1 µl of RNA template and add the following kit reagents to get a total volume of 20 µl:

- cDNA Buffer: 4 µl
- dNTPs: 2 µl
- Oligo dTs: 1 µl
- Enhancer: 1 µl
- Enzyme: 1 µl
- DEPC-water: µl

Step 3: The labelled MCTs are then placed on PCR machine and the protocol is run as: Denaturation @ 42 °C for 30 min; Annealing @ 95 °C for 2 min and Storage/ pause @ 4 °C.

3) **qPCR:** qPCR analysis is done using 100 ng/µl concentration of cDNA, employing Takara Master Mix and quantification is done by normalizing the Ct (cycle threshold) value of gene of interest with internal control such as housekeeping genes (*act-1* or *gpd-1*) as the reference gene.

Step 1: For each reaction, the reaction mixture contains following reagents to get a total volume of 25 µl:

- SYBR green: 12.5 µl
- DEPC water: 6.5 µl
- Forward primer: 2.5 µl
- Reverse primer: 2.5 µl
- cDNA template: 1 µl

Step 2: The contents are vortexed and centrifuged ad then placed on qPCR machine.

Step 3: The qPCR cycle consists of (1) one cycle of pre-incubation for enzyme activation: 95 °C for 30 s; and (2) 40 cycles of amplification: Denaturation: 95 °C for 5 s; Extension/ amplification: 60 °C for 30 s

Step 4: The Ct values obtained are then processed for normalisation with the control, and fold change in gene expression is quantified using the $2^{-\Delta\Delta Ct}$ method.

Notes:

1. The worms should be homogenised while keeping the sample on ice to avoid degeneration of RNA due to mechanical heat.

3.4.4 Reporter strains fluorescence imaging

With the help of fluorescently tagged proteins the expression of histone deacetylases can easily be quantified using a fluorescent microscope and the intensity can be calculated using ImageJ software.

Materials:

Reagents:

1. *C. elegans* strains
2. M9 buffer
3. Sodium azide (Sigma-Aldrich; 71289)

Equipment:

1. Sterile pipettes, tips, glass slides and cover slips
2. Microcentrifuge tubes (MCT)
3. Centrifuge (Eppendorf centrifuge 5702)
4. Fluorescent microscope (Carl Zeiss Axio Imager.M2)

Protocol:

Step 1: Culture the respected strains and age-synchronise them via embryo isolation.

Step 2: Wash the worms using M9 buffer three times to remove excess of adherent bacteria. The worms are then immobilized using 100 mM (10 μl) sodium azide for 5 min.

Step 3: Approximately 20 μl of the sample is placed onto the glass slides, and images are taken using the fluorescence-specific filter on a fluorescent microscope.

Step 4: The fluorescence intensity of the fluorescent protein expression is quantified using ImageJ software.

Notes:

1. Washing should be gentle to avoid debris while imaging.

4. Concluding remarks

Histones are important basic proteins that regulate epigenetic expression. to study the histone modifications, *C. elegans* offers a very good model system, since it is quick and easy to culture and being transparent in nature it makes easier to visualise the expression of fluorescently-tagged proteins with the body. Thus, it provides one of the best models to study in-vivo functions of histones. Besides deacetylation, other modifications of histones such as acetylation, methylation, phosphorylation, ubiquitinylation, sumoylation can also be studied by following the above-mentioned protocols.

Acknowledgements

Authors thank CSIR-CDRI for providing the infrastructure and resources.

Author contribution

A.N. (Aamir Nazir) conceived the idea and edited the manuscript. A. Naseer (Anam Naseer) wrote and edited the manuscript.

Conflict of interest

The authors declare no conflict of interest.

References

Annunziato, B. A. T., College, B., & Education, N. (2008). DNA packaging: Nucleosomes and chromatin Aa the nucleosome. *The Unit of Chromatin Chromatin Is Coiled into Higher-Order Structures,* 1–4.

Caenorhabditis Genetics Center (CGC) – Official Resource for C. elegans Strains and Genetic Reagents https://cgc.umn.edu/. (no.date).

Dna, E., & Packaged, I. (2002). Chromosomal DNA and its packaging in the chromatin fiber eucaryotic DNA is packaged into a set of chromosomes chromosomes contain long strings of genes.

King, J., Patel, M., & Chandrasekaran, S. (2021). Metabolism, HDACs, and HDAC inhibitors: A systems biology perspective.

Kristy, J. W., Hiroshi, Q., & Guy, M. B. (2013). NIH public. *Access,* 1–10. https://doi.org/10.1007/978-1-61779-343-1.

Mareninov, K. S., Diaz, M. F. P., Yong, W. H., & Angeles, L. (2020). HHS public. *Access,* 299–311. https://doi.org/10.1007/978-1-4939-8935-5.

Mariño-ramírez, L., et al. (2014). Histone structure and nucleosome stability. *Expert Review of Proteomics,* 9450.

Naseer, A., & Nazir, A. (2023). Genetics and epigenetics of aging and age - associated diseases.

Naseer, A., Toppo, P., Akbar, M., & Nazir, A. (2025). Knockdown of mitochondrial sirtuin sir-2. 2 reduces alpha-synuclein clearance and impairs energy homeostasis in a model of ageing. https://doi.org/10.1242/dmm.052197.

Seto, E., & Yoshida, M. (no.date). Erasers of histone acetylation: The histone deacetylase enzymes.

Talbert, P. B., & Henikoff, S. (2021). Cell science at a glance. *Histone Variants at a Glance,* 1–10. https://doi.org/10.1242/jcs.244749.

Wu, Q., et al. (2025). Histone deacetylase HDA-5 regulates lipid metabolism through H4K5 and H4K8 acetylation in Caenorhabditis elegans. *Journal of Biological Chemistry, 301,* 110893.

CHAPTER SIXTEEN

Protocols for in vitro analysis of liquid-liquid phase separation of proteins

Niharika Nag[a] and **Timir Tripathi**[b,*]
[a]Molecular and Structural Biophysics Laboratory, Department of Biochemistry, School of Life Sciences, North-Eastern Hill University, Shillong, India
[b]Molecular and Structural Biophysics Laboratory, Department of Zoology, School of Life Sciences, North-Eastern Hill University, Shillong, India
*Corresponding author. e-mail address: timir.tripathi@gmail.com

Contents

Abstract

Liquid-liquid phase separation (LLPS) plays a central role in the organization of biomolecular condensates that regulate diverse cellular processes. *In vitro* LLPS assays provide a versatile platform for studying the molecular determinants that govern condensate formation, morphology, dynamics, and modulation. This chapter presents a detailed protocol for performing in vitro LLPS assays, particularly using polyethylene glycol (PEG) and sodium salt as key modulators of phase behavior. The described procedure enables systematic evaluation of protein condensation under defined physicochemical conditions, thereby facilitating insights into the interplay among macromolecular crowding, ionic strength, and protein concentration. The protocol includes guidelines for sample preparation, phase separation induction, microscopy-based visualization, and quantitative analysis of condensate properties. Collectively, this toolkit provides a robust framework for studying the biophysical principles underlying LLPS and for assessing the effect of environmental or molecular perturbations on protein phase behavior.

Methods in Enzymology, Volume 734
ISSN 0076-6879, https://doi.org/10.1016/bs.mie.2026.05.059

1. Introduction

Intracellular organization of biological macromolecules has, traditionally, been associated with membrane-bound compartments. Such compartmentalization not only provides spatial separation and structural organization but also creates microenvironments that facilitate enzymatic activities and regulate biochemical reactions. In recent years, however, the concept of liquid-liquid phase separation (LLPS) as another means of intracellular organization has gained traction (Antifeeva et al., 2022; Subedi, Uversky, & Tripathi, 2025). LLPS is a process in which a homogeneous solution spontaneously undergoes "separation" or "demixing" to form two co-existing "phases", i.e., the dense phase and the dilute phase (Alberti, Gladfelter, & Mittag, 2019). The process of LLPS is heavily dependent on inter- and intramolecular interactions, as well as finely tuned by the physicochemical parameters of their environment (Banani, Lee, Hyman, & Rosen, 2017; Ruff, Roberts, Chilkoti, & Pappu, 2018; Sasidharan, Nag, Tripathi, & Saudagar, 2023). Unlike traditional membrane-bound compartments, condensates formed via LLPS are dynamic, reversible, and capable of rapidly exchanging components with their surroundings. This allows cells to respond quickly to environmental or signaling cues, altering condensate composition, size, and material properties to modulate function. A few examples of biomolecular condensates formed by LLPS include stress granules, nucleoli, Cajal bodies, P-bodies, etc. Though the transient and reversible nature of these condensates enables cells to dynamically regulate processes such as RNA metabolism, chromatin remodeling, stress response, signal transduction, etc., when dysregulated, the same interactions can drive pathological aggregation, contributing to conditions ranging from neurodegenerative diseases to cancer and viral infections (Wang et al., 2021; Zhang, Jin, Xu, Guo, & Wang, 2024, Deka et al., 2026).

In the past several years, intrinsically disordered regions (IDRs) present in proteins have been found to be key drivers in the process of LLPS. They do so by providing several weak multivalent interactions, such as electrostatic, hydrophobic, and π–π stacking, that promote dynamic molecular clustering (Borcherds, Bremer, Borgia, & Mittag, 2021). Unlike structured proteins that fold into stable three-dimensional conformations, intrinsically disordered proteins (IDPs) or proteins containing extensive IDRs lack a fixed tertiary structure under physiological conditions. This allows them to be present in a broad conformational ensemble and engage in transient, multivalent interactions. These features also make IDPs highly sensitive to changes in their environment, like small variations in ionic strength, pH, temperature, or the presence of crowding agents, etc. (Fonin, Darling, Kuznetsova, Turoverov, & Uversky, 2018; Moses, Ginell, Holehouse, & Sukenik, 2023; Moses et al., 2020). Furthermore, post-translational modifications (PTMs) such as phosphorylation, methylation, or acetylation serve as regulatory switches that dynamically control condensate assembly and dissolution (Clarke, Thibault, Salapa, & Levin, 2021; Kanaan, Hamel, Grabinski, & Combs, 2020; Keating, Bademosi, San Gil, & Walker, 2023; Luo, Wu, & Li, 2021). However, it is now evident that intrinsic disorder is neither absolutely required nor sufficient for condensate formation, as folded domains, RNA molecules, membranes, and other cellular factors can help in the formation of biomolecular condensates (Tripathi & Gupta, 2026).

In vitro approaches provide an elegant approach to studying LLPS in a controlled environment. We are now able to study LLPS by varying key parameters, such as protein concentration, ionic strength, pH, crowding agents, PTMs, and the presence of nucleic acids, to uncover the principles

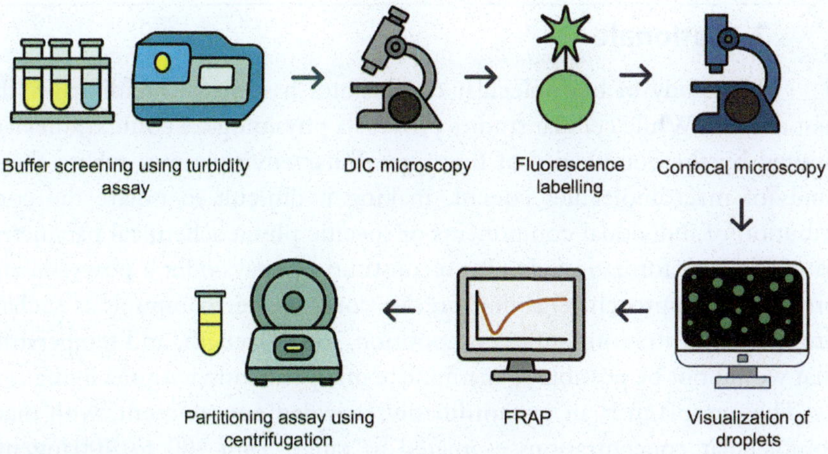

Fig. 1 Workflow of the *in vitro* methodology used to investigate protein phase separation. The experimental workflow begins with buffer screening using a turbidity assay to identify conditions that promote phase separation, which is quantified by measuring absorbance at 350 nm using a spectrophotometer. Differential interference contrast (DIC) microscopy is then used to visualize liquid droplet formation under these conditions. The protein is fluorescently labeled and examined using confocal microscopy to confirm phase separation and droplet morphology. Fluorescence recovery after photobleaching (FRAP) is performed to assess the dynamic exchange of molecules within the condensates. Finally, a partitioning assay using centrifugation is performed to determine the critical protein concentration in the dilute phase.

that govern condensate formation and stability (Alberti et al., 2025). *In vitro* LLPS assays also enable quantitative analysis of droplet size, number, sphericity, and material properties (such as fusion behavior and internal dynamics). Recent methodological advancements now allow the combination of imaging, spectroscopy, and biophysical measurements to provide a detailed understanding of the material state and molecular exchange within condensates. These methods are used not only for IDPs but also for folded proteins, protein-RNA complexes, and viral factors, expanding our understanding of how different molecules assemble into biomolecular condensates.

Given this versatility, *in vitro* LLPS assays have become essential tools for studying condensate morphology, dynamics, and modulation. In this protocol, we describe the procedure for a general *in vitro* LLPS assay, focusing on polyethylene glycol (PEG) and sodium salt concentration-dependent protein condensate experiments that provide a comprehensive toolkit for studying LLPS under defined conditions (Fig. 1).

2. Rationale

The study of biomolecular condensates has grown rapidly over the past decade. While cellular studies provide a physiological context, they are limited by the complexity of the intracellular environment, where thousands of macromolecules coexist, making it difficult to isolate the contribution of individual components or specific physicochemical parameters to condensate formation. *In vitro* reconstitution assays offer a powerful and preliminary approach, enabling precise control over components such as protein concentration, buffer composition, ionic strength, and temperature that would not be possible to manipulate independently within a living cell.

The cytoplasm is an extraordinarily crowded environment, with macromolecular concentrations estimated to range from 100 to 400 mg/mL (Molines et al., 2022; Yu et al., 2016; Zimmerman & Trach, 1991). PEG mimics this excluded volume effect, effectively lowering the thermodynamic threshold for phase separation. Similarly, sodium salt concentration modulates electrostatic interactions between charged residues, which are known to be central to the phase behavior of IDRs, providing a systematic means to probe how ionic strength shapes the condensate landscape. Together, these two modulators allow the construction of a simple yet informative phase diagram that maps the boundaries of condensate formation as a function of crowding and ionic conditions. The turbidity assay, microscopy, and centrifugation-based partitioning described in this protocol provide separate lines of evidence for LLPS; turbidity offers a rapid, quantitative screen; brightfield and confocal microscopy confirm droplet morphology, density, and liquid-like properties; and the partitioning assay yields the critical saturation concentration (C_{sat}), a thermodynamically meaningful parameter that characterizes the phase boundary. Collectively, these assays form a reliable and accessible toolkit for any researcher seeking to characterize the condensate-forming propensity of a protein of interest under defined, reproducible conditions.

3. Methodology

3.1 Buffer screening using turbidity assay

3.1.1 Equipment

1. Shaker Incubator
2. Spectrophotometer
3. Sterilized filters (pore size 0.2 μm)
4. Cuvettes

3.1.2 Reagents

1. Protein of interest
2. Protein storage buffer (Buffer A): 20 mM phosphate, 50 mM NaCl, pH 7.4, filtered using sterilized filters (pore size 0.2 μm).
3. Buffer B: 20 mM phosphate, pH 7.4, filtered using sterilized filters (pore size 0.2 μm).
4. Poly(ethylene glycol) solution: 40 % Poly(ethylene glycol)−8000 (Sigma-Aldrich) in Buffer B
5. 2 M NaCl solution

3.1.3 Procedure

3.1.3.1 PEG screening

1. Prepare fresh protein samples (*see* **Note 1**) in Buffer A at final PEG concentrations of 0 %, 5 %, and 10 %, while keeping the protein concentration (around 10−20 μM) and all other parameters constant (*see* **Note 2**). Include PEG blanks.
2. Mix the samples gently using a micropipette, and avoid the formation of bubbles (*see* **Note 3**).
3. Incubate the samples for 30 min at 37 °C with an rpm of 180−200.

4. Take an appropriate amount of sample and read the absorbance at 350 nm.
5. Subtract the PEG blanks for each sample to obtain the final values. Samples showing $A_{350} > 0.15$ can be considered for further analysis (*see* **Note 4**).
6. Advance the top PEG condition(s) to subsequent NaCl-dependent screening.

3.1.3.2 NaCl screening

1. Prepare Buffer B with varying NaCl concentrations (0, 50, 100, 150, 500, and 1000 mM NaCl).
2. Mix protein into each buffer, keeping the final protein concentration constant across all conditions. Include buffer-only blanks.
3. Mix the samples gently using a micropipette, and avoid the formation of bubbles.
4. Incubate the samples for 30 min at 37 °C with an rpm of 180–200.
5. Take an appropriate amount of sample and read the absorbance at 350 nm.
6. Subtract the buffer blanks for each sample to obtain the final values. Samples showing $A_{350} > 0.15$ can be considered for further analysis.

3.2 Microscopy (brightfield/DIC)

3.2.1 Equipment

1. Shaker Incubator
2. Brightfield Microscope
3. Microscope slide: 1″ × 3″ (1 mm–1.2 mm)
4. Cover slip: 18 mm

3.2.2 Reagents

1. Protein of interest
2. Buffer B: 20 mM phosphate, pH 7.4, filtered using sterilized filters (pore size 0.2 μm).
3. Poly(ethylene glycol) solution: 40% Poly(ethylene glycol)–8000 (Sigma-Aldrich) in Buffer B
4. 2 M NaCl solution

3.2.3 Procedure

1. Choose conditions that met the turbidity threshold in the screens performed above. Prepare fresh mixtures with the same incubation settings as before.

2. Gently load 5–10 μL of the sample onto a standard microscope slide, and drop a coverslip over the drop of sample (*see* **Note 5–6**). Let it stand for 1–2 min.
3. Use an inverted microscope in brightfield (DIC) at RT with a 60X objective.
4. Focus on the mid-plane of the sample. Acquire images from ≥5 randomly selected fields, avoiding edges and obvious debris (*see* **Note 7**).
5. Confirm spherical droplets, absence of irregular clumps, and presence of fusion/coalescence events over short observations (1–5 min) (*see* **Note 8**).
6. Save images as TIFF with embedded scale bars. Analyze images in ImageJ to obtain droplet count, area, etc.

3.3 Confocal microscopy

3.3.1 Fluorescent labeling of protein

3.3.1.1 Equipment

1. Laboratory rocker
2. Cold room
3. Centrifuge

3.3.1.2 Reagents

1. Invitrogen Alexa Fluor C5 maleimide 488
2. Dithiothreitol (DTT)/ Tris(2-carboxyethyl)phosphine (TCEP)
3. Dimethyl sulfoxide (DMSO)
4. Protein of interest
5. Protein storage buffer (Buffer A): 20 mM phosphate, 50 mM NaCl, pH 7.4
6. Amicon Ultra Centrifugal Filter, 30 kDa MWCO (Millipore)

3.3.1.3 Procedure

1. Prepare fresh protein samples in Buffer A. Add a 10-fold molar excess of a reducing agent such as DTT or TCEP to ensure free thiols, and remove excess reagent by dialysis (*see* **Note 9**).
2. Dissolve the dye in dimethylsulfoxide (DMSO) at a concentration of 1–10 mM. Using this stock solution, mix the protein with the dye in a 1:5 molar ratio (protein:dye).
3. Keep the solution in the dark and gently rock at 4 °C overnight (*see* **Note 10**).
4. Remove all free dye, using a 10–30 kDa MWCO spin concentrator, or a gel filtration column (*see* **Note 11**).

5. Measure absorbance at 280 nm (protein) and 495 nm (Alexa Fluor C5 maleimide 488) to calculate the degree of labeling (DOL) using: $\frac{[\text{Dye}]}{[\text{Protein}]}$ (*see* **Note 12**).
6. Aliquot labeled protein, protected from light, and store at −20 °C

3.3.2 Confocal microscopy

3.3.2.1 Equipment

1. Shaker Incubator
2. Confocal Microscope
3. Microscope slide: 1″ × 3″ (1 mm−1.2 mm)
4. Cover slip: 18 mm

3.3.2.2 Reagents

1. Protein of interest
2. Buffer B: 20 mM phosphate, pH 7.4, filtered using sterilized filters (pore size 0.2 μm).
3. Poly(ethylene glycol) solution: 40 % Poly(ethylene glycol)−8000 (Sigma-Aldrich) in Buffer B
4. 2 M NaCl solution

3.3.2.3 Procedure

Using the conditions that tested positive for LLPS in the turbidity assays and microscopy, we proceed to visualize droplets using confocal microscopy (Fig. 2).

1. Use Buffer B to prepare droplet-formation buffers with the desired PEG or NaCl concentration.
2. Add labeled and unlabeled protein to each droplet formation buffer to achieve a final protein concentration of 20 μM, with ~1 % of the protein fluorescently labeled.

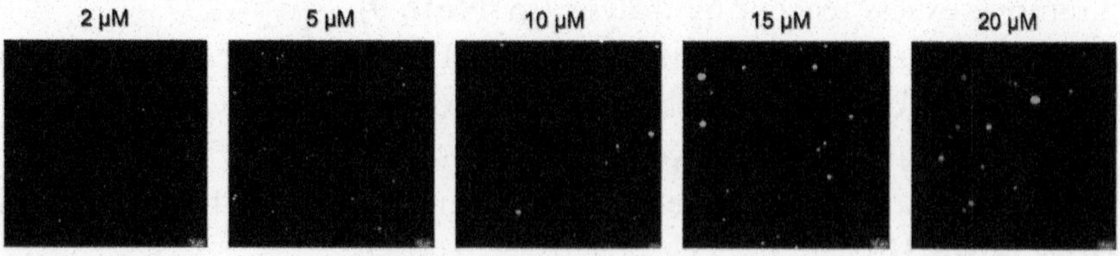

Fig. 2 Confocal images showing the effect of increasing concentrations of tau protein labeled with C5-maleimide. Tau protein solutions at 2, 5, 10, 15, and 20 μM were incubated in phosphate buffer containing 10 % PEG and 50 mM NaCl. Scale bars represent 10 μm in all images.

3. Thoroughly mix each sample, then incubate for 30 min at 37 °C with an rpm of 180–200.
4. Gently load 5–10 μL of the sample onto a standard microscope slide, and drop a coverslip over the drop of sample. Let it stand for 1–2 min.
5. Image droplets using a Leica SP8 confocal microscope (*see* **Note 13**). Excite Alexa Fluor 488-labeled protein with a 488-nm laser and acquire images at a resolution of 1024 × 1024 pixels using a 63 × oil immersion objective.

3.3.3 Fluorescence recovery after photobleaching (FRAP)

3.3.3.1 Equipment

1. Shaker Incubator
2. Confocal Microscope
3. Microscope slide: 1″ × 3″ (1 mm–1.2 mm)
4. Cover slip: 18 mm

3.3.3.2 Reagents

1. Protein of interest
2. Buffer B: 20 mM phosphate, pH 7.4, filtered using sterilized filters (pore size 0.2 μm).
3. Poly(ethylene glycol) solution: 40 % Poly(ethylene glycol)–8000 (Sigma-Aldrich) in Buffer B
4. 2 M NaCl solution

3.3.3.3 Procedure

1. Prepare droplet samples as described for confocal microscopy using 20 μM total protein with ~1 % fluorescently labeled protein.
2. Gently load 5–10 μL of the sample onto a standard microscope slide, and drop a coverslip over the drop of sample. Let it stand for 1–2 min.
3. Visualize the droplets using a Leica SP8 confocal microscope, excited with a 488-nm laser and a 63 × oil-immersion objective.
4. Select a single droplet using the steps mentioned above and photobleach a small region (~1–2 μm) using the 488 nm laser at high intensity (*see* **Note 14**).
5. Measure the recovery at defined intervals for 120 s (*see* **Note 15–16**).

3.4 Partitioning assay using centrifugation

3.4.1 Equipment

1. Shaker Incubator
2. Centrifuge

3.4.2 Reagents

1. Protein of interest
2. Buffer B: 20 mM phosphate, pH 7.4, filtered using sterilized filters (pore size 0.2 μm).
3. Poly(ethylene glycol) solution: 40 % Poly(ethylene glycol)−8000 (Sigma-Aldrich) in Buffer B
4. 2 M NaCl solution

3.4.3 Procedure

1. Prepare a series of total protein concentrations spanning below and above the expected phase-separating conditions, with the desired PEG or NaCl concentrations. Mix gently and incubate to allow droplet formation.
2. Transfer samples to low-binding microcentrifuge tubes (*see* **Note 17**) and centrifuge at 5000 RCF for 10 min at RT to separate a dense-phase pellet from the dilute-phase supernatant.
3. Carefully withdraw the supernatant without disturbing the pellet.
4. Measure A280 of the supernatant using a cuvette or microplate reader with matched blanks. Convert A_{280} to concentration using the protein's extinction coefficient.
5. Measure the concentration of the supernatant; this corresponds to the critical concentration (C_{sat}). This value should remain constant even as the total protein concentration is increased (*see* **Note 18**).

4. Conclusion

The protocol we have described in this chapter provides a systematic and accessible framework for studying LLPS *in vitro*, using PEG-mediated crowding and ionic-strength modulation as the primary drivers of condensate formation. While the methods described here have been validated using the microtubule-associated protein tau, which is a well-characterized IDP known to undergo LLPS and form condensates, the workflow is designed to be broadly applicable. Researchers working with other IDPs, folded proteins with disordered regions, RNA-binding proteins, or multicomponent protein-nucleic acid systems should find the overall workflow easily adaptable to their systems of interest, requiring only appropriate adjustment of buffer conditions, protein concentrations, and crowding agent concentrations.

The strength of this protocol is its step-by-step, multi-method design. Starting with a turbidity screen allows for quick identification of phase-separating conditions before moving to more time- and resource-intensive imaging experiments. Brightfield and confocal microscopy then provide visual confirmation of liquid-like droplets, their size, and density. At the same time, FRAP confirms the dynamic and reversible exchange of molecules within the condensates, which is an important criterion for distinguishing true LLPS from irreversible aggregation. Finally, the centrifugation-based partitioning assay provides a quantitative measure in the form of C_{sat} that allows comparison across different protein constructs, mutants, or experimental conditions.

The approaches outlined here can serve as a foundation for more advanced investigations. The basic assay can be extended to include PTM mimics, disease-associated mutations, or small-molecule inhibitors/modifiers, helping uncover regulatory mechanisms and identify condensate-modulating compounds with potential therapeutic relevance. Integration with techniques such as single-molecule fluorescence, single-molecule fluorescence resonance energy transfer (smFRET), nuclear magnetic resonance (NMR), or atomic force microscopy (AFM) will further deepen mechanistic insight into condensate structure and dynamics. As the field moves toward understanding how condensates are dysregulated in disease and how they might be targeted therapeutically, robust, reproducible *in vitro* methods, such as the one described here, will remain valuable tools for the study of biomolecular condensates.

5. Notes

1. Protein samples should always be freshly prepared. Repeated freeze–thaw cycles or pre-aggregated protein can lead to abnormally high turbidity values.
2. Always prepare fresh buffers to prevent any changes in concentration.
3. Gentle mixing with a micropipette is sufficient to homogenize samples. Avoid vortexing or bubble formation, as PEG is viscous and can trap bubbles.
4. $A_{350} > 0.15$ generally indicates detectable droplet formation; however, these conditions need to be validated microscopically.
5. Microscope slides and coverslips should be thoroughly cleaned before use. Surface contamination can distort droplet morphology or cause nonspecific adhesion.

6. If droplets adhere strongly to glass or lose sphericity, pretreat slides with hydrophilic or hydrophobic coatings, such as pluronic or Sigmacote, to modulate wetting.
7. Image samples within ~5 min of loading to minimize evaporation, which can alter droplet size or concentration.
8. Droplets displaying fusion or coalescence within minutes are indicative of liquid-like behavior, distinguishing them from amorphous aggregates.
9. Treatment with DTT or TCEP ensures that reactive cysteine thiols are reduced and free for labeling.
10. All labeling steps should be carried out in the dark to prevent photobleaching of the fluorophore.
11. Unreacted dye must be completely removed by ultrafiltration or gel filtration to avoid background fluorescence during imaging. The selection of the filtration device would depend on the molecular weight of the protein.
12. A degree of labeling (DOL) between 0.5 and 1.0 is generally optimal.
13. Acquisition of z-stacks can help confirm true droplet morphology and exclude surface artifacts.
14. During FRAP, select isolated, spherical droplets to prevent cross-bleaching of neighboring condensates. The bleached region should not exceed 10% of the droplet diameter; larger regions may yield inaccurate recovery kinetics.
15. Use high laser intensity for a short duration (≤1 s) to achieve complete bleaching without causing photodamage.
16. Recovery curves can be fitted to exponential functions to determine mobile fractions and diffusion rates. Incomplete recovery suggests restricted molecular mobility or gelation.
17. Use low-binding microcentrifuge tubes for partitioning assays to minimize protein adsorption to plastic surfaces.
18. The plateau in supernatant concentration (C_{sat}) across increasing total protein inputs represents the critical saturation concentration for LLPS under those conditions.

References

Alberti, S., Arosio, P., Best, R. B., Boeynaems, S., Cai, D., Collepardo-Guevara, R., et al. (2025). Current practices in the study of biomolecular condensates: A community comment. *Nature Communications, 16*(1), 7730.

Alberti, S., Gladfelter, A., & Mittag, T. (2019). Considerations and challenges in studying liquid-liquid phase separation and biomolecular condensates. *Cell, 176*(3), 419–434.

Antifeeva, I. A., Fonin, A. V., Fefilova, A. S., Stepanenko, O. V., Povarova, O. I., Silonov, S. A., et al. (2022). Liquid-liquid phase separation as an organizing principle of intracellular space: overview of the evolution of the cell compartmentalization concept. *Cellular and Molecular Life Sciences, 79*(5), 251.

Banani, S. F., Lee, H. O., Hyman, A. A., & Rosen, M. K. (2017). Biomolecular condensates: Organizers of cellular biochemistry. *Nature Reviews. Molecular Cell Biology, 18*(5), 285–298.

Borcherds, W., Bremer, A., Borgia, M. B., & Mittag, T. (2021). How do intrinsically disordered protein regions encode a driving force for liquid-liquid phase separation? *Current Opinion in Structural Biology, 67*, 41–50.

Clarke, J. P., Thibault, P. A., Salapa, H. E., & Levin, M. C. (2021). A comprehensive analysis of the role of hnRNP A1 function and dysfunction in the pathogenesis of neurodegenerative disease. *Frontiers in Molecular Biosciences, 8*, 659610.

Deka, N., Nag, N., & Tripathi, T. (2026). Intrinsically disordered proteins and liquid–liquid phase separation in drug discovery. *Drug Discovery Today, 31*(2), 104603.

Fonin, A. V., Darling, A. L., Kuznetsova, I. M., Turoverov, K. K., & Uversky, V. N. (2018). Intrinsically disordered proteins in crowded milieu: When chaos prevails within the cellular gumbo. *Cellular and Molecular Life Sciences: CMLS, 75*(21), 3907–3929.

Kanaan, N. M., Hamel, C., Grabinski, T., & Combs, B. (2020). Liquid-liquid phase separation induces pathogenic tau conformations in vitro. *Nature Communications, 11*(1), 2809.

Keating, S. S., Bademosi, A. T., San Gil, R., & Walker, A. K. (2023). Aggregation-prone TDP-43 sequesters and drives pathological transitions of free nuclear TDP-43. *Cellular and Molecular Life Sciences: CMLS, 80*(4), 95.

Luo, Y.-Y., Wu, J.-J., & Li, Y.-M. (2021). Regulation of liquid–liquid phase separation with focus on post-translational modifications. *Chemical Communications, 57*(98), 13275–13287. https://doi.org/10.1039/D1CC05266G].

Molines, A. T., Lemière, J., Gazzola, M., Steinmark, I. E., Edrington, C. H., Hsu, C.-T., et al. (2022). Physical properties of the cytoplasm modulate the rates of microtubule polymerization and depolymerization. *Developmental Cell, 57*(4), 466–479.e466.

Moses, D., Ginell, G. M., Holehouse, A. S., & Sukenik, S. (2023). Intrinsically disordered regions are poised to act as sensors of cellular chemistry. *Trends in Biochemical Sciences, 48*(12), 1019–1034.

Moses, D., Yu, F., Ginell, G. M., Shamoon, N. M., Koenig, P. S., Holehouse, A. S., et al. (2020). Revealing the hidden sensitivity of intrinsically disordered proteins to their chemical environment. *The Journal of Physical Chemistry Letters, 11*(23), 10131–10136.

Ruff, K. M., Roberts, S., Chilkoti, A., & Pappu, R. V. (2018). Advances in understanding stimulus-responsive phase behavior of intrinsically disordered protein polymers. *Journal of Molecular Biology, 430*(23), 4619–4635.

Sasidharan, S., Nag, N., Tripathi, T., & Saudagar, P. (2023). Chapter 12 - Interactions and interplay of MLOs with classical membrane-bound organelles. In V. N. Uversky (Ed.). *Droplets of life* (pp. 375–395). Academic Press.

Subedi, S., Uversky, V. N., & Tripathi, T. (2025). Chapter 11 - Liquid–liquid phase separation, biomolecular condensates, and membraneless organelles: A novel blueprint of intracellular organization. In T. Tripathi, & V. N. Uversky (Eds.). *The three functional states of proteins* (pp. 177–195). Academic Press.

Tripathi, T., & Gupta, M. N. (2026). Intrinsic disorder in membraneless organelles: necessary, sufficient, or neither? *Critical Insights in Biophysics, 1*(1), 2610923.

Wang, B., Zhang, L., Dai, T., Qin, Z., Lu, H., Zhang, L., et al. (2021). Liquid–liquid phase separation in human health and diseases. *Signal Transduction and Targeted Therapy, 6*(1), 290.

Yu, I., Mori, T., Ando, T., Harada, R., Jung, J., Sugita, Y., et al. (2016). Biomolecular interactions modulate macromolecular structure and dynamics in atomistic model of a bacterial cytoplasm. *eLife, 5*, e19274.

Zhang, Y., Jin, C., Xu, X., Guo, J., & Wang, L. (2024). The role of liquid-liquid phase separation in the disease pathogenesis and drug development. *Biomedicine & Pharmacotherapy, 180*, 117448.

Zimmerman, S. B., & Trach, S. O. (1991). Estimation of macromolecule concentrations and excluded volume effects for the cytoplasm of Escherichia coli. *Journal of Molecular Biology, 222*(3), 599–620.

CHAPTER SEVENTEEN

Methodological approach to investigate peroxiredoxin 1 deacetylation by histone deacetylase 6

Sonali Kumar[a,b], Ozasvi R. Shanker[b,c], Jyotirmoy Banerjee[d], and Aparna Banerjee Dixit[b,*]

[a]Department of Biotechnology, Lingaya's Vidyapeeth, Faridabad, Haryana, India
[b]Dr. B. R. Ambedkar Centre for Biomedical Research, University of Delhi, Delhi, India
[c]All India Institute of Ayurveda (AIIA), Government of India, New Delhi, India
[d]Department of Biophysics, All India Institute of Medical Sciences, New Delhi, India
*Corresponding author. e-mail address: aparnabanerjeedixit@gmail.com

Contents

Abstract

Post-translational acetylation of cytoplasmic proteins has emerged as a critical regulatory mechanism in neurological disease. Peroxiredoxin 1 (Prdx1), a key antioxidant enzyme, undergoes reversible lysine acetylation that modulates its enzymatic activity.

Methods in Enzymology, Volume 734
ISSN 0076-6879, https://doi.org/10.1016/bs.mie.2026.05.040

Histone Deacetylase 6 (HDAC6), a predominantly cytoplasmic deacetylase, has been identified as a regulator of Prdx1 acetylation, linking deacetylase activity to redox homeostasis. Accurate detection of Prdx1 deacetylation requires methodological strategies capable of preserving endogenous acetylation states and selectively enriching acetylated protein fractions.

This chapter describes a reproducible workflow for assessing HDAC6-mediated Prdx1 deacetylation in cortical tissue which integrates optimized protein extraction under acetylation-preserving conditions, acetyl-lysine–based co-immunoprecipitation, and immunoblot detection of Prdx1. Parallel assessment of total Prdx1 and established HDAC6 substrates enables normalization and validation of deacetylase activity. Quantitative densitometric analysis provides comparative evaluation of acetylation levels across experimental conditions.

This methodology offers a practical and translationally applicable approach for investigating non-histone deacetylation mechanisms and can be adapted to other HDAC6-regulated substrates in neurodegenerative research.

1. Introduction

Oxidative stress (OS), which results from an imbalance between the production of reactive oxygen species (ROS) and the body's antioxidant mechanisms, is now being recognized as an important factor in the dysfunction and progression of neurodegenerative and epilepsy-related disorders. An accumulation of excess ROS leads to oxidative damage to proteins, lipids, and nucleic acids, resulting in altered neuronal function, injury, and the development of epilepsy. In focal cortical dysplasia (FCD), which is a leading cause of drug-resistant epilepsy, oxidative stress has been found to be an important pathological mechanism, although the molecular mechanisms that control redox imbalance are not yet fully understood (Arena et al., 2018; Kumar et al., 2025; Zimmer et al., 2019).

Among the intracellular antioxidant mechanisms, the peroxiredoxin (Prdx) family has a pivotal role in redox homeostasis through the reduction of hydrogen peroxide and organic peroxides. Peroxiredoxin 1 (Prdx1) is a prominent cytosolic antioxidant enzyme in the brain and acts as a crucial first-line defense mechanism against ROS. Notably, the activity of Prdx1 is not fixed and is subject to dynamic regulation by post-translational modifications, such as acetylation, which directly affects its peroxidase activity and efficiency in counteracting oxidative stress (Szeliga, 2020). Abnormalities in Prdx1 acetylation have been shown to contribute to compromised antioxidant function and susceptibility to oxidative stress in neurological disorders.

Histone deacetylase 6 (HDAC6) is a mainly cytoplasmic deacetylase with multiple non-histone substrates and plays a crucial role in various cellular processes, such as cytoskeletal regulation, protein aggregation, mitochondrial regulation, and redox regulation (Kumar et al., 2023). Unlike other nuclear HDACs, HDAC6 primarily functions in the post-translational modification of cytoplasmic proteins, making it a central regulator of neuronal homeostasis (Chang et al., 2021; Kumar et al., 2020). Prdx1 has been shown to be a non-histone substrate of HDAC6, thus establishing a direct link between deacetylase activity and antioxidant regulation (Kumar et al., 2025; Leng et al., 2018).

Experimental studies have shown that the enhanced activity of HDAC6 results in the increased deacetylation of Prdx1, leading to decreased efficiency in scavenging ROS and allowing the accumulation of oxidative stress (Choi et al., 2017; Kumar et al., 2025; Leng et al., 2018;). This process has been shown to contribute to neurodegenerative diseases such as Alzheimer's disease, in which the pharmacological inhibition of HDAC6 has been shown to reverse the acetylation of Prdx1, resulting in the improvement of mitochondrial function and the reduction of oxidative stress (Choi et al., 2017).

This idea has recently been expanded to epilepsy-related cortical malformations, and it has been shown that HDAC6 expression and activity are strongly increased in cortical tissue of patients with FCD type II and in the prenatal BCNU model of FCD in rodents (Kumar et al., 2025). This is associated with decreased acetylation of Prdx1, increased ROS, and oxidative stress in dysplastic cortex, suggesting that HDAC6-mediated Prdx1 deacetylation contributes to redox dysregulation and neuronal hyperexcitability (Kumar et al., 2025). Furthermore, pharmacological inhibition of HDAC6 activity by the selective inhibitor Tubastatin A was demonstrated to normalize Prdx1 acetylation, decrease oxidative stress, and strongly reduce seizure susceptibility in vivo (Kumar et al., 2025).

Notwithstanding the increasing appreciation of the role of the HDAC6-Prdx1 pathway in redox biology, there are still limited approaches to investigate Prdx1 deacetylation directly. The precise determination of Prdx1 acetylation, HDAC6 expression, and oxidative stress in a single study setting is technically challenging, especially if the results are to be applicable to human disease. Overcoming these difficulties requires a comprehensive approach that can address post-translational modifications of antioxidant enzymes.

However, despite these improvements, the standardized methodological descriptions for studying Prdx1 deacetylation mediated by HDAC6 are still relatively underdeveloped in the literature. The differences in lysis conditions, antibody choice, immunoprecipitation methods, and normalization techniques can cause significant variability in the detection of acetylated Prdx1. Thus, a clear and protocol-driven roadmap is needed to ensure reproducibility and proper interpretation of Prdx1 acetylation status in experimental systems.

This chapter exclusively describes the methodological strategies for the detection and quantification of Prdx1 deacetylation mediated by HDAC6. Based on an experimentally validated protocol that has been previously employed in human cortical tissue and a rodent model (Kumar et al., 2025), this chapter describes the optimized strategies for protein extraction, acetyl-lysine-based co-immunoprecipitation, immunoblot-based detection of Prdx1 acetylation, and validation of HDAC6 deacetylase activity using well-established non-histone substrates. By strictly focusing on these methodological strategies, this chapter offers a useful and reproducible roadmap for studying HDAC6-mediated regulation of Prdx1 acetylation in neurological studies.

2. Experimental considerations for detecting peroxiredoxin 1 deacetylation by HDAC6

To precisely detect the deacetylation of Peroxiredoxin 1 (Prdx1), it is necessary to carefully design the experiment because of the dynamic and reversible nature of lysine acetylation. In contrast to the measurement of total protein expression, the acetylation level is very sensitive to sample processing, lysis conditions, antibody specificity, and normalization method. In order to avoid the detection of Prdx1 acetylation changes being due to technical issues rather than actual HDAC6-dependent deacetylation, it is important to be meticulous in the experimental design (Kumar et al., 2025; Leng et al., 2018).

2.1 Selection of biological material and experimental system

Prdx1 deacetylation can be studied in human tissue, neural cultures, or animal models, as long as there is sufficient protein yield and quality. In disease-relevant models, cortical tissue has direct translational relevance but also introduces additional complexity in terms of tissue heterogeneity and post-mortem or surgical accessibility. Experimental models should be chosen to ensure that there is a detectable level of endogenous Prdx1 and HDAC6 expression, as overexpression models can skew acetylation dynamics (Kumar et al., 2025).

Regardless of the biological model, it is essential to process the tissue quickly and under conditions that prevent post-lysis deacetylation. All tissue procurement and homogenization procedures should be done at low temperatures, and deacetylase activity should be immediately inhibited during protein extraction.

2.2 Preservation of protein acetylation during lysis

The detection of acetylated Prdx1 is particularly susceptible to loss during protein extraction due to continued enzymatic deacetylation after cell lysis. To prevent artificial deacetylation, lysis buffers must be supplemented with a broad-spectrum deacetylase inhibitor cocktail, including inhibitors effective against class IIb deacetylases such as HDAC6 (Leng et al., 2018). Protease inhibitors alone are insufficient for preserving acetylation status.

Non-denaturing lysis conditions are preferred, as they maintain protein–protein interactions required for efficient immunoprecipitation. Harsh detergents or high ionic strength buffers can disrupt antibody–epitope recognition and reduce recovery of acetylated Prdx1. Optimization of buffer composition is therefore essential to balance protein solubilization with preservation of post-translational modifications (Kumar et al., 2025).

2.3 Rationale for Acetyl-lysine–based co-immunoprecipitation

Direct detection of Prdx1 acetylation by immunoblotting is limited by the availability and specificity of site-specific acetyl-Prdx1 antibodies. As an alternative, acetyl-lysine–based co-immunoprecipitation followed by immunoblot detection of Prdx1 provides a robust and widely applicable approach for assessing global Prdx1 acetylation status (Kumar et al., 2025; Leng et al., 2018).

In this strategy, acetylated proteins are enriched using a pan–acetyl-lysine antibody, after which Prdx1 is detected within the immunoprecipitated fraction by immunoblotting. This method allows selective assessment of acetylated Prdx1 independently of total Prdx1 levels and avoids confounding effects arising from changes in protein expression. Inclusion of appropriate IgG controls and input samples is essential to confirm specificity and efficiency of immunoprecipitation.

2.4 Normalization and interpretation of Prdx1 acetylation levels

Normalization of Prdx1 acetylation quantification is necessary to prevent misinterpretation. Given the fact that the regulation of acetylation, but not the total Prdx1 protein, is independent, the levels of acetylated Prdx1 must be analyzed in relation to total Prdx1 protein expression, rather than housekeeping proteins alone (Kumar et al., 2025; Leng et al., 2018). Simultaneous immunoblot analysis of total Prdx1 in the input lysate is thus necessary for each experimental condition.

Moreover, normalization of total protein input during the immunoprecipitation procedure is essential to normalize differences in pull-down efficiency. The absence of proper normalization may result in the generation of misleading information on the degree of deacetylation.

2.5 Validation of HDAC6-dependent deacetylation

To attribute changes in Prdx1 acetylation specifically to HDAC6 activity, validation controls are essential. Measurement of acetylation levels of established HDAC6 substrates, such as α-tubulin, provides a reliable functional readout of HDAC6 deacetylase activity within the same samples (Kumar et al., 2025). Reduced acetylation of α-tubulin supports increased HDAC6 activity, whereas restoration of acetylation following HDAC6 inhibition confirms effective enzymatic blockade.

Pharmacological inhibition using selective HDAC6 inhibitors serves as an additional methodological control to demonstrate reversibility and specificity of Prdx1 deacetylation. Restoration of Prdx1 acetylation under these conditions validates both the biological relevance of HDAC6 and the robustness of the detection strategy (Kumar et al., 2025).

2.6 Reproducibility and technical limitations

Due to the sensitivity of the acetylation assay, it is highly recommended that the experiment be conducted using biological replicates and technical replicates. This is because there may be variability in the performance of the antibodies, the efficiency of binding to the beads, and the conditions of lysis. However, it is important to note that the acetyl-lysine-based enrichment approach does not detect the specific acetylation sites on Prdx1; therefore, the approach is based on the global acetylation (Leng et al., 2018).

Despite the limitations, this methodological framework can be used as a reliable and reproducible approach for the detection of Prdx1 deacetylation by HDAC6.

3. Methodology

3.1 Preparation of cortical protein lysates

3.1.1 Equipment

1. Refrigerated centrifuge
2. Homogenizer
3. Ice bucket
4. Microcentrifuge tubes

3.1.2 Reagents

1. IP lysis buffer (RIPA buffer)
2. Protease inhibitor cocktail
3. BCA protein assay kit

3.1.3 Procedure

1. Retrieve frozen cortical tissue and keep on ice throughout processing.
2. Add ice-cold lysis buffer supplemented with protease inhibitors.
3. Maintain buffer volume at ~10 × tissue weight.
4. Homogenize tissue thoroughly until complete disruption.
5. Incubate lysate on ice for 20 min.
6. Centrifuge at 12,000 rpm for 5 min at 4 °C.
7. Carefully collect supernatant without disturbing pellet.
8. Quantify protein concentration using BCA assay.

3.1.4 Notes

- Perform all steps at **4 °C** to prevent deacetylation.
- Process samples immediately or store at −80 °C.

3.2 Co-Immunoprecipitation of acetylated proteins

3.2.1 Equipment

1. End-over-end rotator
2. Cold room or 4 °C incubator

3.2.2 Reagents

1. Anti-acetylated lysine antibody
2. Protein A/G Sepharose beads

3.2.3 Procedure

1. Incubate antibody with Protein A/G beads for 2 h at 4 °C with gentle rotation.

2. Add equal amounts of protein lysate to antibody–bead complexes.
3. Incubate overnight at 4 °C with continuous rotation.

3.2.4 Notes

- Ensure **equal protein input** across samples for accurate comparison.
- Use sufficient antibody to maximize enrichment efficiency.

3.3 Washing and elution of immunoprecipitated proteins

3.3.1 Equipment

1. Microcentrifuge
2. Pipettes

3.3.2 Reagents

1. Ice-cold PBS
2. SDS sample buffer

3.3.3 Procedure

1. Centrifuge samples to pellet beads.
2. Wash beads three times with ice-cold PBS.
3. Resuspend gently during washing.
4. Remove supernatant completely after final wash.
5. Add SDS sample buffer directly to beads.
6. Boil samples at 100 °C for 10 min to elute proteins.

3.3.4 Notes

- Avoid vigorous pipetting to prevent bead loss.
- Insufficient washing increases background signal.

3.4 SDS-PAGE and immunoblotting

3.4.1 Equipment

1. SDS-PAGE electrophoresis system
2. PVDF membranes
3. Chemiluminescence imaging system

3.4.2 Reagents

1. 10 % SDS-PAGE gel
2. Blocking solution (3 % BSA)
3. Primary antibody (anti-Prdx1)
4. HRP-conjugated secondary antibody
5. ECL detection reagent

3.4.3 Procedure

1. Load eluted samples onto 10 % SDS-PAGE gel.
2. Perform electrophoresis to separate proteins.
3. Transfer proteins onto PVDF membrane (2 h).
4. Block membrane with 3 % BSA for 1 h.
5. Incubate overnight with primary antibody at 4 °C.
6. Wash membrane with TBST.
7. Incubate with secondary antibody for 1 h at room temperature.
8. Develop signal using chemiluminescence reagent.
9. Capture images using imaging system.

3.4.4 Notes

- Membranes can be stored in TBST at 4 °C if needed.
- Avoid overexposure during imaging.

3.5 Assessment of total Prdx1 (input control)

3.5.1 Procedure

1. Load 50 μg of total protein lysate on SDS-PAGE.
2. Perform western blot using anti-Prdx1 antibody.
3. Normalize protein levels using β-actin.

3.5.2 Notes

- Ensures observed changes are due to acetylation, not protein expression differences.

3.6 Data analysis and quantification

3.6.1 Procedure

1. Quantify band intensity using ImageJ software.
2. Measure integrated density of acetylated Prdx1 bands.
3. Normalize to total Prdx1 levels.
4. Express results relative to control.
5. Perform statistical analysis:
   - Student's *t*-test (two groups)
   - One-way ANOVA (multiple groups)
6. Present data as mean ± SEM.

3.6.2 Notes

Significance threshold set at $\mathbf{p < 0.05}$.

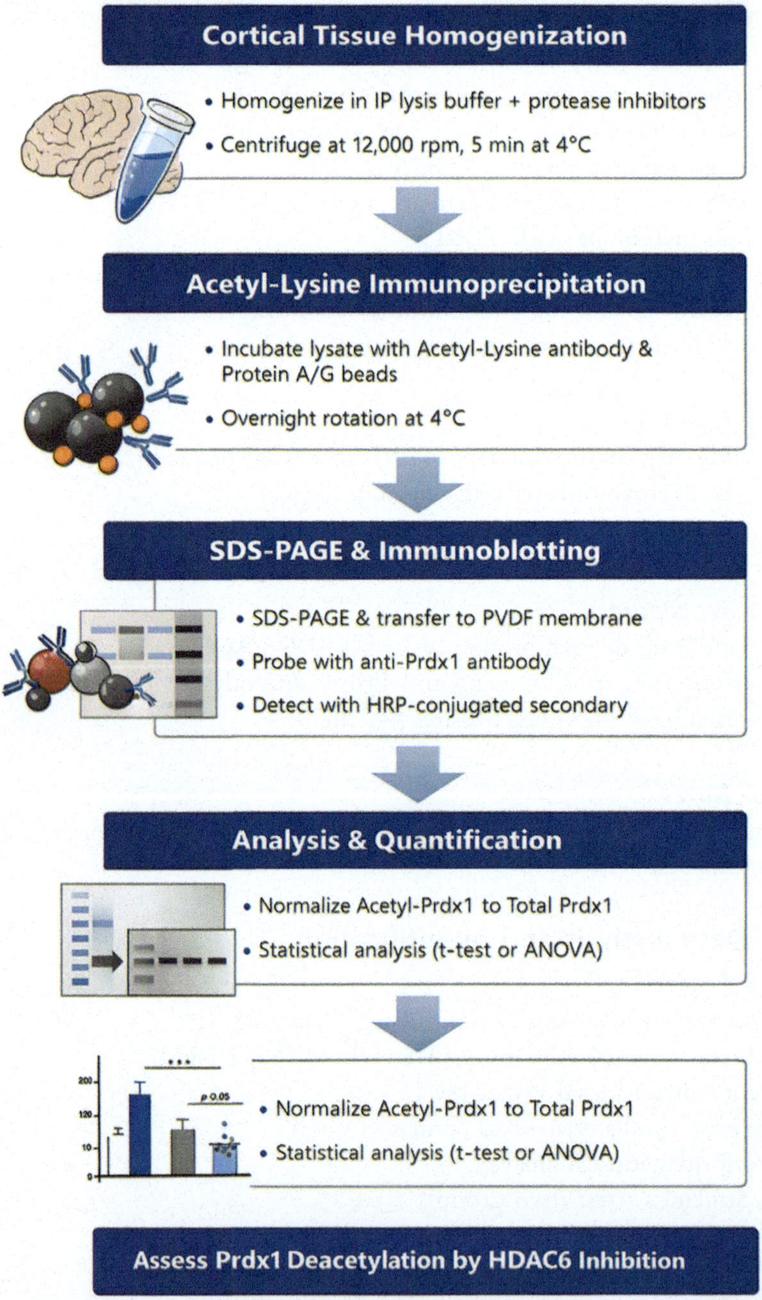

Fig. 1 Workflow for detection of Peroxiredoxin 1 (Prdx1) deacetylation by HDAC6 using acetyl-lysine–based co-immunoprecipitation. Schematic representation of the experimental protocol used to assess Prdx1 acetylation status.

3.7 Expected outcome of assay

1. Detection of a distinct Prdx1 band in acetylated fraction.
2. Reduced acetylation signal indicates increased deacetylation.
3. Restoration of signal confirms HDAC6-dependent regulation.

3.8 Safety and compliance

1. Follow institutional ethical guidelines for human/animal tissue use.
2. Maintain biosafety protocols while handling biological samples.
3. Handle chemiluminescent reagents with care.

3.8.1 Notes

- Use PPE at all times.
- Dispose of biological waste as per regulations.

4. Conclusion

Protein acetylation assessment is vital to the comprehension of the mechanisms of post-translational regulation in neurobiology. The methods described in this chapter outline a strong and reproducible model to measure the deacetylation of Peroxiredoxin 1 (Prdx1) by the enzyme Histone Deacetylase 6 (HDAC6) in the cortex. By using the acetyl-lysine-based co-immunoprecipitation assay and the immunoblot assay to measure Prdx1 levels, the results could be normalized to the total protein levels to selectively measure post-translational regulation without the involvement of transcriptional or translational regulation.

A major advantage of this workflow resides in its ability to be adapted to endogenous protein systems without the need to utilize overexpression models, thus maintaining the physiological relevance of the findings. The addition of parallel control groups, such as the examination of total levels of Prdx1, as well as the verification of the activity of HDAC6 by using well-characterized substrates, adds to the confidence of interpretation, as well as the ability to exclude the possibility of technical misinterpretation of the findings. If carried out correctly, with strict temperature control as well as optimized antibody conditions, this protocol enables the comparative analysis of the acetylation status of Prdx1 in different experimental groups.

Although this protocol does not allow the detailed analysis of the acetylation status of the protein, it allows the development of a base upon which more complex analyses, such as mass spectrometry-based analysis of the acetylome or mutational analysis, can be carried out. As the field begins to realize the importance of non-histone acetylation in the pathogenesis of neurological diseases, the need for transparent descriptions of methodologies, such as the one presented in this protocol, will be of the utmost importance.

In summary, this protocol provides a useful and highly relevant tool to study the regulation of Prdx1 acetylation in an HDAC6-dependent manner and can be readily modified to study the regulation of other cytoplasmically localized proteins that are targets of deacetylation activity. This protocol will help to further advance the mechanistic understanding of post-translational regulation in neurodegenerative and epilepsy-associated conditions (Fig. 1).

References

Arena, A., Zimmer, T. S., Van Scheppingen, J., Korotkov, A., Anink, J. J., Mühlebner, A., ... Aronica, E. (2018). Oxidative stress and inflammation in a spectrum of epileptogenic cortical malformations: Molecular insights into their interdependence. *Brain Pathology (Zurich, Switzerland), 29*(3), 351–365. https://doi.org/10.1111/bpa.12661.

Chang, P., Li, H., Hu, H., Li, Y., & Wang, T. (2021). The role of HDAC6 in autophagy and NLRP3 inflammasome. *Frontiers in Immunology, 12*. https://doi.org/10.3389/fimmu.2021.763831.

Choi, H., Kim, H. J., Kim, J., Kim, S., Yang, J., Lee, W., ... Mook-Jung, I. (2017). Increased acetylation of peroxiredoxin1 by HDAC6 inhibition leads to recovery of Aβ-induced impaired axonal transport. *Molecular Neurodegeneration, 12*(1), https://doi.org/10.1186/s13024-017-0164-1.

Kumar, S., Attrish, D., Srivastava, A., Banerjee, J., Tripathi, M., Chandra, P. S., & Dixit, A. B. (2020). Non-histone substrates of histone deacetylases as potential therapeutic targets in epilepsy. *Expert Opinion on Therapeutic Targets, 25*(1), 75–85. https://doi.org/10.1080/14728222.2021.1860016.

Kumar, S., Shanker, O. R., Banerjee, J., Tripathi, M., Chandra, P. S., & Dixit, A. B. (2023). Epigenetics in epilepsy. *Progress in Molecular Biology and Translational Science, 198*, 249–269. https://doi.org/10.1016/bs.pmbts.2023.01.005.

Kumar, S., Shanker, O. R., Parambath, S. D., Banerjee, J., Tripathi, M., Chandra, P. S., ... Dixit, A. B. (2025). HDAC6 inhibition enhances peroxiredoxin 1 acetylation to mitigate oxidative stress and seizure activity in focal cortical dysplasia. *Experimental Neurology, 392*, 115367. https://doi.org/10.1016/j.expneurol.2025.115367.

Leng, Y., Wu, Y., Lei, S., Zhou, B., Qiu, Z., Wang, K., & Xia, Z. (2018). Inhibition of HDAC6 activity alleviates myocardial ischemia/reperfusion injury in diabetic rats: Potential role of peroxiredoxin 1 acetylation and redox regulation. *Oxidative Medicine and Cellular Longevity, 2018*(1), https://doi.org/10.1155/2018/9494052.

Szeliga, M. (2020). Peroxiredoxins in neurodegenerative diseases. *Antioxidants, 9*(12), 1203. https://doi.org/10.3390/antiox9121203.

Zimmer, T. S., Ciriminna, G., Arena, A., Anink, J. J., Korotkov, A., Jansen, F. E., ... Aronica, E. (2019). Chronic activation of anti- oxidant pathways and iron accumulation in epileptogenic malformations. *Neuropathology and Applied Neurobiology, 46*(6), 546–563. https://doi.org/10.1111/nan.12596.

Printed and bound by CPI Group (UK) Ltd, Croydon, CR0 4YY
17/09/2026
02218552-0004